# Essentials of Cell and Molecular Biology

**E.D.P. De Robertis, M.D.**

University of Buenos Aires
Buenos Aires, Argentina

**E.M.F. De Robertis, Jr., M.D., Ph.D.**

University of Basel
Basel, Switzerland

**SAUNDERS COLLEGE PUBLISHING**

Philadelphia   New York   Chicago
San Francisco   Montreal   Toronto
London   Sydney   Tokyo   Mexico City
Rio de Janeiro   Madrid

*Address orders to:*
383 Madison Avenue
New York, N.Y. 10017
*Address editorial correspondence to:*
West Washington Square
Philadelphia, Pa. 19105

This book was set in Aster by Hampton Graphics, Inc.
The editors were Michael Brown, David Milley and Janis Moore, and Amy Satran.
The art director and cover designer was Nancy E. J. Grossman.
The text designer was Barbara Bert.
The new art was drawn by Marion Krupp.
The production manager was Tom O'Connor.
The printer was Von Hoffman.

**LIBRARY OF CONGRESS
CATALOG CARD NO.: 80-53913**

DeRobertis, Eduardo, & Eduardo
M. DeRobertis, Jr.
    Essentials of cell and
    molecular biology.
Philadelphia, Pa.: Saunders College
400 p.
8103          801010

ESSENTIALS OF CELL & MOLECULAR BIOLOGY                    ISBN 0-03-057713-6

1234    032    987654321

**CBS COLLEGE PUBLISHING**
Saunders College Publishing
Holt, Rinehart and Winston
The Dryden Press

To Nelly and Ana

# Preface

In recent years, cell and molecular biology have made spectacular advances that constitute the basic pillars for the understanding of all biological and medical sciences. The study of cells as the structural and functional units of all living organisms viewed under the optical microscope has expanded into the study of the ultrastructural and macromolecular organization of cell components revealed by the electron microscope.

Furthermore, the role played by molecules such as lipids, proteins, and nucleic acids in cell functions has been one of the main targets of modern biology. This knowledge has reached a climax with the unraveling of the genetic code and of the mechanisms by which genetic information is transcribed and finally translated into specific proteins.

This book is an elementary account of how these two levels of organization — cellular and molecular — are intimately integrated. Although *Essentials* is derived from our more extensive treatment of the subject in *Cell and Molecular Biology* (7th edition, 1980), it really is a completely new book aimed at a much wider audience. It is intended primarily as a textbook for advanced high school and junior college students who need a short, elementary, but modern treatment of cell and molecular biology. This will serve as a basis for more advanced studies such as cytogenetics or cell physiology, or the more applied fields of medical, agronomical, and veterinary sciences.

The material has been organized to facilitate teaching, proceeding from simple to more complex matters. For example, the first two chapters give a general overview of the cell and its main structural and molecular components, most of which are dealt with in more detail in subsequent chapters. In many instances the text and the figures refer to the original experiments from which the new knowledge has been acquired. The message we wish to convey to the student is that the pursuit of science is an unfinished task; we are now seeing only the tip of the iceberg. Much remains to be discovered about the organization and function of the "building blocks" of all living forms on earth.

E. De Robertis
E. M. De Robertis, Jr.

# Acknowledgments

A book that attempts to translate into elementary terms the new concepts of cell and molecular biology could be possible only with the unselfish collaboration of many who contributed to the progress of this challenging field of science.

We would like to express our gratitude to all the colleagues who have contributed their comments and criticisms and those who have generously allowed us to reproduce their illustrations. We are especially indebted to Amy Satran, who edited the entire manuscript and made many suggestions to simplify the text. To Michael Brown and all the staff of Saunders College Publishing we are indebted for the excellent work they have done to make this book more attractive and useful to students.

# Contents

# chapter 1

# The Cell: Structural Organization

## 1-1 INTRODUCTION AND HISTORY OF CELL AND MOLECULAR BIOLOGY

The study of the living world shows that evolution has produced an immense diversity of forms. There are about four million different species of bacteria, protozoa, plants, and animals which differ in their morphology, function, and behavior. However, we now know that when living organisms are studied at the cellular and molecular levels there is a unique master plan of organization. The scope of *cell and molecular biology* is precisely that unifying organizational plan—in other words, the analysis of the cells and molecules that constitute the building blocks of all forms of life.

Ancient philosophers and naturalists, particularly Aristotle in antiquity and Paracelsus in the Renaissance, arrived at the conclusion that "All animals and plants, however complicated, are constituted of a few elements which are repeated in each of them." They were referring to the macroscopic structures of an organism, such as the roots, leaves, and flowers common to different plants, or segments and organs that are repeated in the animal kingdom. Many centuries later, the invention of magnifying lenses led to the discovery of the microscopic world. It was learned that a single cell could constitute an entire organism, as in the protozoa, or that it could be one of many cells grouped and differentiated into tissues and organs to form a multicellular organism.

The cell is thus a fundamental structural and functional unit of living organisms, just as the atom is the fundamental unit in chemical structures (Fig. 1–1). If cellular organization is destroyed by mechanical or other means, cellular function is likewise destroyed. Although some vital functions may persist (such as enzymatic activity), the cell becomes disorganized and dies.

Biochemical studies have demonstrated that living matter itself is composed of the same elements that comprise the inorganic world, although fundamental differences can exist in their organization. In the nonliving world there is a continuous tendency toward a thermodynamic equilibrium with a random distribution of matter and energy, whereas in a living organism there is a high degree of structure and function maintained by energy transformations based on the constant input and output of matter and energy. Biochemists have isolated from the complex mixture of cell constituents not only inorganic components, but also much more complex molecules such as proteins, fats, polysaccharides, and nucleic acids. Such biochemical studies have also demonstrated the underlying unity of the entire living world. Today it is known that the biochemical machinery in all organisms has essentially the same structure and function, and that all living organisms share the same genetic code.

## Levels of Organization and Instrumental Resolving Power

Modern studies of living matter demonstrate that a series of integrated *levels of organization* result in the vital manifestations of the organism. The concept of levels of organization as developed by Needham and others implies that in the entire universe—in both the nonliving and living worlds—there are such various levels of different complexity that "The laws or rules that are encountered at one level may not appear at lower levels."

Table 1–1 shows the limits that separate the study of biological systems at particular dimension levels. The boundaries between levels of organization are imposed artificially by the *resolving power* of the instruments employed. The human eye cannot resolve (discriminate between) two points separated by less than 0.1 mm (100 $\mu$m). Most cells are much smaller than this and must be studied under the full resolving power of the *light microscope* (0.2 $\mu$m). Most cellular components are smaller still, and require the resolution of the

resolving power—ability of an optical system to distinguish images of objects within its field of view

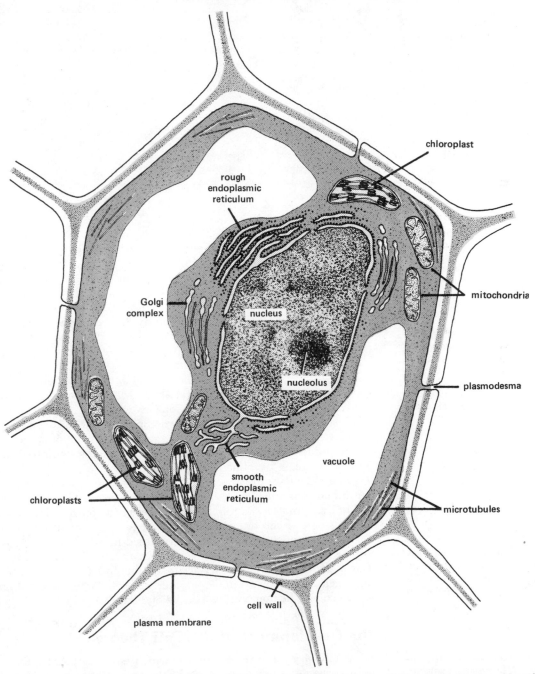

**Figure 1-1**  Diagram of a plant cell showing cell components including the cell wall, the plasmodesmata, chloroplasts, and vacuoles.

**TABLE 1-1  VARIOUS LEVELS OF BIOLOGICAL STRUCTURE**

| Dimension | Field | Structures | Method |
|---|---|---|---|
| 0.1 mm (100 $\mu$m) and larger | anatomy | organs | eye and simple lenses |
| 100 $\mu$m to 10 $\mu$m | histology | tissues } | various types of light microscopes, x-ray microscopy |
| 10 $\mu$m to 0.2 $\mu$m (200 nm) | cytology | cells, bacteria } | |
| 200 nm to 1 nm | submicroscopic morphology ultrastructure | cell components, viruses | polarization microscopy, electron microscopy |
| smaller than 1 nm | molecular and atomic structure | arrangement of atoms | x-ray diffraction |

**electron microscope (EM)—** instrument which uses a focused beam of electrons to produce an enlarged image of an object

*electron microscope* (Fig. 1–2). With this important tool, direct information can be obtained about structures ranging from 0.4 to 200 nm, thus extending our field of observation to the world of macromolecules. Results obtained by the application of electron microscopy have changed the field of cytology so much that a large part of the present book is devoted to discussions of the achievements made possible by this technique. Finally, in recent years great advances have been made in the detailed *x-ray diffraction* analysis of the molecular configuration of proteins, nucleic acids, and larger molecular complexes such as certain viruses.

**x-ray diffraction—scattering** of x-rays by the atoms of a crystal in a manner such that the resulting diffraction pattern provides information about the structure and/or identity of the substance

In Figure 1–3 the sizes of different cells, bacteria, viruses, and molecules are indicated on a logarithmic scale and compared with the wavelengths of various radiations, as well as with the limits of resolution of the eye, the light microscope, and the electron microscope. Note that the light microscope introduces a 500-fold increase in resolution over the eye, and the electron microscope provides a 500-fold increase over the light microscope.

Table 1–2 shows the general relationships among some of the linear dimensions and weights used in different fields of chemical analysis of living matter. Familiarity with these relationships is essential to the study of cell and molecular biology. The weight of the important components of the cell is expressed in picograms (1 pg = 1 $\mu\mu$g or $10^{-12}$ g), and that of molecules in *daltons*. The dalton is the unit of molecular weight (MW); one dalton equals the weight of a hydrogen atom. For example, a water molecule weighs 18 daltons; a molecule of hemoglobin weighs 64,500 daltons.

**dalton—unit of mass equal to** 1/16 that of an oxygen atom, or approximately that of a hydrogen atom

## The Development of the Cell Theory

**cell theory—theory that all** living organisms are composed of cells and cell products

The establishment of the *cell theory*, which essentially states that all living organisms are composed of cells and cell products, was the result of many investigations that started in the 17th century with the development of optical lenses and their combination in the compound microscope (Gr., *mikros*, small + *skopein*, to see, to look). The term *cell* (Gr., *kytos*, cell; L., *cella*, hollow space) was first used by

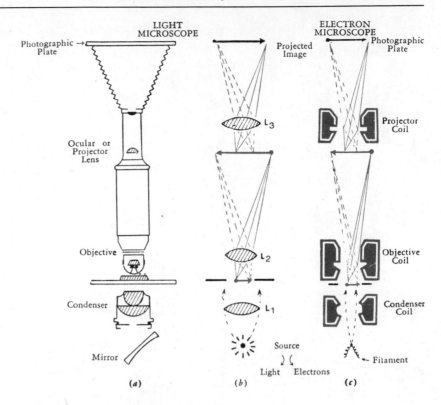

**Figure 1-2** Comparison between the optical microscope and the electron microscope. (See description in text.) (From G. Thompson.)

Robert Hooke (1655) to describe his investigations on "the texture of cork by means of magnifying lenses." In these observations, repeated by Grew and Malpighi on different plants, only the cavities ("utricles" or "vesicles") of the cellulose wall were recognized. In the same century and at the beginning of the next, Leeuwenhoek (1674) discovered free cells, as opposed to the "walled in" cells of Hooke and Grew. Leeuwenhoek observed some organization within cells, particularly the nucleus in some red blood cells. For more than a century afterward, these observations were all that was known about the cell.

At the beginning of the 19th century, several discoveries were made about the structure of plant and animal tissues. These finally led the botanist Schleiden (1838) and the zoologist Schwann (1839) to establish the cell theory in a more definite form. After the discovery of the nucleus by Brown (1831) and the description of the cell content as *protoplasm*, the concept of the cell became that of a mass of protoplasm limited in space by a cell membrane and possessing a nucleus. The protoplasm surrounding the nucleus became known as the *cytoplasm* to distinguish it from the *karyoplasm*, the protoplasm of the nucleus.

A major expansion of the cell theory was expressed by Virchow in 1855 in his famous aphorism, *"Omnis cellulae e cellula"* (i.e., all cells arise from preexisting cells), which established cell division as the central phenomenon in the reproduction of organisms. Years

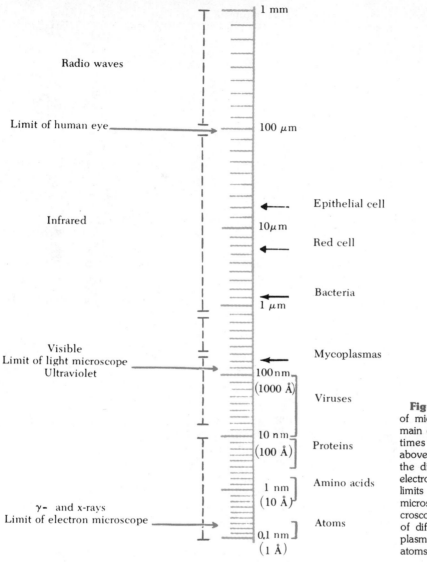

**Figure 1-3** Logarithmic scale of microscopic dimensions. Each main division represents a size ten times smaller than the division above. **To the left,** the position of the different wavelengths of the electromagnetic spectrum and the limits of the human eye, the light microscope, and the electron microscope. **To the right,** the sizes of different cells, bacteria, mycoplasmas, viruses, molecules, and atoms. (Modified from M. Bessis.)

**mitosis**—process occurring in the nucleus during cell division, whereby the genetic material is precisely duplicated and two new chromosome sets identical to the original are generated

later it was shown that cells ensure continuity between one generation and another by the mechanism of *mitosis* (Flemming, 1880) and the precise partitioning of the *chromosomes* (Waldeyer, 1890).

Another important discovery was that the development of an embryo starts with the fusion of two nuclei, one coming from an egg and the other from a sperm cell introduced during *fertilization* (Hertwig, 1875). Before the end of the century it was established that gametes (egg and sperm cells) are formed by a reductional division, later called *meiosis*, by which the number of chromosomes of a species remains constant from one generation to another.

All these discoveries led to the modern version of the cell theory, which states that (1) cells are the morphological and physiological

TABLE 1–2  RELATIONSHIPS BETWEEN LINEAR DIMENSIONS AND WEIGHTS IN CYTOCHEMISTRY*

| Linear Dimension | Weight | Terminology |
|---|---|---|
| 1 cm | 1 g | conventional biochemistry |
| 1 mm | 1 mg or $10^{-3}$ g | microchemistry |
| 100 $\mu$m | 1 $\mu$g or $10^{-6}$ g | histochemistry ⎫ |
| 1 $\mu$m | 1 $\mu\mu$g or 1 picogram or $10^{-12}$ gm | cytochemistry ⎭ ultramicrochemistry |

*From A. Engström and J. B. Finean, *Biological Ultrastructure.* New York, Academic Press, 1958. Copyright 1958, Academic Press, Inc.

units of all living organisms, (2) the properties of a given organism depend on those of its individual cells, (3) cells originate only from other cells, and continuity is maintained through the genetic material, and (4) the smallest unit of life is the cell.

## The Development of Submicroscopic and Molecular Biology

The rapid development of cell and molecular biology in the present century can be attributed to two main factors: (1) the increased resolving power provided by electron microscopy and x-ray diffraction, and (2) the convergence of the field with other branches of biological research, especially genetics, physiology, and biochemistry. The convergence with genetics will be illustrative.

The fundamental laws of heredity were discovered by Gregor Mendel in 1865, but at that time the cytological changes that take place in the sex cells were not sufficiently known to permit an interpretation of his findings. For this and other reasons, little attention was paid to Mendel's work until the botanists Correns, Tschermack, and De Vries independently rediscovered Mendel's laws in 1901. At that time cytology had advanced enough that the mechanism of distribution of the hereditary units postulated by Mendel could be understood and explained. The chromosome theory of heredity was finally established by Morgan and his collaborators, who assigned to the *genes* (Johanssen), or hereditary units, specific *loci* within the chromosomes. From this convergence of cytology and genetics, the study of *cytogenetics* has originated (see Chapter 9).

Within the past decade the study of genetics has also become linked to biochemistry, concerning events that take place at the molecular level. Although Miescher (1871) had already isolated "nuclein"—a substance now known as *deoxyribonucleic acid* or DNA—from white blood cells, its importance as genetic material was not recognized until the 1950s. Until that time the nuclear proteins had been considered the major constituents of genes. This concept was modified when nucleic acids were identified as the bearers of genetic information, especially in the light of fundamental work

**meiosis**—process occurring during the formation of gametes and involving a reduction division, whereby each daughter cell receives one of each pair of homologous chromosomes, thus reducing the number of chromosomes in each cell to one half

performed by Watson and Crick (1953), who proposed the *double helix model* of DNA. This model showed clearly how genes could duplicate and be transmitted from cell to cell (see Chapters 2 and 8). Since then the advances in molecular biology have been extraordinary. Another climax was reached with the deciphering of the *genetic code* (Nirenberg, Ochoa) and the discovery of the molecular mechanisms by which genes are transcribed and subsequently expressed as proteins. At the present time, the main focus of research in molecular biology is on the mechanisms that regulate gene expression.

## 1-2 MAIN TECHNIQUES USED TO STUDY CELL ORGANIZATION

### Microscopy

Our knowledge about the complex organization of cells is relatively recent. It has been obtained in the last 25 years mainly through the use of the electron microscope (Fig. 1–2), whose enormous resolving power has permitted the study of the ultrastructure or submicroscopic morphology of the cell. Most of this knowledge has been gathered by the observation of thin sections (20–50 nm thick) which are fixed and stained with "electron stains" such as osmium tetroxide (Fig. 1–4).

Another important technique involves *freeze-fracture* and *freeze-etching*. After a microscopic specimen is frozen and fractured, the fracture surface is submitted to a vacuum to remove surface ice and produce some degree of etching. Then a *replica* of the surface is made by depositing on it, by evaporation, a thin layer of carbon and other atoms. This replica is separated from the section and observed in the electron microscope. Figure 1–5 shows an onion root cell in which the fracture passes through the nuclear envelope and shows the nuclear pores, as well as several other structural details in the cytoplasm.

*Transmission electron microscopy* is based on the dispersion of electrons impinging upon and passing through the specimen, so that the image reflects the absence of those electrons that have been dispersed (Fig. 1–2). *Scanning electron microscopy* uses the electrons emitted back from the surface of the specimen to produce an image (Fig. 1–6). Figure 1–7, for example, shows a three-dimensional view of a cultured cell in which the details of the surface membrane are apparent.

### Cell Fractionation

Cell fractionation methods involve the *homogenization* or destruction of cell boundaries by different mechanical or chemical procedures, followed by the separation of subcellular fractions according to mass, surface, and specific gravity. The various fractions are then analyzed by biochemical or microchemical methods. As shown in Figure 1–8, the first step is the disruption of the cells by grinding

*Text continued on page 14*

---

**freeze-fracture/freeze etching**—technique by which a microscopic specimen is frozen, fractured by a sharp blow, and submitted to a vacuum to remove surface ice from the fracture faces

**replica**—duplicate of a microscopic specimen produced by depositing, by evaporation, a layer of heavy metal and a layer of carbon on its surface

**transmission electron microscope (TEM)**—instrument in which the image is formed by electrons passing through the specimen; most often used to study internal cell structure

**scanning electron microscope (SEM)**—instrument in which the image is formed by electrons reflected back from the specimen; most often used to study the surfaces of cells and organisms

**cell fractionation**—technique by which subcellular components are separated and purified on the basis of their physical properties

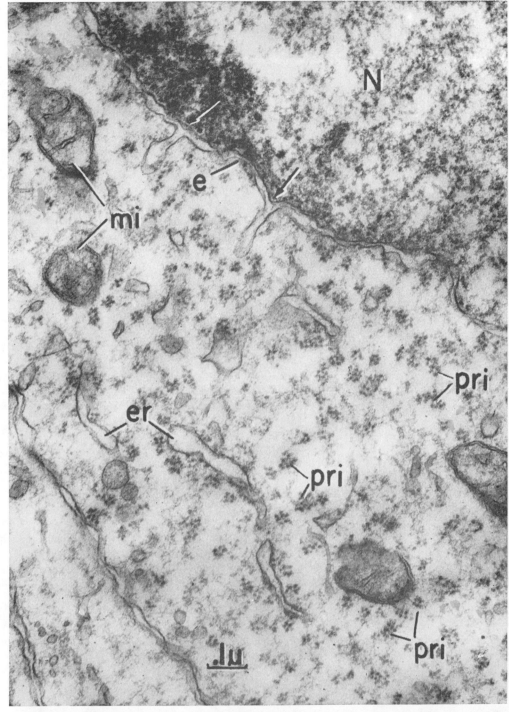

**Figure 1-4** Electron micrograph of a neuroblast of the cerebral cortex of a rat embryo, showing the cytoplasm rich in matrix with numerous ribosomes and little development of the vacuolar system. *e*, nuclear envelope sending projections into the cytoplasm (arrows); *er*, endoplasmic reticulum; *mi*, mitochondria; *N*, nucleus; *pri*, polyribosomes (groups of ribosomes). ×45,000. (From E. De Robertis.)

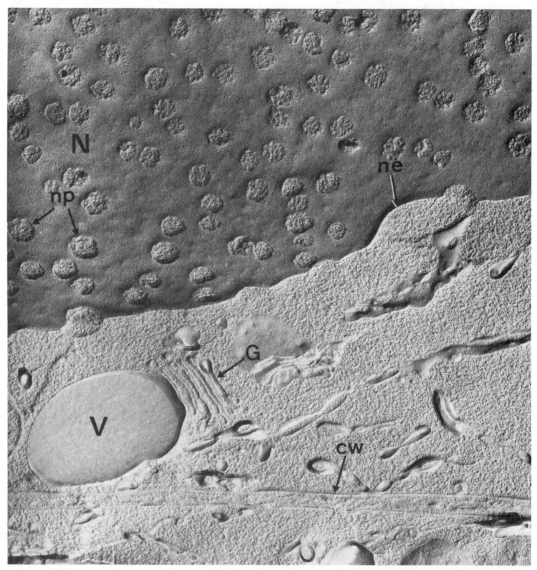

**Figure 1-5**   Onion root cell which has been submitted to the freeze-etching technique. The upper part of the figure corresponds to the nucleus *(N)* and shows the nuclear pore complexes *(np)* and the nuclear envelope *(ne)*. In the cytoplasm, a Golgi complex *(G)* and a large vacuole *(V)* are observed. *cw*, cell wall. ×75,000. (Courtesy of D. Branton.)

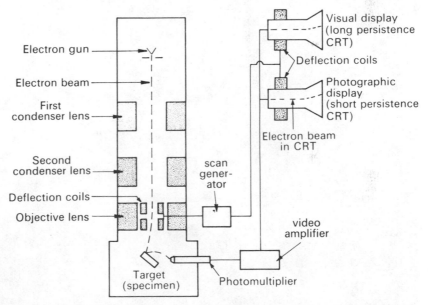

**Figure 1-6**  Diagram of the scanning electron microscope. The electron beam is swept back and forth across the target specimen. The secondary electrons that are scattered reach a photomultiplier and are channeled to cathrode ray tubes (CRT) in which the image can be observed or photographed. (From R. D. Dyson, *Essentials of Cell Biology.* Allyn and Bacon, Boston, 1978.)

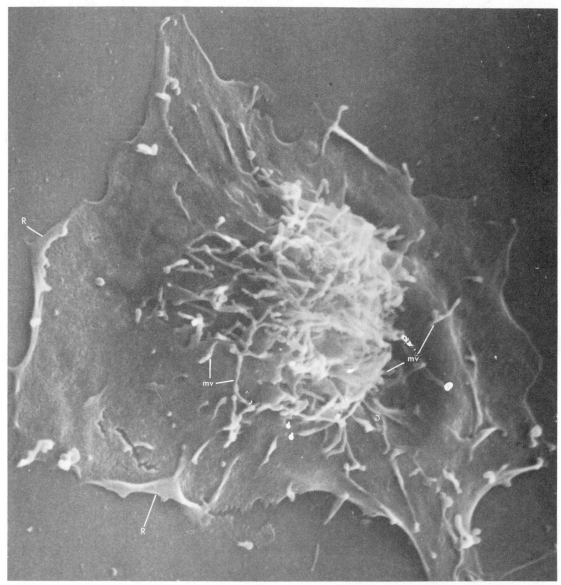

**Figure 1-7**   Scanning electron micrograph of a cultured cell that is in the process of respreading after cytokinesis. Notice the ruffling edge (R) and the numerous microvilli (mv). ×3260. (Courtesy of R. D. Goldman.)

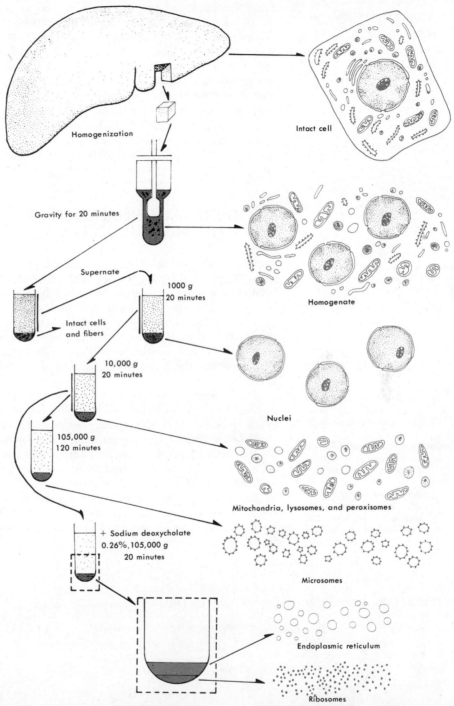

**Figure 1-8**  Diagram showing the various steps used in the technique of differential centrifugation. A piece of liver tissue is homogenized and then subjected to a series of centrifugations of increasing gravitational force, as indicated on the left side of the figure. On the right side are diagrams of the various subfractions as they appear in the electron microscope. (Modified from W. Bloom and D. W. Fawcett, *A Textbook of Histology*. 10th Ed. Saunders, Philadelphia, 1975.)

them in a homogenizer. The heavy unbroken parts are sedimented out of the mixture, and the suspension is then submitted to a series of centrifugations at different gravitational forces. This procedure separates the nuclei, mitochondria, and microsomes (e.g., sections of the endoplasmic reticulum). The final soluble portion consists of the *cytosol*, which includes some of the components of the cytoskeleton as well as smaller components such as enzymes and other soluble proteins. These four main fractions (nuclear, mitochondrial, microsomal, and soluble) can be fractionated further by the use of suitable gradients of sucrose, cesium chloride, and other media. In this way it is possible to obtain pure samples of the various cellular components for biochemical studies.

## 1–3 GENERAL ORGANIZATION OF PROKARYOTIC CELLS

At the beginning of this chapter we mentioned that life is manifested as millions of different species that have their own special morphology and contain specific genetic information. The species can be arranged into progressively more inclusive groups of organisms— genera, orders, families—up to the level of the classic kingdoms, plants and animals. One of the most recent classification schemes, that of Whittaker, postulates a division into the five kingdoms monera, protista, fungi, plantae, and animalia, with their corresponding subdivisions.

This complexity is simplified by examining living forms at the cellular level. Cells are identified as being of one of two recognizable types, prokaryotic or eukaryotic. Table 1–3 shows that only the monera (i.e., bacteria and blue-green algae) are prokaryotic cells, while all the other kingdoms consist of organisms made up of eukaryotic cells. The main difference between these two cell types is that prokaryotic cells (Gr., *karyon*, nucleus) lack a nuclear envelope. The prokaryotic chromosome occupies a space in the cell called a *nucleoid*, and is in direct contact with the rest of the protoplasm. Eukaryotic cells have a *true nucleus* with an elaborate nuclear envelope, through which the nucleocytoplasmic interchanges take place.

**prokaryote**—organism (usually unicellular) that lacks a true nucleus

**eukaryote**—organism (uni- or multicellular) whose cells contain a true nucleus

**nucleoid**—nuclear region within a prokaryotic cell, visibly distinct from the cytoplasm but not isolated by a membrane

---

### TABLE 1–3 CLASSIFICATION OF LIVING ORGANISMS AND CELLS

| Kingdom | Monera | Protista | Fungi | Plantae | Animalia |
|---|---|---|---|---|---|
| Representative organisms | bacteria blue-green algae | protozoa chrysophytes | slime molds true fungi | green algae red algae brown algae bryophytes tracheophytes | metazoa |
| Cell classification | prokaryotes | ←———————— eukaryotes ————————→ | | | |

Table 1-4 compares the structural organization in prokaryotes and eukaryotes, illustrating both the differences and similarities that exist between these two cell types.

From an evolutionary viewpoint, prokaryotes are considered to be ancestors of eukaryotes. Fossils three billion years old contain evidence of prokaryotes alone, whereas eukaryotes probably appeared one billion years ago. In spite of the differences between prokaryotes and eukaryotes, there are considerable homologies in their molecular organization and function. We shall see, for example, that all living organisms employ the same genetic code and similar machinery for protein synthesis.

## Cell Organization and the Energy Cycle

The sun is the ultimate source of energy for living organisms. The energy carried by photons of light is trapped by the pigment *chlorophyll*, present in the chloroplasts of green plants, and accumulates as chemical energy within the different foodstuffs consumed by other organisms.

All cells and organisms can be grouped into another two main classes, this time based on their mechanism of extracting energy for

**TABLE 1-4   COMPARISON OF CELL ORGANIZATION IN PROKARYOTES AND EUKARYOTES**

|  | Prokaryotic Cells | Eukaryotic Cells |
|---|---|---|
|  | Bacteria, blue-green algae, and mycoplasmas | Protozoa, other algae, metaphyta, and metazoa |
| Nuclear envelope | absent | present |
| DNA | naked | combined with proteins |
| Chromosomes | single | multiple |
| Nucleolus | absent | present |
| Division | amitosis | mitosis or meiosis |
| Ribosomes | 70S (50S + 30S)* | 80S (60S + 40S)* |
| Endomembranes | absent | present |
| Mitochondria | respiratory and photosynthetic enzymes in the plasma membrane | present |
| Chloroplast | absent | present in plant cells |
| Cell wall | noncellulosic | cellulosic, only in plants |
| Exocytosis and endocytosis | absent | present |
| Locomotion | single fibril, flagellum | cilia and flagella |

*S refers to the Svedberg sedimentation unit, which is a function of the size and shape of molecules.

autotroph—organism capable of utilizing $CO_2$ as the principal carbon source for its metabolic processes

photosynthesis—process by which autotrophs transform $CO_2$ and $H_2O$ into carbohydrates with the aid of light

heterotroph—organism requiring complex nitrogenous and carbon compounds, including those produced by autotrophs, for its metabolic processes

aerobic respiration—process requiring $O_2$, by which heterotrophs generate energy from ingested organic molecules

their own metabolism. Those of the first class, called *autotrophs* (e.g., green plants), use the process of *photosynthesis* to transform $CO_2$ and $H_2O$ into the elementary organic molecule glucose, from which more complex molecules are produced. The second class of cells, called *heterotrophs* (e.g., animal cells), obtains energy from the various carbohydrates, fats, and proteins synthesized by autotrophic organisms. The energy contained in these organic molecules is released primarily by combustion with $O_2$ from the atmosphere (i.e., oxidation) in a process called *aerobic respiration*. The release of $H_2O$ and $CO_2$ by heterotrophic organisms completes this energy cycle (Fig. 1–9).

These interrelated energy cycles have been maintained throughout evolution. Within the prokaryotic monera there are some species that are autotrophs and others that are heterotrophs. For example, photosynthetic bacteria and blue-green algae are autotrophs. Heterotrophic bacteria absorb soluble nutrients from the medium. Fungi and animals are heterotrophs, while all plants (with the exception of a few species) are autotrophs.

## *Escherichia coli (E. coli):* The Most Studied Prokaryote

*Escherichia coli (E. coli)*— easily cultured prokaryotic bacterial organism used frequently for in vitro biological studies

Although this book deals mainly with the more complex eukaryotic cells, it is important to know that much of our present knowledge of molecular biology stems from the study of viruses and bacteria. A bacterial cell such as *Escherichia coli (E. coli)* is easily cultured in an aqueous solution containing glucose and various inorganic ions. In this medium, at 37°C, the cell mass doubles and divides in about 60 minutes. This time—the *generation time*—can be reduced to 20 minutes if purines and pyrimidines (the precursors of nucleic acids) as well as amino acids are added to the medium.

As shown in Figure 1–10, one cell of *E. coli* is about 2 $\mu$m long and 0.8 $\mu$m in diameter. It is surrounded by a rigid *cell wall*, 10 nm or more thick, containing protein, polysaccharide, and lipid molecules. Inside the cell wall is the *plasma membrane*, a lipoprotein structure that serves as a molecular barrier to the surrounding medium. This plasma membrane, by controlling the entrance and exit of small molecules and ions, contributes to the establishment of a carefully regulated internal milieu for the protoplasm of the bacterium. It is interesting that the enzymes involved in the oxidation of metabolites, and which constitute the *respiratory chain*, are associated with this plasma membrane. (In eukaryotic cells these enzymes are confined to specialized organelles in the cytoplasm, the mitochondria.)

The bacterial chromosome is a single circular molecule of naked DNA tightly coiled within the nucleoid, which appears in the electron microscope as a lighter region of the protoplasm. It is important to remember that the DNA of *E. coli*, which is about 1 mm long ($10^6$ nm) when uncoiled, contains all the genetic information of the organism. In this case there is sufficient information to code for 2000 to 3000 different proteins.

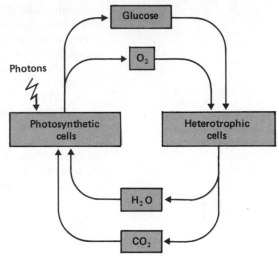

**Figure 1-9**  Simple diagram of the energy cycle and of the interaction between photosynthetic and heterotrophic cells. (After A. L. Lehninger.)

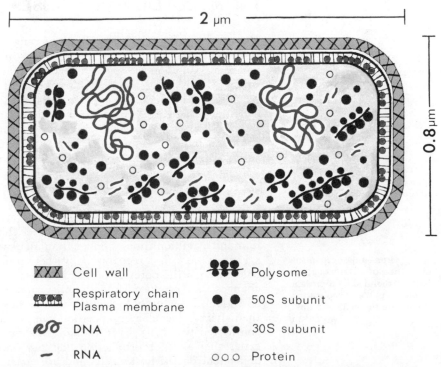

**Figure 1-10**  Diagram of a single cell of the bacterium *Escherichia coli* containing two chromosomes 1 mm long ($10^6$ nm) attached to the cell membrane. 50S and 30S refer to ribosomal subunits.

Figure 1–10 shows two chromosomes because replication of the DNA has occurred and the cell is ready to divide. Notice the attachment of the DNA to the plasma membrane. This anchoring may assist in the separation of the two nucleoids after DNA replication. In fact, the growth of the intervening membrane may accomplish this separation.

In addition to a chromosome, certain bacteria contain a small extrachromosomal circular DNA called a *plasmid*. A plasmid may confer resistance to one or more antibiotics upon the bacterial cell. As will be mentioned in the chapters on molecular biology, plasmids can be separated and reincorporated; genes (specific pieces of DNA) can be inserted into plasmids, which are then transplanted into bacteria using the techniques of *genetic engineering* (see Chapter 11).

Surrounding the DNA in the dark region of the protoplasm are 20,000 to 30,000 particles, about 25 nm in diameter, called *ribosomes*. These particles are composed of *ribonucleic acid (RNA)* and proteins, and are the sites of protein synthesis. Ribosomes consist of a large and a small subunit and exist in groups called *polyribosomes* or *polysomes*. The remainder of the cell is filled with water, various RNAs, protein molecules (including enzymes), and various smaller molecules.

## 1–4 MYCOPLASMAS, VIRUSES, AND VIROIDS

Most prokaryotic cells are small, in the range of 1 to 10 $\mu$m, although some blue-green algae may reach 60 $\mu$m in diameter. From what has been said about *E. coli*, it is evident that there must be a minimum size limit for a cell. It must be large enough (1) to have a plasma membrane, (2) to contain sufficient genetic material to encode the various RNAs involved in protein synthesis, and (3) to contain the biosynthetic machinery that performs this synthesis.

**mycoplasma—simplest known cellular organism, intermediate in size between largest viruses and smallest bacteria**

Among living organisms that have the smallest mass, the best suited for study are small bacteria called *mycoplasmas* (see Fig. 1–3), which produce infectious diseases in animals including humans and which can be cultured in vitro like any bacteria. These agents range in diameter from 0.1 to 0.25 $\mu$m; thus they correspond in size to some of the large viruses. These microbes are of biological interest because each is a thousand times smaller than the average bacterium and a million times smaller than a eukaryotic cell.

**virus—microorganism consisting of a DNA or RNA core surrounded by a protein coat; capable of replicating only within the cells of a host organism**

*Viruses* were first recognized by their property of being able to pass through the pores of porcelain filters and by the pathological changes they produced in cells. All viruses can now be visualized and identified morphologically with the electron microscope, and their macromolecular organization can be studied. Viruses are not considered true cells. Even though they share some cellular properties, such as autoreproduction, heredity, and mutation, viruses are dependent on the host's cells and are considered obligatory parasites.

Outside a living host cell viruses are inactive and may even be crystallized. Upon their introduction into the cell they become active and reproduce. From the point of view of their genetic constitution there are two kinds of viruses: those in which the chromosome is a molecule of RNA, such as tobacco mosaic virus, and those in which the chromosome is a DNA molecule, such as the bacterial viruses or bacteriophages. Viruses use their own genetic program for reproduction but rely on the biosynthetic machinery of the host (i.e., ribosomes, transfer RNA, enzymes) to produce the proteins of the coat, also called the *capsid*.

**bacteriophage—virus that attacks bacterial cells**

**caspid—protein coat surrounding the nucleic acid core of a virus**

Viruses range in size between 30 and 300 nm, and their structures show different degrees of complexity. They are produced within a cell by a process of *macromolecular assembly*, which signifies that their components may be synthesized in different parts of the host cell and then assembled in a coordinated manner. Tobacco mosaic virus (TMV), for example, is a cylindrical particle measuring approximately 16 × 300 nm. This cylinder contains a single-stranded helical RNA molecule 6500 nucleotides in length. Associated with this RNA helix and forming a protein coat are 2130 identical protein subunits (Fig. 1–11). It has been discovered that it is possible to dissociate the RNA and the protein subunits and reassociate them later in order to reconstitute active virus particles. The RNA molecule appears to influence the assembly of the protein subunits.

**Figure 1-11** Diagram of the molecular organization of the tobacco mosaic virus. In the center there is a spiral of RNA which is associated with protein subunits. There is one protein monomer for every three bases in the RNA chain. (From D. L. Caspar and A. Klug, *Cold Spring Harbor Symp. Quant. Biol.*, 27:1, 1962.)

0                                        10 nm

icosahedral symmetry—
characteristic polyhedral
symmetry of the viral capsid
giving the most energetically
stable configuration of its
protein subunits

It has been observed that many viruses display *icosahedral symmetry*. According to Caspar and Klug, this symmetry is due to the fact that the assembly of the protein subunits or *capsomeres* in this configuration enables the capsid of the virus to exist at a state of minimum energy. Such icosahedral symmetry has been found in a virus as small as $\phi$X174, which has only 12 capsomeres, and in one as large as adenovirus, which may have as many as 252 capsomeres.

The bacteriophages are viruses which infect bacterial cells by injecting their own DNA. An interesting example is the small bacteriophage $\phi$29 of *Bacillus subtilis*, which has the elaborate macromolecular structure shown in Figure 1–12.

viroid—virus-like particle
consisting of a single RNA
molecule without a protein
coat

*Viroids* are even simpler organisms than viruses. They are infectious agents that attack plant cells, and consist of a single RNA molecule that is not covered by a capsid of protein. Figure 1–13 shows a viroid that produces a disease in potatoes. Its circular RNA molecule consists of 359 nucleotides, and is able to multiply and infect other plants.

To conclude this short discussion about viruses let us compare them with true cells. We have defined a living cell as having the following characteristics: (1) A *specific genetic program* that permits the reproduction of new cells of the same type. (2) A *cell membrane* that establishes a boundary regulating all exchanges of matter and energy. (3) A *metabolic machinery* that can use energy trapped by the cell or obtained from foodstuffs. (4) A *biosynthetic machinery* for the synthesis of proteins. Viruses have only the first of these characteristics and lack all the others. For this reason they are not generally considered to be living organisms, despite the fact that they do contain a set of genetic blueprints from which a new virus can be made.

## 1–5  GENERAL ORGANIZATION OF EUKARYOTIC CELLS

Having studied the organization of prokaryotic cells it is useful to have another look at Table 1–4, in which the main points of comparison with eukaryotic cells are summarized. When comparing the organization of *E. coli* (Fig. 1–10) with that of a plant cell (Fig. 1–1) or an animal cell (Fig. 1–14), one is struck by the relative complexity of the eukaryotes. In a nondividing eukaryotic cell the *nucleus* exists as a separate compartment surrounded and limited by the nuclear envelope. Another and generally larger compartment is represented by the *cytoplasm*, and finally there is the *cell membrane* with its multiple infoldings and differentiations. Each of these three main components or compartments of the cell contains several subcomponents or subcompartments. Table 1–5 can be used as a guide to this complex organization since it lists the main morphological features of a eukaryotic cell and indicates the main functions of each. It will become apparent that this is an oversimplification when each of these components is discussed individually in later chapters of this book.

nucleus—membrane-enclosed
cell organelle containing the
genetic material and various
proteins involved in its
replication and transcription

cytoplasm—protoplasmic
material of the cell not
including the nucleus and its
contents

*Text continued on page 24.*

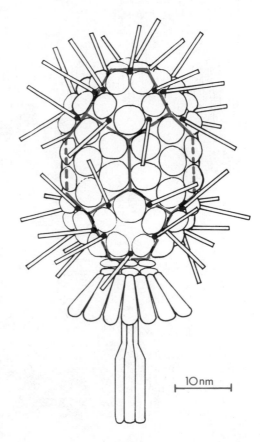

10nm

**Figure 1-12**   Diagram of the small bacteriophage φ29 of *B. subtilis* having a total of 172 protein molecules of which 145 integrate the capsid. (Courtesy of C. Vásquez.)

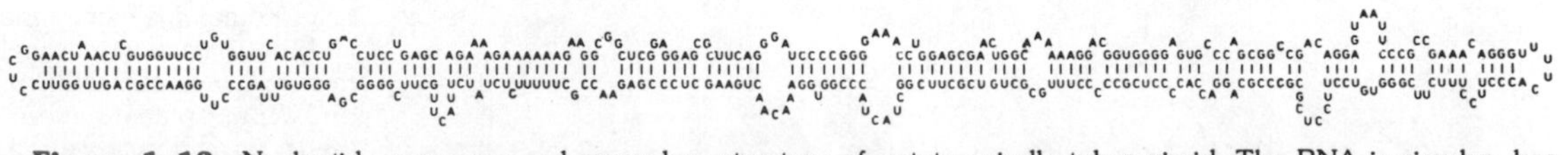

**Figure 1-13**   Nucleotide sequence and secondary structure of potato spindle tuber viroid. The RNA is circular, has only 359 nucleotides, and does not contain AUG codons (the signal for the start of protein synthesis). Viroids are the simplest infectious agents known and are not covered with protein. (From J. J. Gross et al., *Nature, 273*:203, 1978.)

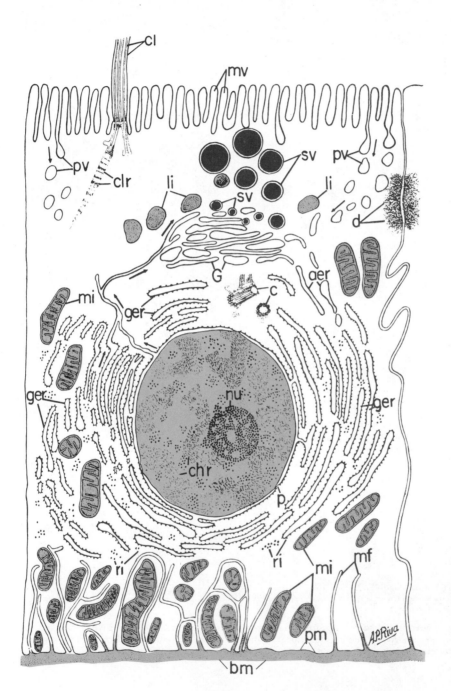

**Figure 1-14** General diagram of the ultrastructure of an idealized animal cell. *aer,* agranular endoplasmic reticulum; *bm,* basal membrane; *c,* centriole; *chr,* chromosome; *cl,* cilium; *clr,* cilium root; *d,* desmosome; *G,* Golgi complex; *ger,* granular endoplasmic reticulum; *li,* lysosome; *mf,* membrane fold; *mi,* mitochondria; *mv,* microvilli; *nu,* nucleolus; *p,* pore; *pm,* plasma membrane; *pv,* pinocytic vesicle; *ri,* ribosome; *sv,* secretion vesicle. (From E. De Robertis and A. Pellegrino de Iraldi.)

**TABLE 1–5  GENERAL ORGANIZATION OF THE EUKARYOTIC CELL**

| Main Components or Compartments | Subcomponents or Subcompartments | Main Function |
|---|---|---|
| Cell membrane | cell wall<br>cell coat<br>plasma membrane | protection<br>cell interactions<br>permeability, endo- and exocytosis |
| Nucleus | chromatin and chromosomes<br><br>nucleolus<br><br>nucleoplasm | genetic information system<br>synthesis of ribosomes |
| Cytoplasm<br>  Matrix, cytosol<br>  Cytoskeleton | soluble enzymes<br>microfilaments<br>microtubules<br><br>ribosomes | glycolysis<br>cell motility<br>cell shape and motility<br>protein synthesis |
| Endomembrane system | nuclear envelope<br>endoplasmic reticulum, rough and smooth<br><br>Golgi complex | nuclear permeability<br>synthesis and transport of materials, secretion<br>secretion |
| Membrane organelles | mitochondria<br>chloroplasts<br>lysosomes<br>peroxisomes | cell respiration<br>photosynthesis<br>digestion<br>peroxidation |
| Microtubular organelles | centrioles and spindle<br>basal bodies, cilia, and flagella | cell division<br>cell motility |

## Morphological Diversity of Eukaryotic Cells

The cells of a multicellular organism vary in shape and structure and are differentiated according to their specific function in the various tissues and organs. This functional specialization causes cells to acquire special characteristics, but a common pattern of organization persists in all cells.

Some cells, such as amebae and leukocytes, change their shape frequently. Others, such as nerve cells and most plant cells, always have a typical shape which is more or less fixed and specific for each cell type. The shape of a cell depends mainly on its functional adaptations and partly on the surface tension and viscosity of the protoplasm, the mechanical action exerted by adjoining cells, and the rigidity of the cell membrane. The orientation of microtubules within the cytosol also has a great influence on the shape of many cells (see Chapter 4).

The size of different cells ranges within broad limits. Some plant and animal cells are visible to the naked eye. Most cells, however, are visible only with a microscope, since they are only a few micrometers in diameter (Fig. 1–3). The smallest animal cells have a diameter of 4 $\mu$m.

In general, the volume of a cell is fairly constant for a particular cell type and is independent of the size of the organism. For example, kidney or liver cells are about the same size in the bull, horse, and mouse; the difference in the total mass of the organ depends on the number, not the volume, of cells.

## The Cell Membrane

plasma membrane—lipoprotein structure serving as a selective barrier between the cell's cytoplasm and its immediate environment

The structure that separates the cell content from the external environment is the *plasma membrane*. This is a thin film (6–10 nm thick) consisting of a continuous *lipid bilayer* with proteins intercalated in or adherent to both surfaces. The plasma membrane can be resolved only with the electron microscope, which reveals its numerous infoldings and differentiations as well as the different types of junctions that establish connections with neighboring cells (Fig. 1–9). The main function of the plasma membrane is to control selectively the entrance and exit of materials. This includes the entrance of water and large molecules by the process of *endocytosis* and the exit of cell products by *exocytosis* (Table 1–5). The plasma membrane is covered and reinforced by the *cell wall* in plant cells and by the *cell coat* in animal cells.

cell wall—rigid exoskeletal structure enclosing and protecting the contents of most plant and some bacterial cells

The *cell wall* is responsible for the rigidity of most plant tissues. This structure consists mainly of cellulose fibers into which other substances (such as lignin, the main component of timber) may be incorporated. As shown in Figure 1–1, there are tunnels running through the cell wall called *plasmodesmata*, which allow communication with the other cells in a tissue.

In most animal cells the plasma membrane is covered by a *cell coat* made of glycoproteins, glycolipids, and polysaccharides that may extend the thickness of the cell membrane and continue far beyond it. The cell coat has several functions in addition to that of protection. It is involved in molecular recognition between cells, contains enzymes and antigens, and is fundamental in the association of cells in a tissue (see Chapter 3).

## The Nucleus and the Cell Cycle

The shape of the nucleus is sometimes related to that of the cell, but it may be completely irregular. In spheroidal, cuboidal, or polyhedral cells, the nucleus is generally a spheroid. In cylindrical, prismatic, or fusiform cells, it tends to be an ellipsoid.

By 1905 Boveri had already noted that, in sea urchin larvae, the size of the nucleus was proportional to the chromosome number. In oocytes the nucleus (often called the *germinal vesicle*) is very active and may attain a large volume. In general it may be said that each somatic nucleus has a specific size that depends partly on its DNA content and mainly on its protein content, and that size is related to functional activity during the period of nondivision.

Almost all cells are *mononucleate*, but *binucleate* cells (some liver and cartilage cells) and *polynucleate* cells also exist. In the *syncytia*, which are large protoplasmic masses not divided into cellular territories, the nuclei may be extremely numerous. Such is the case with striated muscle fibers and certain algae, which may contain several hundred nuclei.

In general, every cell has two major periods in its life cycle: *interphase* (nondivision) and *division* (which produces two daughter cells). This cycle is repeated at each cell generation, but the length of the cycle varies considerably in different types of cells. The essential function of the nucleus is to store and make available to the cell the information present in its DNA molecule(s). This molecule duplicates during a special period of interphase called the *S phase* (or synthetic phase) in preparation for cell division (Fig. 1–15). During interphase the genetic information is also *transcribed* into different RNA molecules (messenger, ribosomal, and transfer RNAs) which, after passing into the cytoplasm, will *translate* the genetic information by facilitating the synthesis of specific proteins. Thus, the essential roles of the nucleus are (1) the storage of genetic information, (2) DNA duplication, and (3) transcription.

**syncytium—large protoplasmic mass not divided into distinct cells**

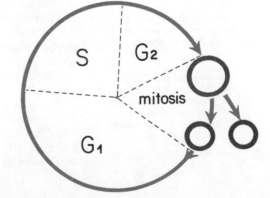

**Figure 1-15**  Diagram of the life cycle of a cell indicating the mitotic and the interphase periods. Interphase consists of the $G_1$, S, and $G_2$ phases. Duplication of DNA takes place during the synthetic or S phase.

In fixed and stained material the structure of the nucleus is distinguished by its complexity. The following structures are generally recognized in the interphase nucleus (Fig. 1–16): (1) A *nuclear envelope* composed of two membranes and perforated at intervals by the *nuclear pores*. (2) The *nucleoplasm* (or *nuclear sap*) that fills most of the nuclear space. This material represents regions of uncondensed *chromatin* (a substance consisting mainly of DNA and protein), the dispersed form assumed by the chromosomes in the nondividing cell. These regions correspond to the *euchromatin* (Gr., *eu*, true). (3) The *chromocenters* that (along with the twisted filaments of chromatin) represent parts of the chromosomes which remain condensed at interphase. These condensed regions of *heterochromatin* are frequently found near the nuclear envelope and are also attached to the *nucleolus*. (4) The *nucleoli*, which are generally spheroidal and very active in protein synthesis. The nucleoli are either single or multiple, and their role is to assemble the RNA molecules and numerous proteins that make up the ribosome before this organelle passes into the cytoplasm.

During cell division the nucleus undergoes a series of complex but remarkably regular and constant changes in which the nuclear envelope and the nucleolus disappear and the chromatin becomes condensed into dark-staining bodies, the *chromosomes* (Gr., *chroma*, color + *soma*, body). Chromosomes are always present in the nucleus even though they are not usually visible during interphase. A considerable amount is known about the structure of the chromatin fiber. Within the nucleus the DNA molecule is associated with basic proteins—the *histones*—to form very fine granular structures of about 8.5 nm, called *nucleosomes* (or nuclear bodies). Strings of nucleosomes constitute the thicker chromatin fiber, which has a mean diameter of 20 nm (Fig. 1–17). Prior to cell division, as mentioned above, the chromatin fibers become coiled more tightly and can be seen under the light microscope as chromosomes (Fig. 1–18).

## The Ultrastructure of the Cytoplasm

The cytoplasmic compartment of the cell has a very complex structural organization. Examination of this compartment in the electron microscope reveals a strikingly prodigious network of membranes. The *endomembrane system* pervades the ground cytoplasm, dividing it into numerous sections and subsections. This system is so polymorphous that it is difficult to describe and categorize. The cytoplasm is generally considered to have two parts, one contained within this membrane system and the other, the *cytoplasmic matrix* proper, remaining outside.

The most important constituents of the cytoplasm are in the matrix (ground cytoplasm) which lies outside the endomembrane system. This matrix constitutes the true internal milieu of the cell and contains all the principal structures involved in cell shape and movement, protein synthesis, and metabolic activity.

---

nuclear envelope—flattened sac or double membrane surrounding the nucleoplasm and genetic material

nucleolus—clearly defined structure within the nucleus, consisting of chromatin and large amounts of RNA; site of ribosome formation and rRNA synthesis

chromosome—extremely compact linear molecule consisting of chromatin strands made up of DNA, RNA, and protein

endomembrane system—polymorphous network of channels within the cell cytoplasm

cytoplasmic matrix—that portion of the cell contents partitioned by but not contained within the endomembrane system

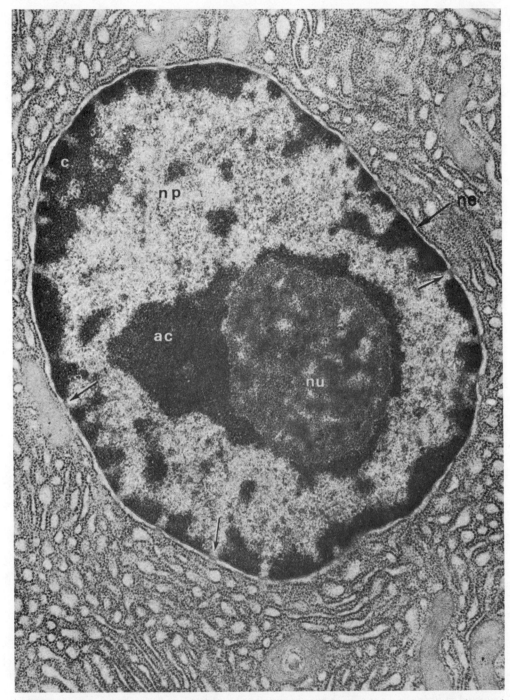

**Figure 1-16**  Electron micrograph of the nucleus from a mouse pancreatic cell. Staining with uranyl acetate enhances mainly the DNA-containing parts of the cell. Arrows in the interchromatin channels point to nuclear pores; c, chromatin; ac, chromatin associated with the nucleolus *(nu); np,* nucleoplasm; *ne,* nuclear envelope. ×24,000. (Courtesy of J. André.)

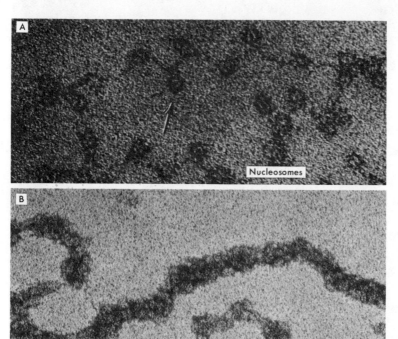

Nucleosomes

20 to 30 nm fiber

**Figure 1-17** Chromatin from chicken erythrocyte nuclei, negative staining. **A,** the chromatin has stretched during spreading, and the nucleosomes can be seen as beads on a string of DNA. ×500,000. **B,** the structure of the 20 to 30 nm fiber has been preserved, and the nucleosomes can be visualized closely packed within it. ×250,000. (Courtesy of A. L. Olins.)

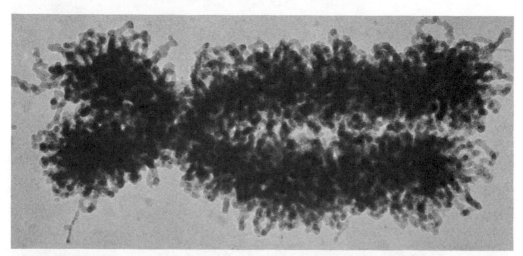

**Figure 1-18** Electron micrograph of a whole-mounted chromosome 12 showing the two chromatids composed of fibrils 30 nm thick. The two chromatids are joined only at the centromere. (Courtesy of E. J. DuPraw.)

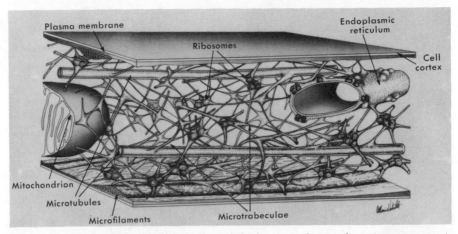

**Figure 1-19**   An idealized model of cell cytoskeletal structure showing the various components contained in the cytoplasmic matrix. Below the plasma membrane there are bundles of actin microfilaments. A microtrabecular lattice pervades the cytosol and is in contact with the endoplasmic reticulum, the microtubules, and ribosomes. ×90,000. [Courtesy of K. R. Porter. From K. R. Porter, *In* Cell Motility, Vol. 1, p. 1. (Goldman, R. D., et al., eds.) Cold Spring Harbor Laboratory, Cold Spring Harbor, New York, 1976.]

## THE CYTOSKELETON: MICROTUBULES AND MICROFILAMENTS.

The microtubules and a variety of microfilaments and microtrabeculae present in the cytoplasmic matrix constitute a kind of dynamic and spongy cytoskeleton, providing a framework for soluble proteins, enzymes, and ribosomes (Fig. 1–19). This cytoskeleton is involved in the maintenance of cell shape, and it has a role in cell mobility and the colloidal changes that the cytoplasm may undergo.

*Microtubules* are thin, rather rigid tubular structures about 25 nm in diameter, the walls of which are made of 13 individual filaments. The main protein comprising microtubules is *tubulin*. Cytoplasmic microtubules can change rapidly by a process of polymerization or depolymerization of the tubulin subunits (Fig. 1–20). Microtubules facilitate changes of cell shape as well as the displacement of macromolecules and organelles throughout the cytoplasm. Microtubules can produce dynamic structures related to cell division, such as asters and the spindle, and even more complex organelles such as centrioles, basal bodies, cilia, and flagella.

*Microfilaments* are among the smallest structures observed with the electron microscope. There are two main kinds of filaments, thin ones (4–6 nm) composed mainly of the protein *actin*, and thicker ones (10 nm) about whose composition less is known, although their role is mainly mechanical. The thin microfilaments, in addition to the contractile protein actin, also contain *myosin* and other proteins related to the process of contraction. It is now recognized that essentially all the manifestations of cell motility in nonmuscle cells involve the interaction of actin and myosin in microfilaments.

## THE ENDOMEMBRANE SYSTEM: NUCLEAR ENVELOPE, ENDOPLASMIC RETICULUM, GOLGI COMPLEX.

Figure 1–14 illustrates the possible continuities and functional interconnections of these different portions of the cytoplasmic membrane system.

cytoskeleton—complex cytoplasmic network of microtubules and microfilaments located predominantly near the cell membrane and thought to be responsible for most aspects of cell shape and movement

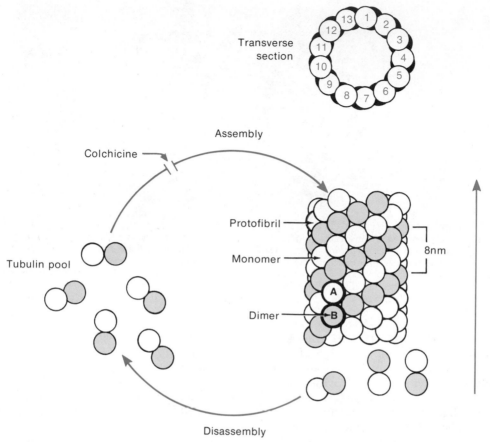

**Figure 1-20** Diagram of a microtubule and the process of assembly and disassembly of tubulin. **Top,** transverse section of a microtubule showing 13 protofibrils. **Bottom,** longitudinal view showing the helicoidal protofibrils made of tubulin dimers and the polarization of the microtubule. Observe that the microtubule is being disassembled at the bottom while being simultaneously assembled at the top. Colchicine, by blocking the assembly process, produces depolymerization of the microtubules.

endoplasmic reticulum (ER)—system of membrane-enclosed cytoplasmic channels involved in cellular transport processes; rough endoplasmic reticulum (rER) can be distinguished from smooth (sER) by the presence of ribosomes on its membrane surface

Golgi complex—cytoplasmic membrane system consisting of stacks of flattened sacs and other vesicles; involved in secretion processes

The *nuclear envelope* is made of flattened sacs or *cisternae* which consist of two membranes. These merge at the *pores,* openings which allow the transfer of materials between the nucleus and the cytoplasm. The inner membrane is in contact with the chromatin fibers, while the outer membrane is covered with ribosomes (Fig. 1–16).

The *endoplasmic reticulum (ER)* constitutes the bulk of the endomembrane system (Fig. 1–21). It is made of tubules and flattened sacs. The outer surface of the *rough* or *granular* ER is covered with ribosomes which synthesize protein molecules that are delivered into the lumen, or cavity, of the reticulum. The *smooth* or *agranular* ER is in continuity with the rough and is engaged, along with the rough, in the transport of products within their cavities. Both parts of the ER are involved primarily in secretion.

The *Golgi complex* is a differentiated part of the endomembrane system. It is made up of *dictyosomes,* which are stacks of flattened sacs and vesicles (Fig. 1–21). This complex is involved in the pro-

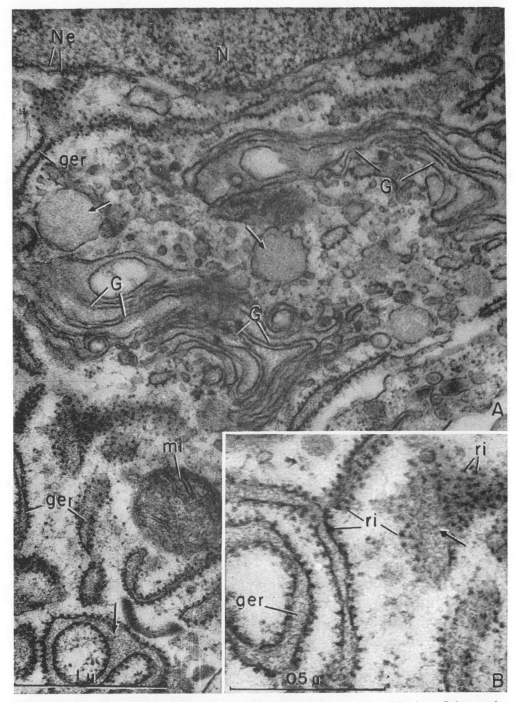

**Figure 1-21**   Electron micrograph of a plasma cell showing near the nucleus *(N)* a large Golgi complex *(G)* formed of flat cisternae and small and large vesicles. Some of the large vesicles (arrows) are filled with material. Surrounding the Golgi complex is abundant granular endoplasmic reticulum *(ger)* having cisternae filled with amorphous material (arrows). *mi,* mitochondrion; *Ne,* nuclear envelope; *ri,* ribosomes. ×48,000; inset ×100,000. (From E. De Robertis and A. Pellegrino de Iraldi.)

cessing and packaging of secretory products that come through the ER and, as secretory vesicles or granules, are released from the cell by exocytosis. The ER and the Golgi complex are also involved in the formation of lysosomes and peroxisomes.

**MEMBRANE ORGANELLES: MITOCHONDRIA, CHLORO-PLASTS, LYSOSOMES, PEROXISOMES.** As indicated in Table 1–4, the cytoplasm contains a variety of organelles (i.e., rather permanent structures with definite functions), several of which are membrane-bound.

*Mitochondria* are present in nearly all eukaryotic cells in the form of cylindrical structures less than 1 $\mu$m in diameter. They consist of a double membrane. The inner membrane is folded into the *cristae* or *crests* (Fig. 1–22). These crests partially subdivide the inner chamber of the mitochondrion, and their inner surface is covered with mushroom-like projections that are related to phosphorylation, a portion of the process of cell respiration. Within the *matrix* of the mitochondria are numerous soluble enzymes that are involved in the *Krebs cycle*, the first step in the aerobic utilization of energy from nutrients.

Plant cells contain a variety of organelles, generally called *plastids*, which are not present in animal cells. Some of them, such as the leukoplasts, are colorless and participate mainly in the storage and metabolism of starch. Other plastids contain various pigments and are collectively called *chromoplasts*. The most important of these are the *chloroplasts*, which contain the green pigment *chlorophyll* (Fig. 1–1). The chloroplast has a double outer membrane, a *stroma* filled with many soluble enzymes, and a complex system of membrane-bound compartments. Chloroplasts are the site of photosynthesis, by which a plant is able to capture light energy and, using $H_2O$ and $CO_2$, to synthesize a variety of compounds that are stored and eventually used as foodstuffs.

Both mitochondria and chloroplasts contain a kind of semiautonomous genetic system with its own DNA. They also contain ribosomes and RNA molecules, and are able to synthesize some of their own proteins.

*Lysosomes* are polymorphous organelles enclosed by a single membrane. They contain a vast array of hydrolytic enzymes (Fig. 1–22). These enzymes are responsible for the digestion of foreign substances that have been incorporated into the cell by endocytosis, and also for the digestion of parts of the cytoplasm. Lysosomes originate from the ER and the Golgi complex.

*Peroxisomes* are also bound by a single membrane. They contain enzymes related to the production and breakdown of peroxides ($H_2O_2$), and they fulfill a protective function in the cell.

**MICROTUBULAR ORGANELLES: CENTRIOLES AND BASAL BODIES, CILIA AND FLAGELLA.** Microtubules are the main components of the *asters* and the *spindle*, which integrate the mitotic apparatus during cell division. These organelles appear and disappear at the appropriate stages of mitosis (prophase and telophase, respectively; see Chapter 9).

---

**mitochondrion—membrane-bound organelle that generates chemical energy in the form of ATP for use in cellular metabolic processes**

**plastid—any of several membrane-enclosed, self-replicating cytoplasmic organelles found in plant cells and functioning in metabolic processes**

**lysosome—one of a diverse group of single-membrane organelles containing many kinds of hydrolytic enzymes; involved in intracellular digestion**

**microtubule—hollow cylindrical structure made up of longitudinal fibrils and thought to assist in intracellular movements; also component of the cytoskeleton and numerous organelles**

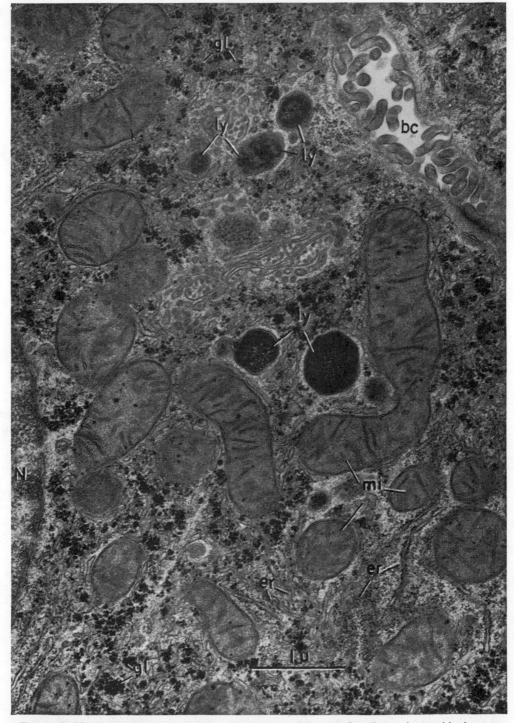

**Figure 1-22**  Peripheral region of a liver cell, showing a biliary capillary *(bc)* and several bodies interpreted as lysosomes *(ly)*. *er*, endoplasmic reticulum; *gl*, glycogen; *mi*, mitochondria; *N*, nucleus. ×31,000. (Courtesy of K. R. Porter.)

*Centrioles* are cylindrical structures of about $0.2 \times 0.5$ μm. They are open at both ends, and their walls contain nine groups of microtubular triplets arranged in a circle. Centrioles are generally double and disposed at right angles to one another. During mitosis, centrioles migrate to the poles of animal cells and are apparently involved in the formation of the spindle; plant cells lack centrioles, but the spindle is formed without their aid.

*Basal bodies* or *kinetosomes* are structurally similar to the centrioles but are located at the base of cilia and flagella. *Cilia* are short (3–10 μm) processes that extend into the surrounding medium. *Flagella* are longer than cilia but have the same diameter (0.5 μm) and structure. The axis of these organelles, the *axoneme*, is surrounded by the plasma membrane and has a characteristic microtubular structure consisting of nine doublets in a circle and two in the center. This 9 + 2 organization differs slightly from that of the centrioles and basal bodies, which have nine triplets in a circle and none in the center (see Chapter 4). Cilia are used for locomotion in isolated cells, such as certain protozoa, or to move particles in the medium, as in air passages or the oviduct. Flagella are generally used for locomotion of cells, such as the spermatozoon.

# SUMMARY

## 1-1 INTRODUCTION AND HISTORY OF CELL AND MOLECULAR BIOLOGY

Although there are about four million different species of living organisms, at the cellular and molecular levels there is a unique master plan of organization common to all. All organisms have the cell as their basic structural and functional unit, all have essentially the same biochemical machinery, and all share the same genetic code.

The boundaries between the various levels of organization are imposed artificially by the resolving power of the instruments used to study them. The limit of resolution of the human eye is about 100 μm, of the light microscope 0.2 μm, and of the electron microscope ~0.4 nm (an increase of approximately 500-fold in each of the latter two cases). The weight of cellular components is expressed in picograms, and that of molecules in daltons.

Modern cell theory states that (1) cells are the morphological and physiological units of all living organisms, (2) the properties of any organism depend on those of its individual cells, (3) cells originate from other cells, and continuity is maintained through the genetic material, and (4) the smallest unit of life is the cell.

The rapid development of cell and molecular biology in the present century can be attributed to (1) the increased resolving power provided by electron microscopy and x-ray diffraction, and (2) its convergence with other branches of biological research, such as genetics and biochemistry. Current research is focused mainly on the mechanisms that regulate gene expression.

## 1-2 MAIN TECHNIQUES USED TO STUDY CELL ORGANIZATION

The enormous resolving power of the electron microscope has provided a wealth of knowledge about cell ultrastructure and submicroscopic morphology. Samples are generally examined as either thin sections, fixed and selectively stained, or replicas, produced by freeze-fracture, freeze-etching, and the deposition of a layer of atoms on the sample surface. Transmission electron microscopy uses the electrons passing through the specimen to create an image; scanning electron microscopy uses the electrons emitted back from the surface of the specimen to create an image.

Cell fractionation methods involve the homogenization of cell boundaries by physical or chemical procedures, followed by the separation of the nuclear, mitochondrial, microsomal, and soluble fractions according to their mass and specific gravity. Series of centrifugations at different gravitational forces and in different media yield pure samples of the various cellular components for further study.

## 1–3 GENERAL ORGANIZATION OF PROKARYOTIC CELLS

Cells are identified as being of one of two types, prokaryotic or eukaryotic. The main difference between these types is that prokaryotic cells lack a nuclear envelope; their DNA occupies a space in the cell called a nucleoid. From an evolutionary standpoint, prokaryotes are considered to be ancestors of eukaryotes.

All organisms can be identified as either autotrophs or heterotrophs. Autotrophs can synthesize their own nutrients by the process of photosynthesis, using light energy from the sun to transform $CO_2$ and $H_2O$ into glucose; heterotrophs must obtain their nutrients from among the products of autotrophs, extracting energy from them by the process of aerobic respiration.

*Escherichia coli* is a common and representative prokaryote. It is easily cultured, has a short generation time, and is well suited to laboratory studies.

## 1–4 MYCOPLASMAS, VIRUSES, AND VIROIDS

Among cellular organisms having the smallest mass, the mycoplasmas (0.25–0.1 $\mu$m) are the best suited for study.

Viruses are not considered to be true cells. Even though they share such cellular properties as heredity and mutation, they are dependent on the host's cells and are regarded as obligatory parasites. Viruses may contain either DNA or RNA, and they rely on the biosynthetic machinery of the host to produce capsid proteins from their genetic information. The capsids of many viruses display icosahedral symmetry, a configuration of the protein subunits that allows them to exist at a state of minimum energy. The entire virus particle is generally produced within the cell by a process of macromolecular assembly.

Viroids are even simpler than viruses, consisting of a single RNA molecule that is not covered by a protein capsid.

## 1–5 GENERAL ORGANIZATION OF EUKARYOTIC CELLS

Eukaryotes are generally more complex than prokaryotes, consisting of three main components (which are subdivided further): the nucleus, the cytoplasm, and the cell membrane. The shape, size, and volume of a eukaryotic cell can all vary considerably, but are ultimately determined by the specific function of the cell.

The cell or plasma membrane is a continuous lipid bilayer with proteins intercalated in or adherent to both of its surfaces. Its main function is to exercise selective control over the passage of materials in and out of the cell.

The essential functions of the nucleus are the storage of genetic information; DNA duplication, which takes place during the S phase of interphase; and DNA transcription into RNA. The major structures recognized in the interphase nucleus are (1) the nuclear envelope, (2) the nucleoplasm, representing uncondensed chromatin, (3) the chromocenters, and (4) the nucleoli, which are involved in the assembly of ribosomes. At the time of cell division, the dispersed chromatin condenses and becomes visible as the chromosome(s).

The cytoplasmic matrix is pervaded by a complex and polymorphic endomembrane system which is involved in the transport of materials within the cell. The principal structures responsible for the other activities of the cell are located within the matrix.

The cell is supported and mobilized by a variety of microtubules and microfilaments made up of the proteins tubulin, actin, and myosin. These structures are generally involved in dynamic processes such as changes of cell shape, the partitioning of chromosomes at cell division, and muscle contraction.

The endomembrane system consists of the nuclear envelope, the smooth and rough endoplasmic reticulum, and the Golgi complex. This system is engaged in the sequestering, packaging, and transport of substances produced either for use within the cell or for export to the medium.

The membrane organelles are generally involved in cellular metabolic processes. Mitochondria and chloroplasts are the sites of energy production; lysosomes are responsible for the digestion of substances within the cell; and peroxisomes fulfill a similar protective function.

Microtubules are the major structural components of several cell organelles including the asters, spindle, and centriole, which are involved in mitosis, and cilia and flagella, which are used for cellular locomotion or to move particles in the surrounding medium.

# STUDY QUESTIONS

- What is the modern version of the cell theory?
- Compare the resolving power of the eye, the light microscope, and the electron microscope.

- What are the major differences between prokaryotes and eukaryotes? Autotrophs and heterotrophs?
- In what respects does a virus resemble a true cell, and in what respects does it differ?
- Describe the structure and function of the plasma membrane.
- What are the functions of the various structures found in the cell nucleus?
- What is the relationship between chromatin and chromosomes?
- Describe the distribution of the endomembrane system. What are the roles of its various parts?
- What are some of the functions of microtubules and microfilaments, and in what structures are they found?
- What is the purpose of the complex inner membrane found in mitochondria and chloroplasts?

# SUGGESTED READINGS

**Claude, A.** (1975) The coming of age of the cell. *Science, 189:*433.

**Giese, A. C.** (1979) *Cell Physiology,* 5th Ed. Saunders, Philadelphia.

**Hess, E. L.** (1970) Origins of molecular biology. *Science, 168:*664.

**Margulis, L.** (1971) Symbiosis and evolution. *Sci. Am., 225:*48.

**Novikoff, A. B., and Holtzman, E.** (1976) *Cells and Organelles,* 2nd Ed. Holt, New York.

**Schwartz, R., and Dayhoff, M.** (1978) Origins of prokaryotes, eukaryotes, mitochondria and chloroplasts. *Science, 199:*395.

**Swanson, C. P., and Webster, P.** (1977) *The Cell,* 4th Ed. Prentice-Hall, Englewood Cliffs, New Jersey.

# The Cell: Molecular Organization

The structure of the cell which is visible with optical and electron microscopes is the result of molecules arranged in a very precise order. Although there is still much to be learned, the general principles of the molecular organization of some cell structures, such as membranes, ribosomes, chromosomes, mitochondria, and chloroplasts, are beginning to emerge. The biology of cells is inseparable from that of molecules because, in the same way that cells are the building blocks of tissues and organisms, molecules are the building blocks of cells.

Cells contain 75 to 85 percent water, 10 to 20 percent protein, and 2 to 3 percent inorganic salts. Among these molecules, the organic compounds derived from the chemistry of carbon atoms stand out as the molecules of life. Numerous cell structures are made up of very large molecules called *macromolecules* or *polymers*, consisting of repeating units (called *monomers*) which are linked together by covalent bonds.

There are three important examples of polymers in living organisms. (1) *Nucleic acids* result from the repetition of four different

**polymer**—large molecule made up of repeating structural subunits; also called a macromolecule

**monomer**—single subunit constituent of a polymer

units called *nucleotides*. The linear sequence of the four nucleotides in the DNA molecule is the ultimate source of genetic information. (2) *Polysaccharides* can be polymers of glucose, forming starch, cellulose, or glycogen, or may also involve the repetition of other molecules to form more complex polysaccharides. (3) *Proteins* or *polypeptides* are made up of some 20 amino acids, present in various proportions, linked together by *peptide bonds*. The order in which these 20 monomers can be linked gives rise to an astounding number of combinations in different protein molecules, and determines not only their specificity but also their biological activity.

## 2-1 NUCLEIC ACIDS

Nucleic acids are macromolecules of the utmost biological importance. All living organisms contain nucleic acids in the form of deoxyribonucleic acid (DNA) and ribonucleic acid (RNA). Some viruses contain only DNA, while others have only RNA.

DNA is the major store of genetic information. This information is copied or *transcribed* into RNA molecules, the nucleotide sequences of which contain the "code" for specific amino acid sequences. Proteins are then synthesized in a process involving the *translation* of the RNA. The series of events just outlined is often referred to as the *central dogma* of molecular biology, and can be summarized in the form:

$$DNA \xrightarrow{\text{transcription}} RNA \xrightarrow{\text{translation}} Protein$$

In higher cells, DNA is localized mainly in the nucleus, within the chromosomes. A small amount of DNA is present in the cytoplasm and contained in mitochondria and chloroplasts. RNA is found both in the nucleus, where it is synthesized, and in the cytoplasm, where the synthesis of proteins takes place (Table 2–1).

DNA (deoxyribonucleic acid)—double-stranded helical molecule in which genetic information is encoded as a sequence of purine and pyrimidine bases, attached pairwise by hydrogen bonds and longitudinally by sugar-phosphate backbones

RNA (ribonucleic aicd)—single-stranded molecule having a nucleotide subunit structure similar to that of DNA; involved in the translation of DNA into protein

transcription—process by which an RNA molecule is polymerized on a DNA template with the aid of various enzymes

translation—process by which a protein is synthesized from amino acids according to specifications encoded in the RNA

## TABLE 2–1  DNA AND RNA: STRUCTURE, REACTIONS, AND ROLE IN THE CELL

|  | Deoxyribonucleic Acid | Ribonucleic Acid |
|---|---|---|
| Localization | primarily in nucleus; also in mitochondria and chloroplasts | in cytoplasm, nucleolus, and chromosomes |
| Pyrimidine bases | cytosine thymine | cytosine uracil |
| Purine bases | adenine guanine | adenine guanine |
| Pentose | deoxyribose | ribose |
| Cytochemical reaction | Feulgen | basophilic dyes with ribonuclease treatment |
| Hydrolyzing enzyme | deoxyribonuclease (DNAse) | ribonuclease (RNAse) |
| Role in cell | genetic information | synthesis of proteins |

# Nucleic Acids: A Pentose, Phosphate, and Four Bases

Nucleic acids consist of a sugar moiety (pentose), nitrogenous bases (purines and pyrimidines), and phosphoric acid. A nucleic acid molecule is a linear polymer in which the monomers *(nucleotides)* are linked together by means of *phosphodiester* "bridges" or bonds (Fig. 2–1). These bonds link the 3' carbon in the pentose of one nucleotide to the 5' carbon in the pentose of the adjacent nucleotide. Thus the backbone of a nucleic acid consists of alternating phosphates and pentoses. The nitrogenous bases are attached to the sugars of this backbone.

As shown in Figure 2–1, the phosphoric acid uses two of its three acid groups in the 3',5' diester links. The remaining group confers on the polynucleotide its acid properties and enables the molecule to form ionic bonds with basic proteins. (We mentioned in Chapter 1 that the DNA of eukaryotic cells is associated with basic proteins called histones, forming a nucleoprotein complex called chromatin.) This free acid group also causes nucleic acids to be highly *basophilic;* i.e., they stain readily with basic dyes.

A mild hydrolysis cleaves the nucleic acid into its component nucleotides. Figure 2–2 shows that nucleotides result from the covalent bonding of a phosphate and a heterocyclic base to the pentose.

*Pentoses* are of two types: *ribose* in RNA, and *deoxyribose* in DNA. The only difference between these two sugars is that there is one less oxygen atom in deoxyribose (Fig. 2–1). A cytochemical reaction specific for the deoxyribose moiety, called the *Feulgen reaction,* can be used to visualize DNA under the microscope.

The *bases* found in nucleic acids are also of two types: *pyrimidines* and *purines.* Pyrimidines have a single heterocyclic ring, whereas purines have two fused rings. In DNA the pyrimidines are *thymine* (T) and *cytosine* (C); the purines are *adenine* (A) and *guanine* (G) (Fig. 2–1). RNA contains *uracil* (U) instead of thymine (Table 2–1).

It is useful to remember that there are two main differences between DNA and RNA: DNA has a deoxyribose and RNA a ribose moiety; DNA contains thymine and RNA uracil. The difference in pyrimidine bases has made it possible for cell biologists to use radioactive thymidine as a specific DNA label, and radioactive uridine to label RNA.

The combination of a base plus a pentose, minus the phosphate, constitutes a *nucleoside.* For example, *adenine* is a purine base; *adenosine* (adenine + ribose) is the corresponding nucleoside; while *adenosine monophosphate* (AMP), *adenosine diphosphate* (ADP), and *adenosine triphosphate* (ATP) are nucleotides (Fig. 2–2).

In addition to functioning as the building blocks of nucleic acids, nucleotides are important because they are used to store and transfer chemical energy. Figure 2–2 shows that the two terminal phosphate bonds of ATP contain high energy. When these bonds are cleaved, the energy released can be used to drive a variety of cellular reactions. The high energy ~ P bond enables the cell to accumulate a large quantity of energy in a very small space and keep it ready for use when needed.

---

**nucleic acid**—macromolecule consisting of repeating units made up of a sugar moiety, a nitrogenous base, and phosphoric acid

**pentose**—5-carbon sugar moiety (deoxyribose in DNA, ribose in RNA)

**pyrimidine**—type of nitrogenous base found in nucleic acids, consisting of a single heterocyclic ring (cytosine, thymine, or uracil)

**purine**—type of nitrogenous base found in nucleic acids, consisting of two fused rings (adenine or guanine)

**nucleoside**—nucleotide minus the phosphoric acid, i.e., the combination of pentose moiety and nitrogenous base

**nucleotide**—monomeric unit of a nucleic acid macromolecule; also used to store and transfer energy

**ATP (adenosine triphosphate)**—nucleotide containing high energy bonds; provides energy for many biochemical cellular processes by undergoing enzymatic hydrolysis

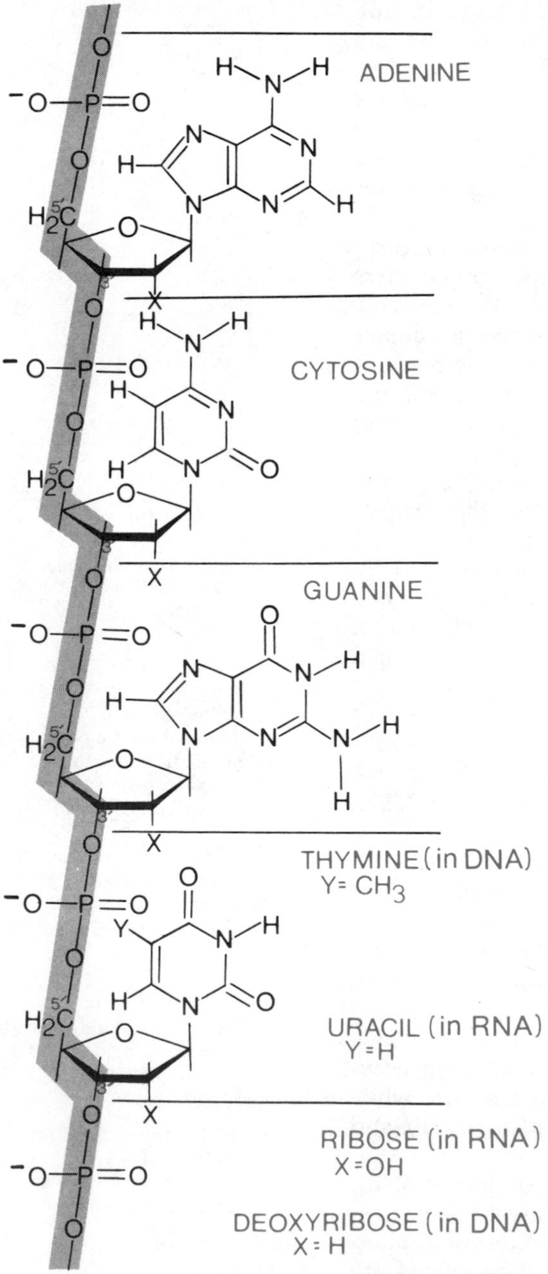

**Figure 2-1**  A segment of a single hypothetical nucleic acid chain showing the nucleotides and their constituent parts. The pentose-phosphate backbone is indicated.

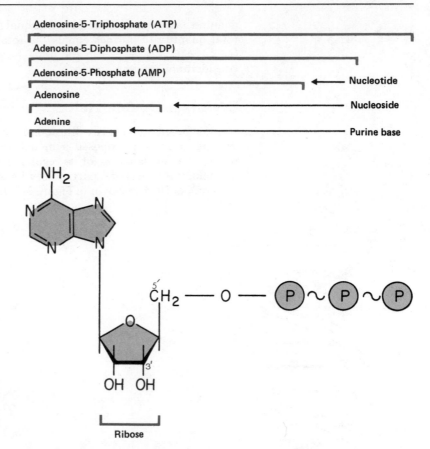

**Figure 2-2** Structure of adenosine triphosphate and its components. Note the presence of two high-energy phosphate bonds.

## DNA Base Composition: A = T and G = C

DNA is present in living organisms as linear molecules of extremely high molecular weight. *E. coli*, for example, has a single circular DNA molecule which weighs about $2.7 \times 10^9$ daltons and has a total length of 1.4 mm. In higher organisms the amount of DNA may be several thousand times larger; for example, the DNA in a single human diploid cell, if fully extended, would have a total length of 1.7 meters.

All the genetic information of a living organism is stored in its linear sequence of the four bases. Therefore, a four-letter alphabet (A, T, G, C) must code for the primary structure (i.e., the number and sequence of the 20 amino acids) of all proteins. One of the most exciting discoveries in molecular biology was the elucidation of this code (Chapter 11). One prelude to this discovery, having direct bearing on the understanding of DNA structure, was the finding that there were predictable regularities in base content. Between 1949 and 1953 Chargaff studied the base composition of DNA in great detail. He found that although the base composition varied from one species to another, in all cases the amount of adenine was equal to the amount of thymine (A = T). The number of cytosine and guanine bases was also found to be equal (C = G). Consequently, the total quantity of purines equals the total quantity of pyrimidines (i.e., A + G = C + T). On the other hand, the AT/GC ratio varies considerably between species.

## DNA Is a Double Helix

In 1953, based on the x-ray diffraction data of Wilkins and Franklin, Watson and Crick proposed a model for the DNA structure that provided an explanation for its regularities in base composition and its biological properties, particularly its duplication in the cell. The structure of DNA is shown in Figure 2–3. It is composed of two right-

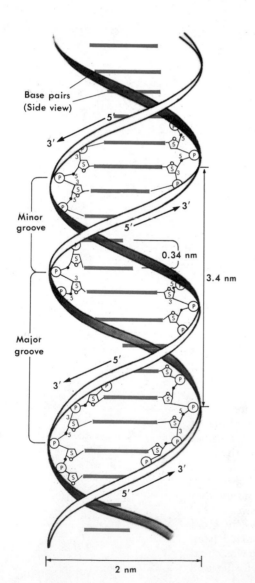

**Figure 2-3**   The DNA double helix. The phosphate-ribose backbones are indicated as ribbons. The base pairs are flat structures stacked one on top of another perpendicular to the long axis of DNA, and they are therefore represented as horizontal lines in this side view. Note that the two strands are antiparallel. The double helix gives one complete turn every ten base pairs (3.4 nm). *P*, phosphate group; *S*, sugar.

handed helical polynucleotide chains that form a *double helix* around the same central axis. The two strands are *antiparallel*, meaning that their 3′,5′ phosphodiester links run in opposite directions. The bases are stacked inside the helix in a plane perpendicular to the helical axis.

The two strands are held together by *hydrogen bonds* established between the pairs of bases. Since there is a fixed distance between the two sugar moieties in the opposite strands, only certain base pairs can fit into the structure. As shown in Figure 2–4, the only two pairs that are possible are AT and CG. It is important to note that two hydrogen bonds are formed between A and T, and three are formed betwen C and G, and that therefore a CG pair is more stable than an AT pair. In addition to hydrogen bonds, hydrophobic interactions established between the stacked bases are important in maintaining the double helical structure.

The *axial sequence* of bases along one polynucleotide chain may

**double helix**—physical configuration typically adopted by the polynucleotide chains of DNA

**antiparallel**—term describing alignment of the two DNA strands in a helix with their 3′,5′ phosphodiester linkages in opposite directions

**hydrogen bond**—weak noncovalent interaction resulting from the sharing of a proton between adjacent electronegative atoms; holds antiparallel DNA strands together by uniting AT and GC base pairs; also found in proteins

**axial sequence**—linear base sequence along one strand of the DNA molecule

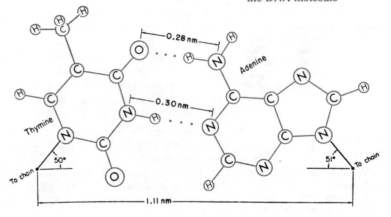

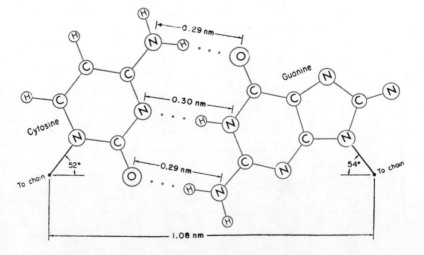

**Figure 2-4**  The two base pairs in DNA. The complementary bases are thymine and adenine (T—A) and cytosine and guanine (C—G). Observe that between T—A there are two, and between C—G three, hydrogen bonds. (From L. Pauling and R. B. Corey.)

vary considerably, but on the other chain the sequence must be *complementary*, as in the following example:

First chain:    $^{5'}$T  G  C  T  G  A  C  G  T$^{3'}$

Second chain:  $_{3'}$A  C  G  A  C  T  G  C  A$_{5'}$

Because of this property, given an order of bases on one chain, the other chain is exactly complementary. During DNA duplication the two chains dissociate and each one serves as a template for the synthesis of a new complementary chain. In this way two double-stranded DNA molecules are produced, each having exactly the same molecular constitution.

## DNA Strands Can Be Separated and Reannealed

Since the structure of the DNA double helix is preserved by weak interactions (i.e., hydrogen bonds and hydrophobic interactions established between the stacked bases), it is possible to separate the two strands by treatments involving heating, for example, or alkaline pH. This separation is called *melting* or *denaturation* of DNA. Since the temperature required to break the GC pairs (having three hydrogen bonds) is higher than that needed to break the AT pairs (having two hydrogen bonds), the temperature at which the DNA strands separate (the *melting point*) depends on the AT/GC ratio.

If the DNA is cooled slowly after denaturation, the complementary strands will base-pair in register, and the *native* (double helical) conformation will be restored. This process is called *renaturation* or *annealing*, and is a consequence of the base-pairing properties of nucleotides.

Renaturation of DNA is a very useful tool in molecular biology. Figure 2–5 shows how DNA renaturation can be used to estimate the size (number of nucleotides) of the genome of a given organism. When DNA is renatured under standardized conditions, a large genome (e.g., calf) takes more time to reanneal than a small genome (e.g., *E. coli* or bacteriophage T$_4$). This is because the individual sequences take longer to find the correct partners (the larger the genome, the more chances there are of incorrect molecular collisions).

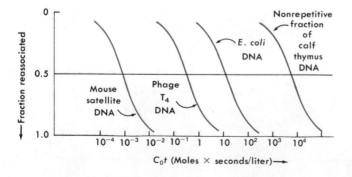

**Figure 2-5** Renaturation kinetics of various DNAs ($C_0$t curves). The velocity of renaturation is measured in the $C_0$t units required to obtain ½ renaturation. $C_0$t (short for *Concentration* × *time*) by convention is the initial concentration of DNA (in moles of nucleotides per liter) in the reaction mixture, multiplied by the time (in seconds) during which reannealing is allowed to proceed. A large genome renatures more slowly than a small one. (Redrawn from the classic paper by R. J. Britten and D. E. Kohne, *Science, 161*:530, 1968.)

Figure 2–5 also shows how renaturation studies have led to the discovery of *repeated sequences* in eukaryotic DNA. When certain DNA sequences are repeated many times, the rate of renaturation will be much faster than for sequences present as single copies. Some sequences (called *satellite* DNAs) can be repeated millions of times in the genome (see Chapter 13).

Single-stranded DNA will also anneal to complementary RNA, resulting in a hybrid molecule in which one strand is DNA and the other is RNA. *Molecular hybridization* is a very powerful method for characterizing RNAs because an RNA molecule will hybridize only to the DNA from which it was transcribed.

molecular hybridization—method of characterizing RNA molecules whereby they are allowed to base-pair with single-stranded DNA, to determine the extent and completeness of complementarity

## RNA Structure: Classes and Conformation

The primary structure of RNA is similar to that of DNA, except for the substitution of ribose for deoxyribose and uracil for thymine, as has already been discussed (see Table 2–1). The base composition of RNA does not follow Chargaff's rules, as described earlier in this section, since RNA molecules consist of only one chain.

As shown in Table 2–2, there are three major classes of ribonucleic acid: *messenger* RNA (mRNA), *transfer* RNA (tRNA), and *ribosomal* RNA (rRNA). All are involved in protein synthesis. mRNA carries the genetic information for the sequence of amino acids; tRNA identifies and transports amino acid molecules to the ribosome, and rRNA represents 50 percent of the mass of ribosomes, the organelles which provide a molecular scaffold for the chemical reactions of polypeptide assembly. The structure and function of these RNAs will be described in detail in the chapters on gene expression.

Although each RNA molecule has only a single polynucleotide chain, RNA is not a simple, smooth, linear structure. RNA molecules have extensive regions of complementarity in which hydrogen bonds between AU and GC pairs are formed between different regions of the same molecule. Figure 2–6 shows that as a result the molecule folds upon itself, forming structures called *hairpin loops*. In the base-paired regions the RNA molecule adopts a helical structure comparable to that of DNA. The compact structure of RNA molecules folded upon themselves has important biological consequences. For example, in bacteriophage MS2 a sequence that indicates the start-

messenger RNA (mRNA)—class of RNA molecules that encode the amino acid sequence of a protein in their nucleotide base sequence and serve as templates for protein synthesis

transfer RNA (tRNA)—class of RNA molecules that identify amino acids in the cytoplasm and transport them to the ribosome

ribosomal RNA (rRNA)—class of RNA molecules comprising the ribosomal subunits

**TABLE 2-2 MAJOR CLASSES OF RIBONUCLEIC ACIDS IN *E. COLI***

| Type | Sedimentation Coefficient | Molecular Weight | Number of Nucleotide Residues | Percent of Total Cell RNA |
|------|---------------------------|------------------|-------------------------------|---------------------------|
| mRNA | 6S to 25S | 25,000 to 1,000,000 | 75 to 3,000 | ~2 |
| tRNA | ~4S | 23,000 to 30,000 | 75 to 90 | 16 |
| rRNA | 5S | ~35,000 | ~100 | |
| | 16S | ~550,000 | ~1,500 | 82 |
| | 23S | ~1,100,000 | ~3,100 | |

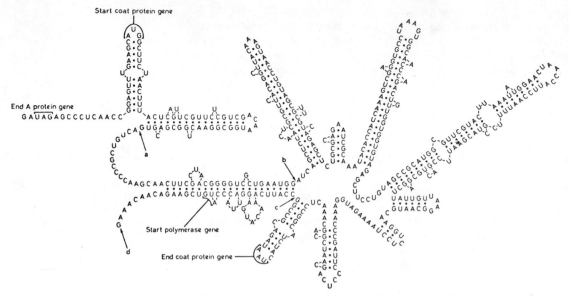

**Figure 2-6** Part of the nucleotide sequence of the RNA bacteriophage MS2, a virus that infects *E. coli*. Note that the molecule folds back on itself, forming *hairpin loops*. The starting codon (AUG) for the coat protein is readily accessible to ribosomes. The start for the polymerase protein is blocked by the secondary structure and is accessible only when the structure is opened by the translation of the coat gene by ribosomes. (From W. Fiers, et al., *Nature, 260*:500, 1976.)

ing site for protein synthesis is in a part of the molecule inaccessible to the ribosomes, and can only be expressed when additional factors unfold the molecule (see Fig. 2–6).

## 2-2 CARBOHYDRATES

carbohydrate—simple or polymerized macromolecule composed of C, H, and O and serving as an energy source or structural component in the cell; also called a saccharide (mono-, di-, or poly-, depending on the number of subunits)

*Carbohydrates*, composed of carbon, hydrogen, and oxygen, are the main source of cellular energy and are also important structural components of cell walls and intercellular materials. Carbohydrates are classified according to the number of monomers they contain.

*Monosaccharides* are simple sugars having the general formula $C_n(H_2O)_n$. Depending on the number of carbon atoms they contain, these are designated trioses, pentoses, or hexoses. The pentoses *ribose* and *deoxyribose* are found in the molecules of nucleic acids. *Glucose*, a hexose, is the primary energy source for the cell. Other important hexoses are *galactose*, *fructose*, and *mannose*.

*Disaccharides* are sugars formed by the condensation of two hexose monomers with the loss of one molecule of water. Their formula is therefore $C_{12}H_{22}O_{11}$. The most important members of this group are *sucrose* (formed by glucose and fructose) and *lactose* (formed by galactose and glucose).

*Polysaccharides* result from the condensation of many hexose monomers, with a corresponding loss of water molecules. Their formula is $(C_6H_{10}O_5)_n$. Upon hydrolysis they yield molecules of simple sugars. The most important polysaccharides in living organisms are

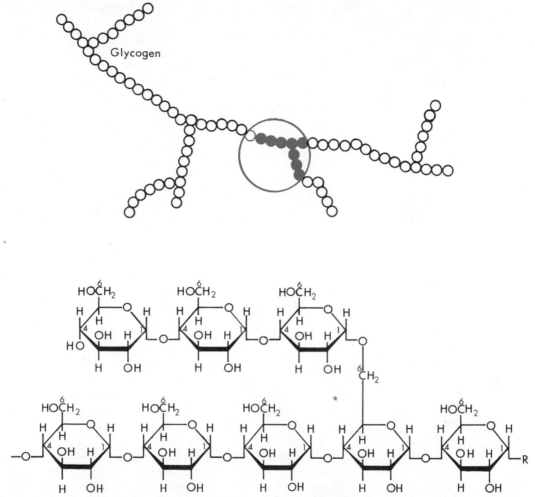

**Figure 2-7**   Glycogen is a branched polymer formed by as many as 30,000 glucose units. The glycosidic bonds are established between carbons 1 and 4 of glucose, except at the branching points, which involve linkages between carbons 1 and 6. The **top** part of the figure shows a low resolution diagram; the circled area is enlarged in the **bottom** part.

*starch* and *glycogen*, which are reserve substances in plant and animal cells, respectively, and *cellulose*, the most important structural element of the plant cell wall. These three substances are all polymers of glucose molecules, but differ in the way they are joined together. Figure 2–7 shows that glycogen is a branched molecule in which the glucose monomers can be joined together by two types of linkages.

*Complex polysaccharides* consist of hexoses plus nitrogen-containing compounds, such as glucosamine, that can also be acetylated or substituted with sulfuric or phosphoric acid. These polymers are important in molecular organization, particularly as intercellular substances. They are often found in combination with proteins or lipids.

## Glycoproteins: Two-Step Carbohydrate Addition

glycoprotein—conjugated protein having a carbohydrate portion, often involved in membrane interaction and recognition

*Glycoproteins* are complexes composed of a protein bound to a carbohydrate group(s). They can be either cellular or secretory. Cellular glycoproteins are present mainly in the cell membrane, with the carbohydrate moiety protruding to the outside of the cell, and they have important functions in membrane interaction and recognition (see Section 3–5). Secreted glycoproteins are produced by many different kinds of cells (e.g., liver, thyroid, plasma). In most cases the protein is linked to the carbohydrate moiety by way of an asparagine (Asn) residue on the former.

A constant feature of asparagine-bound glycoproteins is a pentasaccharide "core" consisting of N-acetylglucosamine (GlcNAc) and mannose, to which are attached two side chains (R and R′) that differ in length and composition in the various proteins.

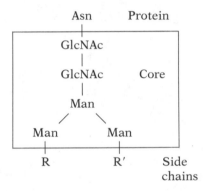

This complex is synthesized in two stages, each one taking place at a different intracellular location. The "core" is added to the protein inside the endoplasmic reticulum as the nascent peptides are being synthesized, while the side chains are added later in the Golgi complex by enzymes called *glycosyltransferases* (see Section 5–2). The oligosaccharide core is initially assembled on a lipid carrier (dolichol-phosphate) and then transferred to the protein molecule. The function of the lipid carrier is to enable the hydrophilic oligosaccharide to traverse the membrane of the endoplasmic reticulum.

## 2–3 LIPIDS

lipid—major structural component of cells, containing long hydrocarbon chains or benzene rings; characterized by solubility in nonpolar organic solvents

The compounds in this group are characterized by their relative insolubility in water and solubility in organic solvents. This general property of lipids can be attributed to the predominance of long linear hydrocarbon chains or benzene rings which are nonpolar and hydrophobic.

# Triglycerides: Three Fatty Acids Bound to Glycerol

*Neutral fats*, often called *triglycerides*, are triesters of fatty acids and glycerol. (An *ester* is a compound produced by the condensation of an acid with an alcohol.) Fatty acids have long hydrocarbon chains with the general formula:

$$COOH$$
$$|$$
$$(CH_2)_n$$
$$|$$
$$CH_3$$

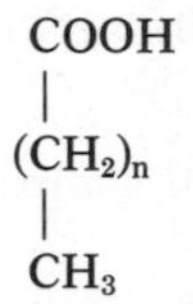

Fatty acids always have an even number of carbons because they are synthesized by joining two-carbon acetyl units. For example, palmitic acid has 16 carbons, and stearic acid has 18 carbons. Sometimes the hydrocarbon chain has double bonds ($-C=C-$), and in such cases the fatty acid is said to be nonsaturated. Double bonds are important because they increase the flexibility of the hydrocarbon chain, and thereby the fluidity of biological membranes.

The carboxyl groups of fatty acids react with the alcohol groups of glycerol in the following way:

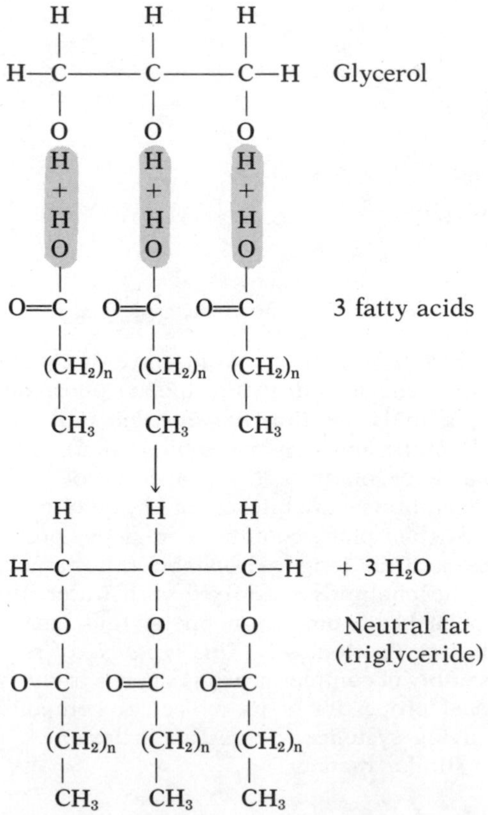

Glycerol

3 fatty acids

+ 3 H₂O

Neutral fat (triglyceride)

triglyceride—triester of fatty acids and glycerol; serves as an energy source and, on a macromolecular level, as a source of insulation; also called a neutral fat

The resulting triglycerides, which accumulate in adipose tissue, are used by organisms for the convenient storage of spare energy. The oxidation state of the long hydrocarbon chains is very low, and they therefore liberate large amounts of energy (about twice as many calories per gram as carbohydrates and proteins) when oxidized to form $CO_2$ and $H_2O$ in cells.

## Phospholipids and Biological Membranes

Not all lipids are reserve substances. We mentioned earlier that some are important structural components of the cell and particularly of cell membranes.

*Phospholipids* have only two fatty acids attached to the glycerol molecule. The third hydroxyl group of glycerol is esterified to phosphoric acid instead of to a fatty acid. This phosphate is also bound to a second alcohol molecule, which can be choline, ethanolamine, inositol, or serine, depending on the type of phospholipid:

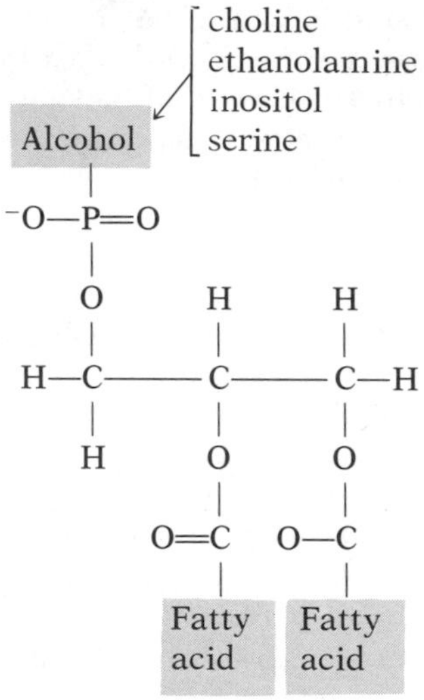

**phospholipid—main constituent of biological membranes, which are made up of bilayers of these molecules**

As shown in Figure 2–8, phospholipids have two long hydrophobic fatty acid "tails" and a hydrophilic (polar) phosphate-containing "head." Phospholipids are thus *amphipathic* molecules (i.e., they contain a hydrophilic and a hydrophobic region), and their configuration accounts for many of the properties of biological membranes. Such membranes are bilayers of phospholipids with the hydrophilic heads (phosphate-containing regions) positioned at the water interface and the long hydrophobic tails arranged in the interior. When phospholipids are mixed with water, the bilayer arrangement—polar heads outside, nonpolar tails inside—is adopted spontaneously (see Section 3–2). This *principle of self-assembly*, in which the assembly of complex structures arises exclusively from the physicochemical properties of its molecular components, is characteristic of living systems. Viruses and ribosomes, for example, assemble in a similar manner.

**hydrophobic—having no affinity for water**

**hydrophilic—having a strong affinity for water**

**amphipathic—term used to describe a molecule having both a hydrophobic and a hydrophilic region**

**principle of self-assembly—principle that the assembly of complex biological structures can arise exclusively from the physical properties of its molecular components**

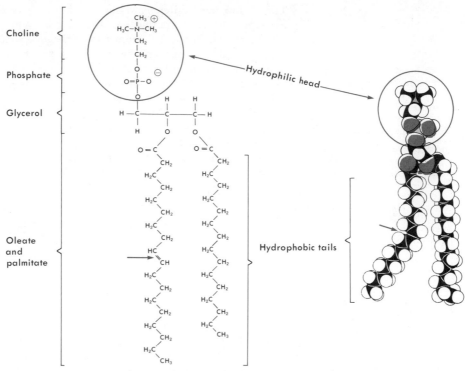

**Figure 2-8**   A membrane phospholipid molecule has a hydrophilic head and two hydrophobic tails. The phospholipid represented here is palmitoyl-oleoyl-phosphatidyl-choline. Note that the double bond in the oleic acid produces a bend in the hydrocarbon chain (indicated by an arrow). Double bonds in the fatty acids increase cell membrane fluidity because unsaturated chains are more flexible (the rotation of the carbon-carbon single bonds on either side of the arrow is enhanced).

# 2-4  PROTEINS

## Proteins: Chains of Amino Acids Linked by Peptide Bonds

The building blocks of proteins are the *amino acids*. An amino acid is an organic acid in which the carbon next to the —COOH group (called an *alpha carbon*) is also bound to an —$NH_2$ group. In addition, the alpha carbon is bound to a *side chain* (R) which is different in each amino acid.

$$H_2N—\overset{\overset{\text{H}}{|}}{\underset{\underset{\text{R}}{|}}{C}}—COOH$$

(side chain)

protein—polymer of amino acids having a structural or enzymatic function in the cell

amino acid—one of about 20 nitrogen-containing organic acids used by the cell for polymerization into protein molecules

amino group—(—$NH_2$)—(basic group found in all amino acids

carboxyl group (—COOH)—acidic group found in all amino acids

alpha carbon—carbon atom adjacent to the carboxyl group of an amino acid

side chain (R)—chemical group attached to the alpha carbon that differs in each amino acid and accounts for the various properties of each one

The amino acids differ from one another only in the side chain; for example, the R in alanine has one carbon, while in leucine it has four carbons. Table 2–3 shows that the properties of the various amino acids depend on the chemical composition of their side chains; for example, lysine and arginine are basic because their side chains contain an extra amino group, and the acidic amino acids (glutamic and aspartic acids) contain an extra carboxyl group.

The condensation of amino acids to form a protein molecule occurs in such a way that the —NH₂ group of an amino acid combines with the —COOH group of the adjoining one, with the simultaneous loss of one molecule of water, forming a *peptide bond*. Many amino acids joined together form a *polypeptide chain*.

**peptide bond**—covalent bond formed when the carboxyl group of one amino acid joins with the amino group of an adjacent one, with the loss of one water molecule

**polypeptide chain**—protein chain consisting of amino acids linked by peptide bonds

## Four Levels of Protein Structure

Four levels of structure are commonly distinguished in proteins.

The *primary structure* is the sequence of amino acids, which form a chain connected by peptide bonds (Fig. 2–9). The amino acid sequence of a protein determines the higher levels of structure of the molecule. The biological importance of the amino acid sequence is exemplified by the human hereditary disease *sickle-cell anemia*, in which profound biological changes are produced by a single amino acid change in the hemoglobin molecule.

The *secondary structure* is the spatial arrangement of amino acids that are close to each other in the peptide chain. Some regions may display a rod-shaped structure, the *α-helix* (called alpha because it was the first structure deduced by Pauling and Corey in the early 1950s). In an *α-helix* the peptide chain is coiled around an imaginary cylinder (Fig. 2–10) and stabilized by hydrogen bonds between the amino group of an amino acid and the carboxyl group of the amino acid situated four residues ahead in the *same* polypeptide chain. In

**primary structure**—linear amino acid sequence of a protein molecule

**secondary structure**—spatial arrangement of neighboring amino acids in a protein molecule

**α-helix**—protein secondary structure in which the polypeptide chain assumes a helical configuration, stabilized by hydrogen bonds between amino acids in the same chain

## TABLE 2-3 THE TWENTY AMINO ACIDS

| Type of Amino Acid | 3-Letter Symbol | 1-Letter Symbol |
|---|---|---|
| *Hydrophobic (Aliphatic Side Chain)* | | |
| Glycine | Gly | G |
| Alanine | Ala | A |
| Valine | Val | V |
| Leucine | Leu | L |
| Isoleucine | Ile | I |
| *Basic (Diamino)* | | |
| Arginine | Arg | R |
| Lysine | Lys | K |
| *Acidic (Dicarboxylic)* | | |
| Glutamic acid | Glu | E |
| Aspartic acid | Asp | D |
| *Amide-Containing* | | |
| Glutamine | Gln | Q |
| Asparagine | Asn | N |
| *Hydroxyl-Containing* | | |
| Threonine | Thr | T |
| Serine | Ser | S |
| *Sulfur-Containing* | | |
| Cysteine | Cys | C |
| Methionine | Met | M |
| *Aromatic* | | |
| Phenylalanine | Phe | F |
| Tyrosine | Tyr | Y |
| *Heterocyclic* | | |
| Tryptophan | Trp | W |
| Proline | Pro | P |
| Histidine | His | H |

a *β-pleated sheet* the amino acids adopt the conformation of a sheet of pleated paper, and the structure is stabilized by hydrogen bonds between the amino and carboxyl groups in *different* polypeptide strands. Other segments of the protein are not highly cross-linked and adopt a *random coil* configuration. This is partly because certain amino acids, such as proline, tend to disrupt helical structure.

The *tertiary structure* is the way in which helical and random coil regions fold with respect to each other. That is, it refers to the three-dimensional relationship of amino acid segments that may be far apart from each other in the linear sequence.

The *quaternary structure* is the arrangement of protein subunits within complex proteins made up of two or more such subunits. For example, the hemoglobin molecule is composed of four polypeptide chains, two designated $\alpha$ and two $\beta$. Separation and association of the subunits may occur spontaneously. Hemoglobin may be broken into two half-molecules (two $\alpha$ and two $\beta$) by urea. When urea is removed they reassemble, forming complete functional molecules.

**β-pleated sheet**—protein secondary structure in which parallel or antiparallel polypeptide chains assume a pleated configuration, stabilized by hydrogen bonds between molecules in adjacent chains

**random coil**—protein secondary structure of apparently unordered nature, typically interspersed with more organized regions

**tertiary structure**—orientation of regions of secondary structure with respect to each other

**quaternary structure**—subunit structure of proteins made up of more than one polypeptide chain

**Figure 2-9** The primary structure of bovine pancreatic ribonuclease. Notice the position of the four disulfide bridges between cysteine residues. (From C. B. Anfinsen.)

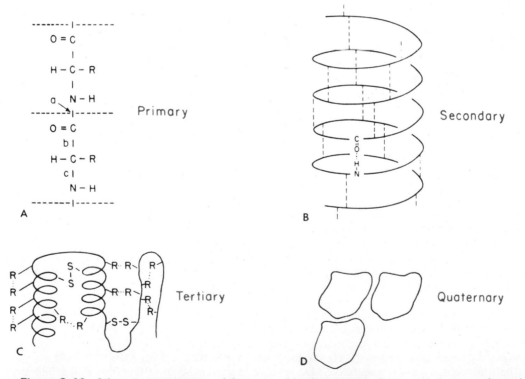

**Figure 2-10** Schematic representation of the structural levels in proteins. Amino acid chains are denoted by R, non-covalent interactions by dotted lines. (Courtesy of Prof. C. Nemethy.)

This binding is highly specific and takes place only between the half-molecules; it is yet another example of the principle of self-assembly (mentioned in the previous section with regard to lipid bilayers). We shall see in the next section of this chapter that most enzymes exist as multi-subunit structures.

## Factors Involved in Determining Protein Structure

The spatial arrangement of a protein molecule is predetermined by its amino acid sequence (primary structure). This can be demonstrated by experiments in which protein *denaturation*, or disruption of tertiary structure, is brought about by high temperatures or other nonphysiological conditions. Denaturation of a protein usually results in the loss of its biological activity.

Figure 2–11 shows that in the case of the enzyme ribonuclease, denaturation can be achieved by treatment wth β-mercaptoethanol and high concentrations of urea. Mercaptoethanol is a reducing agent which can disrupt S—S bridges (disulfide bonds), reducing them to —SH groups, while urea disrupts other weak molecular interactions. The tertiary structure of ribonuclease is maintained by four disulfide bonds which are established between pairs of cysteines (an amino acid that contains an —SH group). After denaturation the enzyme can be *renatured*, or correctly refolded into its natural conformation, by gradually removing the urea and mercaptoethanol, after which the activity of the enzyme is recovered (Fig. 2–11). There are 105 possible combinations in which eight cysteines can pair to produce four disulfide bridges, but only the biologically active conformation is produced after careful renaturation because it is thermodynamically the most stable structure. This is clear evidence that all the information needed to produce the complex folding of a protein molecule is contained in its primary structure.

Several different types of bonds are involved in maintaining the four levels of protein structure. The *covalent bonds* in proteins are of two main types. The first is the *peptide bond* uniting amino acid

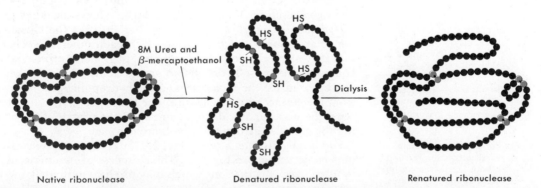

**Figure 2–11** Denaturation and renaturation of ribonuclease. This experiment shows that the information for protein folding is contained in the amino acid sequence (primary structure) of proteins. (Cys residues are indicated with shaded beads. See Fig. 2–9.) Sulfhydryl groups, *SH, HS.*

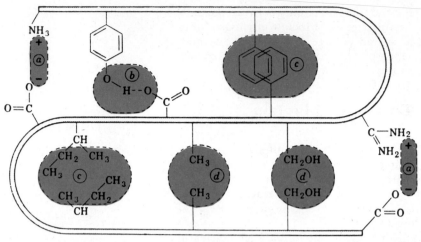

**Figure 2-12** Types of noncovalent bonds that stabilize protein structure. *a*, ionic bonds; *b*, hydrogen bonds; *c*, hydrophobic interactions; *d*, van der Waals interactions. (From C. B. Anfinsen.)

**disulfide bond**—covalent bond formed between —SH groups of two cysteine residues

subunits in the primary sequence. The second is the *disulfide bond* (S—S bridge) which, as we have just seen, is established between the —SH groups of two cysteine residues and is responsible for some aspects of the tertiary structure.

Various kinds of *weak interactions* are important in the establishment of secondary and tertiary structure. These weak bonds are all noncovalent; the main types (illustrated in Fig. 2–12) are as follows:

*Ionic* or *electrostatic bonds* result from the attractive force between ionized groups having opposite charges (Fig. 2–12, *a*).

**ionic (electrostatic) bond**—noncovalent bond resulting from attraction between ionized groups having opposite charges

*Hydrogen bonds* result when an $H^+$ (proton) is shared between two neighboring electronegative atoms. The $H^+$ can be shared between nitrogen or oxygen atoms which are close to each other. Hydrogen bonds have many important biochemical functions. They are essential for the specific pairing between nucleic acid bases, thus providing the main force that holds the two DNA strands together as well as allowing the specific copying of DNA into RNA. Figure 2–4 shows hydrogen bonds in DNA, and Figure 2–12, *b* shows them in a protein.

**hydrophobic interaction**—exclusion of water by nonpolar groups of a macromolecule, causing compaction with hydrophobic groups inside and hydrophilic groups outside

*Hydrophobic interactions* (Fig. 2–12, *c*) involve the clustering of nonpolar groups, which associate with each other in such a way that they are not in contact with water. In globular proteins the side chains of the most hydrophobic amino acids (Table 2–3) tend to aggregate inside the molecule, and the hydrophilic groups protrude from the surface of the structure. The hydrophobic residues tend to repel the water molecules that surround the protein, thereby causing the globular structure to be more compact.

**Van der Waals interaction**—weak attraction created between molecules by mutually induced charge fluctuations

*Van der Waals interactions* (Fig. 2–12, *d*) occur only when two atoms come very close together. The closeness of two molecules can induce charge fluctuations which may produce mutual attraction at very short range.

The essential difference between a covalent and a noncovalent bond is in the amount of energy needed to break the bond. For example, a hydrogen bond requires only 4.5 kcal mole$^{-1}$, as compared with 110 kcal mole$^{-1}$ for the covalent O—H bond in water. Although each individual bond is weak, large numbers of them can produce very stable structures, as in the case of double-stranded DNA. Covalent bonds are generally broken by the intervention of enzymes, whereas noncovalent bonds are easily dissociated by physicochemical forces.

Many cellular proteins exist as complexes of multiple subunits which are held together by weak interactions. For analytical purposes, cell biologists sometimes find it desirable to dissociate them into their component polypeptides. Figure 2–13 shows how several

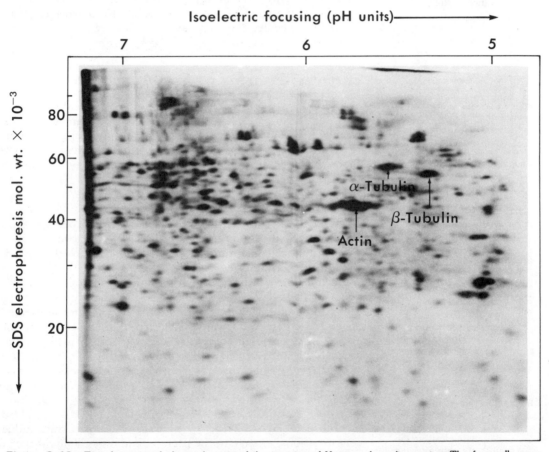

**Figure 2-13** Two-dimensional electrophoresis of the proteins of *Xenopus borealis* oocytes. The frog cells were labeled with $^{35}$S-methionine and homogenized, and the total proteins were separated according to their charge (by isoelectric focusing) in the first dimension and according to their size (by sodium dodecyl sulfate gel electrophoresis) in the second. The polyacrylamide gel was exposed against photographic film (autoradiography). Several hundred radioactive proteins can be seen. The quaternary structure of the polypeptide chains was dissociated using high concentrations of urea (which destroys hydrogen bonds and hydrophobic interactions) and β-mercaptoethanol (which destroys disulfide bridges) in the first dimension and the detergent SDS (which destroys ionic interactions) in the second. As in most other cell types, the most abundant protein is actin (3 to 5 percent of the total cell protein), followed by tubulin (which has α and β subunits, both of which form part of cell microtubules). Notice that the distance migrated by the proteins in the second dimension is proportional to the logarithm of the molecular weight. (From E. M. De Robertis).

hundred polypeptide chains can be separated from one another by the technique of two-dimensional polyacrylamide gel electrophoresis.

## 2-5 ENZYMES AND THEIR REGULATION

**enzyme—protein that acts as a biological catalyst**

**catalyst—substance that facilitates a chemical reaction without being depleted or permanently modified in the process**

Enzymes are biological catalysts. A *catalyst* is a substance that accelerates chemical reactions but is not itself modified in the process, so that it can be used again and again. Enzymes are the largest and most specialized class of protein molecules. More than a thousand different enzymes have been identified; many of them have been obtained in pure, and even crystalline, condition. Enzymes represent one of the most important products of the genes contained in the DNA molecule. The complex network of chemical reactions involved in cell metabolism is directed by enzymes.

### Enzymes: Proteins with an Active Site

**active site—location(s) on an enzyme molecule at which the substrate attaches and is chemically modified**

**substrate—substance acted upon by an enzyme and converted into one or more products**

*Enzymes* (E) are proteins with one or more loci, called *active sites*, to which the *substrate* (S) attaches. A substrate is any substance that is acted upon by enzymes; as a result of this interaction the substrate is chemically modified and converted into one or more products (P). Since this is generally a reversible reaction, it may be expressed as follows:

$$E + S \rightleftharpoons [ES] \rightleftharpoons E + P \tag{1}$$

**enzyme-substrate complex—intermediary molecular complex formed by the enzyme and substrate while the catalyzed reaction is in progress**

where [ES] is an intermediary *enzyme-substrate complex*. Enzymes accelerate the reaction until an equilibrium is reached. They are so efficient that the reaction may proceed up to $10^8$ to $10^{11}$ times faster than in the noncatalyzed condition.

**lock-and-key—term used to indicate the high specificity of enzymes for their respective substrates**

Enzymes have great specificity for their substrates and will frequently not accept related molecules of a slightly different shape. This can be explained by assuming that enzyme and substrate have a *lock-and-key* interaction. As shown in Figure 2–14, the enzyme has an *active site* complementary to the shape of the substrate. If a substrate has a different shape, it will not bind to the enzyme.

**induced fit—phenomenon by which the active site of some enzymes becomes exactly complementary to the substrate only after the substrate has bound to it**

**conformational change—structural alteration caused in an enzyme by the binding of its substrate, usually of a type that facilitates the reaction**

Although we can think of enzymes in terms of locks and keys, this does not mean that the active site is a rigid structure. In some enzymes the active site is precisely complementary to the substrate only *after* the substrate is bound, a phenomenon called *induced fit*. As shown in Figure 2–14, the binding of the substrate induces a conformational change in the protein, and only then will the chemical groups essential for catalysis come in close contact with the substrate. Some enzymes bind preferentially to one of several possible conformations of the substrate; in this way the flexibility of both enzyme and substrate may contribute to catalysis.

The binding of the substrate to the active site involves forces of a noncovalent nature (ionic and hydrogen bonds, van der Waals forces), which are of very short range. This explains why the enzyme-substrate complex can be formed only if the enzyme has a site that is exactly complementary to the shape of the substrate.

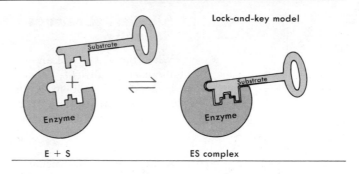

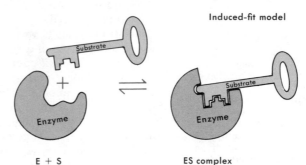

**Figure 2-14**   Substrates interact with the active site in a precise way. Some enzymes have an *induced fit:* the shape of the active site is complementary to the substrate only after the substrate is bound.

## The Cell Is Not Simply a Bag Full of Enzymes

Enzymes catalyze the thousands of chemical reactions that occur in cells. The enzymes of some biochemical pathways are in solution in the cytosol, and the substrates must diffuse freely from one enzyme to the next. However, in other cases the enzymes involved in a chain of reactions are bound to one another and function together as a *multi-enzyme complex*. For example, the seven enzymes that synthesize fatty acids are tightly bound to one another. Similarly, the pyruvate-dehydrogenase complex is formed by three enzymes. Multi-enzyme systems facilitate complex reactions because they limit the distance through which the substrate molecules must diffuse during the sequence of reactions. The substrate is not released from the complex until all the reactions are completed.

The most complex multi-enzyme systems are associated with biological membranes and the ribosomes. For example, the respiratory enzymes necessary for electron transfer are arranged in a precise way in the two-dimensional scaffold provided by the inner membrane of mitochondria in eukaryotes (in prokaryotes they are arranged within the cell membrane). These multi-enzyme systems require this well-defined structure for activity; the enzymes become inactive when removed from the membrane. For this reason, the study of membrane biochemistry is especially difficult.

Enzyme distribution is never random. Some enzymes are packed into lysosomes, and others into secretion granules. Other enzymes, such as the RNA and DNA polymerases, are located in the nucleus and not in the cytoplasm. The mechanisms by which these proteins are segregated into the correct cellular compartments are emphasized throughout this book.

multi-enzyme complex—physical attachment of several enzymes involved in a chain of related or sequential reactions

## Allosteric Enzymes Have Multiple Interacting Subunits

When the velocity of an enzyme reaction is plotted as a function of increasing substrate concentration, many enzymes display the *hyperbolic* curve shown in Figure 2–15. As more substrate is added, more enzyme is found in the ES complex form, and the velocity of appearance of product increases.

$$E + S \rightleftharpoons [ES] \rightarrow P + E \qquad (2)$$

At high substrate concentrations essentially all of the enzyme molecules are in the ES complex form, and the *maximal velocity* of the reaction is achieved.

Not all enzymes have hyperbolic kinetic curves, however; some have kinetic curves of a *sigmoidal* shape (Fig. 2–15). The latter are called *regulatory* or *allosteric* enzymes. They are *oligomers* containing two *(dimer)*, four *(tetramer)*, or more subunits which are able to interact with one another. The sigmoidal shape of the curve results from the fact that the binding of the first substrate molecule enhances the affinity for the binding of the second substrate molecule, and so forth. This change of affinity is due to a change in the conformation of the enzyme molecule. Figure 2–15 shows that there is a region in the curve in which a small increase in S concentration causes a very large increase in enzyme activity. Allosteric enzymes are of great regulatory value, since large changes in activity can be obtained by small changes in S concentration.

Regulatory enzymes are sensitive to the so-called *allosteric modifiers* (i.e., modulators) which act as *inhibitors* or *activators*. Such modifiers do not bind to the active site but rather to a different (allosteric) site. The binding of the modifier induces a conformational change in the active site, resulting in a decreased or increased affinity of the enzyme for the substrate. Figure 2–15 shows that in the presence of an activator the curve tends to be more hyperbolic, while under the influence of an inhibitor it is more sigmoidal. As will be discussed below, allosteric enzymes are of paramount importance in the regulation of cell metabolism.

## Enzyme Regulation at the Genetic and Catalytic Levels

The living cell seldom wastes energy synthesizing or degrading more material than necessary. Therefore, the thousands of chemical reactions that occur inside the cell must be carefully controlled. Enzyme activity is regulated by two major mechanisms: *genetic control* and *control of catalysis* (Fig. 2–16).

*Genetic control* implies a change in the total amount of enzyme molecules. The best known examples of this form of regulation are *enzyme induction* and *repression* in microorganisms in which enzyme synthesis is regulated at the gene level by the indirect action of certain small molecules. In Chapter 13 we will analyze in detail the cases of the *lactose operon* (induction) and the *tryptophan operon* (repression) of *E. coli*. We will see that gene expression is regulated

---

**allosteric regulation—** noncompetitive regulation of enzyme activity by binding of a substance (not necessarily either substrate or product) to a site other than the active site, in such a way as to stabilize the enzyme in an active or inactive state

**oligomer—**polymer having relatively few structural units

**inhibitor—**modifier binding to the allosteric site of an enzyme, resulting in decreased affinity for the substrate

**activator—**modifier binding to the allosteric site of an enzyme, resulting in increased affinity for the substrate

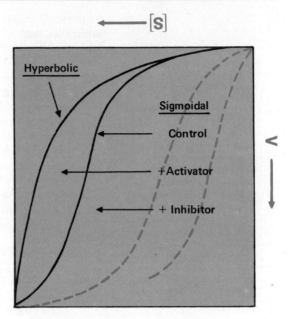

**Figure 2-15** Sigmoidal curve characteristic of a regulatory or allosteric enzyme. Observe the difference with a hyperbolic type of curve. Under the action of an activator the curve becomes more hyperbolic, while an inhibitor makes it more sigmoidal.

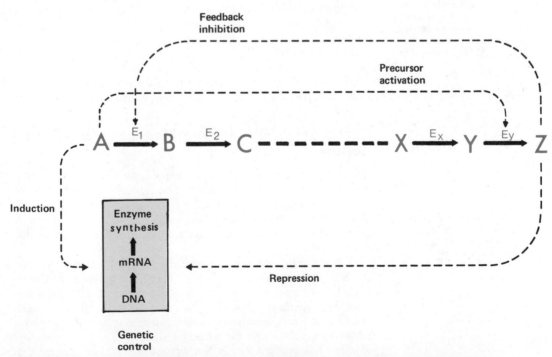

**Figure 2-16**  Enzymes are regulated at the level of catalysis and at the gene level (enzyme synthesis).

In *feedback inhibition* the end-product (Z) of a metabolic chain acts as an allosteric inhibitor of the *first* enzyme of the pathway. In *precursor activation* the first metabolite (A) of a pathway is an allosteric activator of the final enzyme. The *genetic control* mechanisms modify enzyme synthesis according to cellular requirements. In *induction* (e.g., the lactose operon) the presence of a substrate (A) stimulates the synthesis of the enzymes that degrade it, while in *repression* (e.g., the tryptophan operon) the accumulation of the end-product (Z) switches off enzyme production.

by proteins called *repressors*, which bind to the DNA, and that small molecules such as the disaccharide lactose or the amino acid tryptophan can affect gene activity by binding to these repressors. As shown in Figure 2–16, in enzyme *induction* the availability of a substrate (e.g., lactose) induces the synthesis of the enzymes that degrade it, while in enzyme *repression* the accumulation of the end-product of a metabolic chain (e.g., tryptophan) turns off production of the enzymes involved in its synthesis. In both cases, the net result is that enzymes are synthesized only when required.

*Control of catalysis* involves a change in enzyme activity without a change in the total amount of enzyme synthesized. This is frequently produced in regulatory or allosteric enzymes by the action of allosteric activators or inhibitors. Two important mechanisms for this type of control are *feedback inhibition* and *precursor activation* (Fig. 2–16). In *feedback inhibition* the end-product of a metabolic pathway acts as an *allosteric inhibitor* of the first enzyme of this metabolic chain. Thus, when enough product is synthesized, the entire chain can be shut off, and useless accumulation of metabolites is avoided. In *precursor activation* the first metabolite of a biosynthetic pathway acts as the *allosteric activator* of the last enzyme of the sequence (Fig. 2–16).

Another way in which metabolism may be regulated involves enzyme *interconversions*. Some enzymes may exist in two forms (active and inactive) which are interconvertible. Frequently the mechanism of interconversion consists of *phosphorylation*, i.e., the covalent binding of a phosphate group that is provided by ATP.

Genetic control (induction, repression) is generally considered to be a coarse and relatively slow type of regulation, and feedback inhibition a finer, almost instantaneous way of ensuring that enzyme activity is adequate for cellular requirements.

## Cyclic AMP: The Second Messenger in Hormone Action

Hormones are molecules that transfer information from one group of cells to another distant tissue. The molecular bases of hormonal action were poorly understood until 1956, when Sutherland discovered *cyclic adenosine monophosphate* (cAMP), a cyclic nucleotide that has been found to regulate a large number of metabolic processes.

In 3',5' cyclic AMP the phosphate group is covalently bound to the 3' and 5' carbons of the ribose ring. (Normally in AMP only the 5' carbon is bound to the phosphate; see Fig. 2–2.) This cyclic nucleotide is synthesized by *adenylate cyclase*, an enzyme that is tightly bound to the cell membrane. Many hormones modify the activity of adenylate cyclase and therefore produce a change in the intracellular level of cyclic AMP. In Sutherland's model of hormonal action (Fig. 2–17), the hormone is regarded as a *first messenger* that interacts with specific receptor sites located in the outer surface of the cell membrane. This hormone-receptor interaction results in a change in the activity of adenylate cyclase. The active site of this

---

induction—genetic form of enzyme regulation by which synthesis of an enzyme is indirectly enhanced by certain metabolites (inducers)

repression—genetic form of enzyme regulation by which synthesis of an enzyme is indirectly repressed by certain metabolites (repressors)

feedback inhibition—form of enzyme regulation by which the end-product of a metabolic pathway becomes an allosteric inhibitor of the first enzyme that catalyzed its production

precursor activation—form of enzyme regulation by which the first metabolite of a biosynthetic pathway becomes an allosteric activator of the last enzyme of the sequence

cAMP (cyclic adenosine monophosphate)—cyclic nucleotide having a regulatory function in many metabolic processes

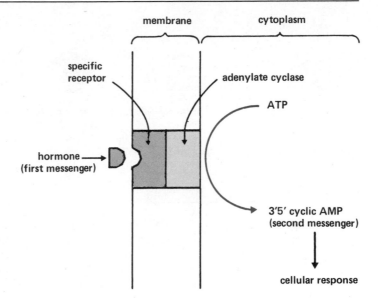

**Figure 2-17** Diagram showing the effect of a hormone (first messenger) upon a specific receptor of the cell membrane and its effect on the enzyme adenylate cyclase.

enzyme is on the inner surface of the membrane and, using ATP as substrate, it produces cyclic AMP in the cytoplasm. This nucleotide is considered a *second messenger* that carries the information to the metabolic machinery of the cell. The effect of cyclic AMP depends on the target organ; for example, an increase in cyclic AMP will produce glycogen degradation in the liver and steroid production in the adrenal cortex.

Many hormones are known to act by way of specific receptors that stimulate adenylate cyclase. Although each hormone has a particular receptor protein, all the receptors can interact with the same adenylate cyclase enzyme. Some cells may have more than one receptor (for example, liver cells respond to both adrenalin and glucagon). As we will see in Section 3–2, the receptors are mobile and not permanently bound to the adenylate cyclase; they can diffuse on the lipid bilayer, thus enabling the enzyme to interact with several receptors.

The main way in which cAMP affects metabolism is by stimulating the activity of *protein kinases*, a group of enzymes that catalyze the phosphorylation of proteins:

protein kinase—type of enzyme whose activity may be stimulated by cAMP and which catalyzes the phosphorylation of proteins

$$\boxed{\text{protein}} + \text{ATP} \xrightarrow{\text{C}} \boxed{\text{protein}} - \textcircled{P} + \text{ADP}$$

Figure 2–18 shows the way in which cAMP activates protein kinase. The enzyme has two subunits, one catalytic and one regulatory, which can bind cAMP. In the absence of cAMP the regulatory and catalytic subunits form a complex that is enzymatically inactive. When cAMP binds to the regulatory subunit it causes a conformational change, and the complex dissociates. The catalytic subunit thus freed is enzymatically active and will phosphorylate other enzymes.

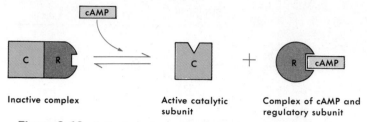

Inactive complex          Active catalytic          Complex of cAMP and
                             subunit                 regulatory subunit

**Figure 2-18**   Activation of a protein kinase by cyclic AMP. The enzyme has a catalytic (C) and a regulatory (R) subunit, which form an inactive complex. Cyclic AMP binds to the regulatory subunit and induces a conformational change. The free catalytic subunit is the active form, which phosphorylates proteins in the reaction:

$$\boxed{\text{Protein}} + \text{ATP} \xrightarrow{\text{C}} \boxed{\text{Protein}}\!-\!\text{P} + \text{ADP}$$

The classic example of regulation produced by cyclic AMP is the degradation and synthesis of glycogen. Epinephrine (adrenalin) and glucagon induce the degradation of glycogen in the liver. These hormones, acting on the corresponding receptor, stimulate adenylate cyclase and raise the cyclic AMP level. Cyclic AMP activates protein kinase and the phosphorylations thus induced elicit the effects shown in Figure 2–19. By means of this *cascade mechanism*, the cell amplifies considerably the initial signal given by the hormones at the membrane level. The end result is that glycogen degradation is stimulated and its synthesis is inhibited. This is the same molecular mechanism by which adrenalin (which is liberated in conditions of alarm or stress) increases the amount of glucose available to the bloodstream.

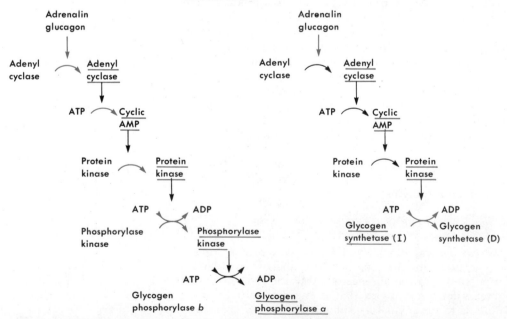

**Figure 2-19**   Hormones regulate glycogen metabolism by inducing a cascade of reactions mediated by cyclic AMP. The active forms are underlined. Note that phosphorylation activates glycogen phosphorylase (thus increasing glycogen degradation), but inhibits glycogen synthetase. The cascade mechanism greatly amplifies the hormonal signal.

Cyclic GMP is a second type of cyclic nucleotide found in cells. Sometimes (although not always) the levels of GMP are in an inverse relationship to those of cAMP (when cAMP increases, cGMP decreases, and vice versa). The precise function of cGMP is not yet understood.

# SUMMARY

## 2-1 NUCLEIC ACIDS

All living organisms contain DNA and RNA. The genetic information contained in the DNA is transcribed into RNA, which in turn is translated into protein; this series of events is often referred to as the central dogma.

Nucleic acids are linear polymers of nucleotides linked together by phosphodiester bonds. The nucleotide monomers result from the covalent bonding of a phosphate and a base to a pentose moiety. The pentose is ribose in RNA and deoxyribose in DNA. The bases found in DNA are thymine (T) and cytosine (C), which are pyrimidines, and adenine (A) and guanine (G), which are purines. RNA contains uracil (U) instead of thymine. A nucleotide without its phosphate group is called a nucleoside. In addition to their role as nucleic acid constituents, nucleotides also have a major role in the storage and transfer of chemical energy.

All the genetic information of a living organism is stored in its linear sequence of the four bases. Although base composition varies from one species to another, the amount of adenine always equals the amount of thymine (A = T), and the amounts of cytosine and guanine are also equal (C = G).

A DNA molecule is composed of two antiparallel polynucleotide chains that form a double helix around a central axis. The bases are stacked inside the helix in a plane perpendicular to its axis, and the two strands are held together by hydrogen bonds established between the base pairs. The only pairs that occur are AT, held together by two hydrogen bonds, and GC, held together by three hydrogen bonds. The latter pair is the more stable. The pairing properties of the bases are such that, whatever the axial sequence on one strand may be, the sequence on the other strand must be exactly complementary to it.

If the two strands of a DNA molecule are denatured (separated) by physical or chemical treatments, they can subsequently reanneal as a consequence of their nucleotide base-pairing properties. Renaturation studies have led to the discovery of repeated sequences in eukaryotic DNA; similar hybridization studies have provided a powerful method for characterizing RNA molecules, which hybridize only to the DNA from which they were transcribed.

The three major classes of RNA are messenger, transfer, and ribosomal, all of which are involved in protein synthesis. Although each RNA molecule consists of only a single polynucleotide chain, the chain folds upon itself to make a more compact structure having such secondary characteristics as hairpin loops. These structures have important biological consequences.

## 2-2 CARBOHYDRATES

Carbohydrates are the main source of cellular energy and are also important structural components. They are classified as mono-, di-, or polysaccharides, depending on the number of monomers they contain. Complex polysaccharides often contain other elements in addition to C, H, and O, and are often found in combination with proteins or lipids.

Glycoproteins consist of a protein bound to a carbohydrate group, and can be either cellular or secretory. A constant feature of glycoproteins is a pentasaccharide core, to which are attached two side chains that differ in the various glycoproteins. The core is assembled on a lipid carrier and transported into the endoplasmic reticulum, where it is added to the protein. The side chains are added later in the Golgi complex.

## 2-3 LIPIDS

Neutral fats or triglycerides are triesters of fatty acids and glycerol. These substances accumulate in adipose tissue and are used for energy storage.

Phospholipids are amphipathic molecules having two nonpolar fatty acid tails and a polar phosphate-containing head. They are the basic constituent of biological membranes, which assemble spontaneously in a bilayer arrangement determined by the hydrophilic and hydrophobic properties of the phospholipids.

## 2-4 PROTEINS

The building blocks of proteins are amino acids, in which the alpha carbon is bound to an amino group

(—NH$_2$), a carboxyl group (—COOH), and a side chain (R). The side chain is different in each amino acid. Peptide bonds between adjacent amino acids involve the amino group of one residue and the carboxyl group of the next.

Proteins are considered to have four levels of structure. The primary structure is the amino acid sequence of the polypeptide chain. The secondary structure is the spatial arrangement of neighboring amino acids; common secondary structures are the $\alpha$-helix, $\beta$-pleated sheet, and random coil. The tertiary structure is the way in which helical and random coil regions fold with respect to each other. The quaternary structure is the arrangement of protein subunits within multi-subunit proteins.

The spatial arrangement of a protein is predetermined by its amino acid sequence. Covalent disulfide bonds contribute to protein folding. In addition to these covalent bonds there are also several types of weak interactions important for secondary and tertiary structure; these include ionic or electrostatic bonds, hydrogen bonds, hydrophobic interactions, and van der Waals interactions.

## 2–5 ENZYMES AND THEIR REGULATION

Enzymes are biological catalysts, which can be present individually in the cytosol or nucleus, located in lysosomes or secretion granules, or bound to one another in a multi-enzyme complex. The most complex multi-enzyme systems are associated with biological membranes or ribosomes.

Enzymes are proteins having one or more active sites to which the substrate attaches in order to be chemically modified. Enzymes have great specificity for their substrates, and they interact in a lock-and-key fashion such that a substrate whose shape differs only slightly will not be able to bind. In some enzymes the active site is precisely complementary to the substrate only after the latter is bound; this phenomenon is called induced fit.

The velocity of an enzyme reaction plotted as a function of increasing substrate concentration may yield a hyperbolic or a sigmoidal curve. Allosteric enzymes have sigmoidal kinetics. The binding of the substrate to the active site produces a conformational change in the enzyme, enhancing its affinity for binding other substrate molecules. Large changes in activity can be obtained by small changes in [S]. Regulatory enzymes are sensitive to allosteric modifiers which act as inhibitors or activators; these modifiers bind to a different (allosteric) site, inducing a conformational change that affects operations at the active site.

Enzyme activity is regulated by genetic control (enzyme induction, repression), control of catalysis (feedback inhibition, precursor activation), or enzyme interconversions. Genetic control is generally considered to be a coarse and relatively slow mechanism, while feedback inhibition is a finer, almost instantaneous form of enzyme regulation. Cyclic AMP is a nucleotide that has been found to regulate a large number of metabolic processes by acting as a "second messenger" in hormone action.

# STUDY QUESTIONS

- Describe the components of DNA and RNA, and explain what physical and chemical properties account for the structure of each molecule.
- How does the structure of DNA account for the mechanisms of DNA replication and RNA transcription?
- Which bond pairs are more stable, AT or GC? Why?
- How and where are glycoproteins synthesized?
- What are the structural differences between a triglyceride and a phospholipid?
- Why do phospholipids spontaneously assemble in bilayers?
- What are the primary, secondary, and tertiary structures of a protein molecule?
- What kinds of interactions maintain protein structure?
- How do the lock-and-key and induced fit theories explain the substrate specificity of enzymes?
- Why do allosteric enzymes have sigmoidal kinetics, and how does this type of behavior help to regulate metabolism?
- What are feedback inhibition and precursor activation of biochemical pathways? What are enzyme induction and repression?
- Describe Sutherland's model of hormone action.

# SUGGESTED READINGS

**Anfinsen, C. B.** (1973) Principles that govern the folding of protein chains. *Science, 181*:223.

**Frieden, E.** (1972) The chemical elements of life. *Sci. Am., 227*:52.

**Koshland, D. E., Jr.** (1973) Protein shape and biological control. *Sci. Am., 229*:52.

**Lehninger, A. L.** (1975) *Biochemistry.* Worth, New York.

**Pastan, I.** (1972) Cyclic AMP. *Sci. Am., 227*:97.

**Perutz, M.** (1978) Hemoglobin structure and respiratory transport. *Sci. Am., 239*:68.

**Stryer, L.** (1975) *Biochemistry.* Freeman, San Francisco.

**Watson, J. D.** (1968) *The Double Helix.* Atheneum, New York.

**Watson, J. D., and Crick, F. H. C.** (1953) Molecular structure of nucleic acid: a structure for deoxy-ribose nucleic acid. *Nature, 171*:737.

# chapter 3

# Molecular Organization and Function of the Cell Surface

The internal milieu of a cell differs from that of its external environment. This difference is maintained throughout the life of the cell by the thin surface membrane, the *cell* or *plasma membrane*, which controls the entrance and exit of molecules and ions. The capacity of the plasma membrane to act as a selective barrier between the cell and the medium is called *permeability*. The plasma membrane is so thin that it cannot be resolved with the light microscope, but in some cells it is covered by thicker protective layers that are within the limits of microscopic resolution. Most plant cells, for example, have a thick cellular wall that covers and protects the

plasma membrane—lipoprotein structure serving as a selective barrier between the cell's cytoplasm and its immediate environment

permeability—susceptibility of the cell to penetration by some molecules but not others, as regulated by the plasma membrane

plasma membrane (Fig. 1–1). Some animal cells are surrounded by a cement-like layer called a cell coat, which generally plays no role in permeability but does have other important functions.

Cell surface research is currently one of the most active areas of cell and molecular biology, since it deals not only with the traffic of materials between the cell and its environment but also with many other functions related to the communication and interaction among cells in a tissue. We will see that at the cell surface there is a mechanism of *molecular recognition* by which one cell is able to recognize similar cells. This mechanism is greatly impaired when normal cells become *transformed* into cancerous cells. One of the defining characteristics of cancer is a failure in the social behavior of cells, resulting in uncontrolled growth and division because the molecular mechanisms of recognition, acting at the cell membrane, are altered. It is hoped that study of this alteration will contribute to our understanding of the differences between normal and cancerous cells.

## 3–1 MOLECULAR ORGANIZATION OF THE PLASMA MEMBRANE

**erythrocyte—a red blood cell**

**ghost—an erythrocyte membrane which has undergone hemolysis and contains only cytoplasm**

The first step in the study of the molecular organization of the cell membrane is its isolation. This is easily achieved with red blood cells *(erythrocytes)*, whose chemical composition is well known. If blood is treated with a *hypotonic solution* (i.e., a solution having a lower osmotic pressure than the blood plasma), the erythrocytes burst and lose their content of hemoglobin. This phenomenon is called *hemolysis*, and the resulting membrane is frequently referred to as a *ghost*.

Figure 3–1 introduces some of the concepts about the molecular organization of the red cell membrane which will be dealt with in more detail later on. Free-floating erythrocytes and a red cell ghost cut in half are represented as seen under the light microscope; there is also a cross-sectional view of the membrane at the molecular level, in which the "fluid mosaic" model is represented. It is important to observe that (1) the membrane is formed by a fairly continuous lipid bilayer into which protein complexes are embedded in a kind of mosaic arrangement; (2) there are other proteins peripheral to the bilayer and disposed on its inner surface; and (3) the molecular asymmetry of the membrane is emphasized further by the oligosaccharide chains that protrude only at its outer surface.

**lipid bilayer—major structural component of the plasma membrane; assembles spontaneously due to the physical properties of its constituent molecules (hydrophobic tails inside, hydrophilic heads outside)**

The composition of the red cell membrane (by mass) is approximately 52 percent proteins, 40 percent lipids, and 8 percent carbohydrates. Many of its constituent molecules are complex glycoproteins and glycolipids (i.e., oligosaccharides bound to proteins and lipids, respectively). Although our discussion will emphasize the red cell membrane, it is important to remember that the plasma membrane in other cells may have a very different chemical composition of proteins, lipids, and carbohydrates.

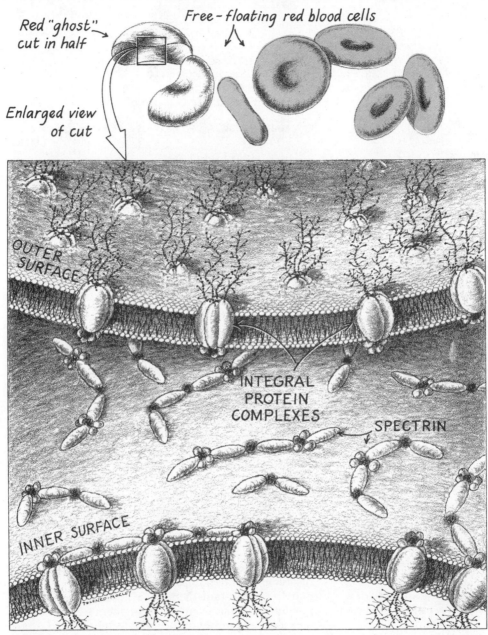

Red "ghost" cut in half

Free-floating red blood cells

Enlarged view of cut

OUTER SURFACE

INTEGRAL PROTEIN COMPLEXES

SPECTRIN

INNER SURFACE

**Figure 3-1**  Diagram showing red blood cells, a red cell ghost cut in half, and a view of the molecular organization of the membrane according to the fluid mosaic model. Integral protein complexes are represented with the oligosaccharide chains sticking out on the outer surface. At the inner surface the peripheral protein spectrin is represented. (Courtesy of G. L. Nicolson, 1978.)

**TABLE 3–1 CRITERIA FOR DISTINGUISHING PERIPHERAL AND INTEGRAL MEMBRANE PROTEINS***

| Property | Peripheral Protein | Integral Protein |
|---|---|---|
| Requirements for dissociation from membrane | mild treatments sufficient: high ionic strength, metal ion chelating agents | hydrophobic bond-breaking agents required: detergents, organic solvents, chaotropic agents |
| Association with lipids when solubilized | usually soluble—free of lipids | usually associated with lipids when solubilized |
| Solubility after dissociation from membrane | soluble and molecularly dispersed in neutral aqueous buffers | usually insoluble or aggregated in neutral aqueous buffers |
| Examples | cytochrome c of mitochondria; spectrin of erythrocytes | most membrane-bound enzymes; histocompatibility antigens; drug and hormone receptors |

*From S. J. Singer, The molecular organization of biological membranes. In *The Structure and Function of Biological Membranes*. (Rothfield, L. I., ed.) Academic, New York, 1971, p. 145.

## Membrane Components: Proteins, Lipids, Carbohydrates

Proteins represent the main component of most biological membranes. They have an important role not only in the mechanical structure of the membrane, but also as carriers or channels for transport; they may also be involved in regulatory or recognition properties. Numerous enzymes, antigens, and various kinds of receptor molecules are also present in plasma membranes.

Membrane proteins have been classified as *integral (intrinsic)* or *peripheral (extrinsic)* according to the degree of their association with the membrane and the methods by which they can be solubilized (Table 3–1). Peripheral proteins can be separated by mild treatment, are soluble in aqueous solutions, and are usually free of lipids. Integral proteins represent more than 70 percent of the two protein types and require drastic procedures for isolation. They are usually insoluble in water solutions, and the presence of detergents is required to maintain them in a nonaggregated form. The study of integral proteins from different membranes has shown that they vary considerably in molecular weight. These proteins may be attached to oligosaccharides, forming glycoproteins.

Table 3–2 lists the main proteins and glycoproteins that have been isolated from the erythrocyte membrane. Among the peripheral proteins are *spectrin* (polypeptides 1 and 2) and *actin* (polypeptide 5). These two proteins associate to form microfilaments, which provide a kind of skeletal support for the membrane. The major integral protein *(polypeptide 3)* spans the thickness of the membrane and has

integral (intrinsic) protein—type of membrane protein that requires drastic treatments for isolation and is usually insoluble in aqueous solutions

peripheral (extrinsic) protein—type of membrane protein that can be isolated by mild treatments and is soluble in aqueous solutions

**TABLE 3–2  THE MAJOR ERYTHROCYTE MEMBRANE POLYPEPTIDES AND GLYCOPROTEINS***

| Component | Mol. Wt. | Percent Stained Protein | Polypeptides per Ghost | Other Designations |
|---|---|---|---|---|
| Polypeptides | | | | |
| 1 | 240,000 | 15.1 | 216,000 | spectrin |
| 2 | 215,000 | 14.7 | 235,000 | tektin A |
| 3 | 88,000 | 24.1 | 940,000 | myosin-like polypeptide |
| 4.1 | 78,000 | 4.2 | 180,000 | |
| 4.2 | 72,000 | 5.0 | 238,000 | |
| 5 | 43,000 | 4.5 | 359,000 | actin-like polypeptide |
| 6 | 35,000 | 5.5 | 540,000 | $G_3PD$ |
| 7 | 29,000 | 3.4 | 403,000 | |
| Glycoproteins | | | | |
| PAS-1 | 55,000 | 6.7 | 500,000 | glycophorin |
| PAS-2 | | | | sialoglycoprotein |

*From L. Steck, *J. Cell Biol.*, 62:1, 1974.

a small amount of carbohydrate on the pole at the outer surface (Fig. 3–2). This protein appears to be involved in the diffusion of anions across the membrane. Another important integral protein is *glycophorin* (PAS-1 and 2 in Table 3–2). Approximately 60 percent of this glycoprotein is carbohydrate. Near the COOH end of the molecule is a very hydrophobic region which interacts with the lipids of the membrane; this end is probably exposed to the interior of the red cell.

It is worth re-emphasizing that every protein in the cell membrane is distributed asymmetrically with respect to the lipid bilayer, as is clearly apparent in Figures 3–1 and 3–2. This asymmetry also applies to the numerous enzymes (more than 30 in some cases) that have been detected in isolated cell membranes.

The lipid bilayer (Fig. 3–2, *A*) of the membrane is interrupted only by the proteins that traverse it. This bilayer consists primarily of *neutral phospholipids* and cholesterol, plus 5 to 20 percent *acidic phospholipids*. Within the bilayer the lipids themselves are also distributed asymmetrically.

We have already seen that the oligosaccharide portion of glycoproteins and glycolipids is asymmetrical, and in all cases these residues are exposed only at the outer surface of the membrane.

# 3–2  MOLECULAR MODELS OF THE PLASMA MEMBRANE

The early theories about the molecular organization of the plasma membrane were based on indirect information. Overton observed in 1902 that liposoluble substances (i.e., those soluble in organic

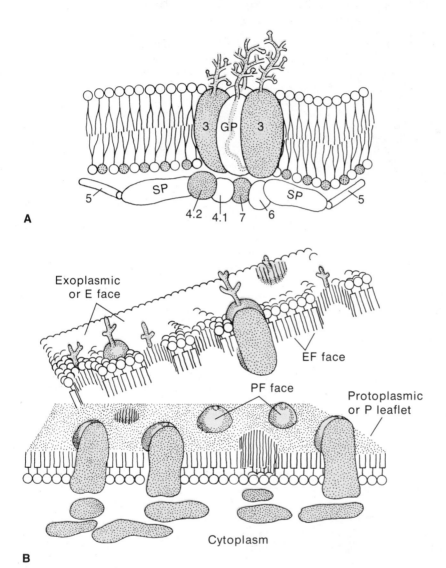

**A**

**B**

Exoplasmic or E face

EF face

PF face

Protoplasmic or P leaflet

Cytoplasm

**Figure 3-2 A,** diagram of the erythrocyte membrane showing the lipid bilayers and the main protein components. The nomenclature used is the same as in Table 3–2. (GP) glycophorin, the main glycoprotein; (band 3) the main intrinsic protein; (SP) spectrin, a typical peripheral protein. The asymmetry of the lipids is indicated. (Courtesy of G. L. Nicolson, redrawn.) **B,** diagram of a membrane submitted to freeze-fracture. Observe that the cleavage plane occurs along the contact between the lipid leaflets. The PF face corresponds to the protoplasmic leaflet and shows more intrinsic proteins that protrude on the surface (see the electron micrograph of Fig. 3–7). The EF face corresponds to the exoplasmic face and may show pits corresponding to the intrinsic proteins in the other leaflet.

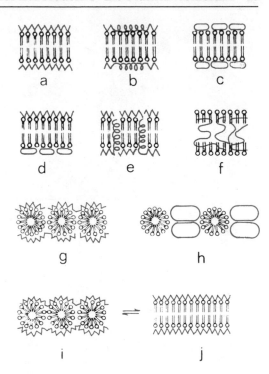

**Figure 3-3**   A variety of molecular models proposed for the plasma membrane; **a-f,** models based on a lipid bilayer structure; **g-j,** models based on globular arrangements. **a,** protein in $\beta$-form; **b,** $\alpha$-helix; **c,** globular protein; **d,** asymmetry in the protein; **e,** partial penetration with protein channels or pores; **f,** protein within the lipid bilayer; **g,** lipid micelles with $\beta$ protein; **h,** lipid micelles with globular protein; **i** and **j,** globular-bilayer transformation. (Courtesy of A. L. Lehninger.)

solvents) penetrate more easily into the cell than those that are water-soluble, and on this basis he postulated that the plasma membrane was composed of a thin layer of lipid. In 1926 Gorter and Grendel measured the lipid content of hemolyzed erythrocytes and found that there was enough lipid to make a double layer of lipid molecules over the entire surface of each cell. Based on these and other properties of the plasma membrane (e.g., high electrical impedance, low interfacial tension), Danielli and Davson (1935) proposed a molecular model in which the membrane was a kind of sandwich, with the lipid bilayer in the middle and protein attached to both surfaces (Fig. 3–3).

## Artificial Model Systems

The Danielli-Davson model was also supported by the study of artificial lipid systems in which *monolayer* and *bilayer* films were formed. Recall that the various lipid molecules described in Chapter 2 are *dipoles*, containing both a polar (hydrophilic) group and a nonpolar (lipophilic) hydrocarbon chain (Fig. 3–4). If a lipid dissolved in certain kinds of solvents is spread on the surface of water, it tends to form a *lipid monolayer* whose thickness depends on the number of carbons in the hydrocarbon chain. The monomolecular film can be deposited on the surface of a glass slide dipped into water, and bilayers or multilayers can be formed by the process of successive dippings.

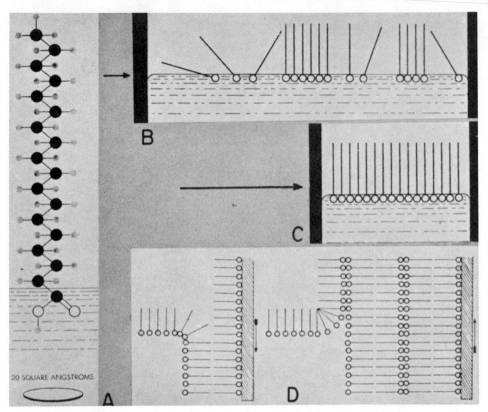

**Figure 3-4**   Diagram of the technique of making monolayer films. **A,** a molecule of stearic acid with the polar group dipped in water. **B,** at low compression, molecules are oriented at different angles or form packed aggregates. **C,** at high compression, molecules are tightly packed and are vertical. Circles represent polar groups and straight lines the nonpolar hydrocarbon chains. **D,** method of building up molecular films at an air-water interface. *Left,* a glass slide previously coated with a monomolecular film of barium stearate (notice the polar groups attached to the glass surface) is dipped in water that has a monomolecular film at the interface. The second monomolecular layer attaches to the first by the nonpolar ends. *Right,* several bimolecular layers of barium stearate have been deposited on the glass slide by successive dips into the water. (A, B, C, from H. E. Ries, Jr.; D, courtesy of D. Waugh.)

Because of their many experimental applications, the *planar* and the *vesicular bilayers* are even more useful than the monolayers just described. By applying a droplet of lipid solubilized in an organic solvent to a small hole in a septum dividing two water chambers, it is possible to produce a lipid bilayer across which many biophysical properties (e.g., electrical resistance, ion permeability) can be studied. It is also possible to introduce into these membranes certain small polypeptides, proteins, and other substances capable of making channels for the passage of ions. These and other artificial systems have proved to be useful models for the study of some permeability mechanisms as well as the protein-lipid interactions within the membrane. Several types of phospholipid-water systems are illustrated in Figure 3–5.

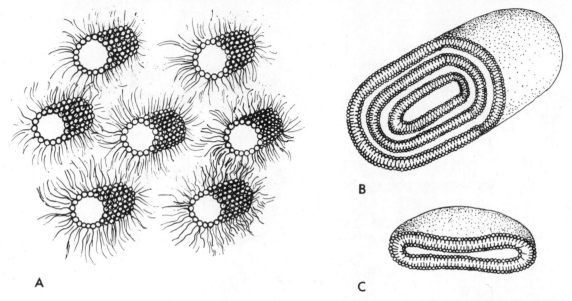

**Figure 3-5** Different types of phospholipid-water systems. **A,** hexagonal phase in a partially hydrated phospholipid. Observe that the water is contained within the cylinders which are limited by polar groups; **B,** multilayered mesophase corresponding to one liposome. Water is contained between lipid bilayers; **C,** phospholipid vesicle, single bilayered structure of 2 nm. (Courtesy of A. D. Bangham and D. A. Haydon, *Br. Med. Bull., 24*:124, 1968.)

## The Unit Membrane Model

The electron microscope revealed, for the first time, that all cells are surrounded by a plasma membrane 6 to 10 nm thick. Early observations revealed a three-layered structure with two outer dense layers of 2.0 nm each and a middle one of about 3.5 nm (Fig. 3–6, **A**). On the basis of these findings, Robertson (1959) postulated the *unit membrane model.* In this model the electron microscopic image was interpreted along the lines of the Danielli-Davson model, with the clear central layer corresponding to the hydrocarbon chains of the lipids and the dense surrounding layers to the proteins on both sides. It was thought that many kinds of intracellular membranes were constructed according to this pattern.

It is now known that the unit membrane model is an oversimplification. It does not, for example, account for the numerous protein molecules that traverse the membrane. In recent years the model has had to be re-evaluated on the basis of the following observations: (1) Extremely thin sections have revealed the existence of fine bridges that cross the bilayer (Fig. 3–6, **C**). (2) Freeze-fracture through the erythrocyte membrane shows that numerous particles are intercalated in the plane of cleavage of the membrane (Fig. 3–7). (3) Fixation methods that avoid the removal of proteins tend to give a more granular appearance to the membrane in a cross-section. According to these findings the unit membrane appears to be somewhat artifactual, perhaps signifying only the appearance of the three-layered structure after preparation for electron microscopy.

unit membrane model—early model of plasma membrane structure, postulating a lipid bilayer sandwiched between two layers of protein

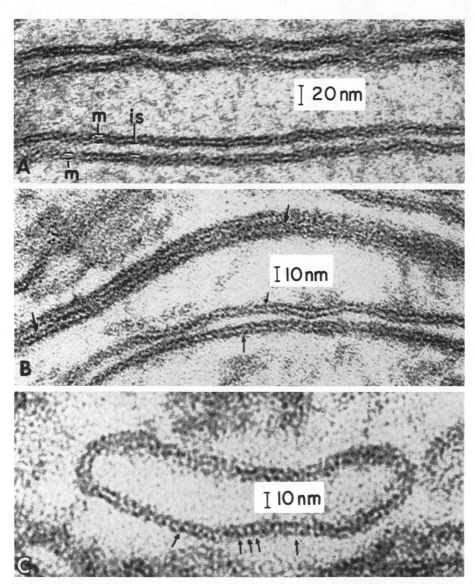

**Figure 3-6  A,** electron micrograph of cell membranes of intestinal cells *(m),* showing the three-layered structure (unit membrane). *is,* intercellular space. ×240,000. **B,** cell membranes in the rat hypothalamus showing the unit membrane structure and, with arrows, some finer details across the membrane. The upper arrows indicate a region in which the two cell membranes are adherent (tight junction) and the intercellular space has disappeared. ×360,000. **C,** the same as **B,** showing fine bridges (arrows) across the unit membrane. ×380,000. (From E. De Robertis.)

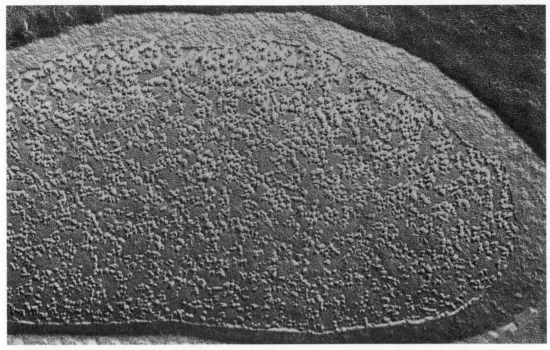

**Figure 3-7**  Electron micrograph of a freeze-fractured and etched red cell ghost. Most of the surface shows particles that are intercalated in the plane of cleavage of the membrane. ×88,000. (Courtesy of D. Branton.)

## The Fluid Mosaic Model

Current knowledge about the molecular organization of biological membranes comes mainly from chemical analyses and from the application of several biophysical techniques. The important concepts that have emerged are summarized in the *fluid mosaic model* of membrane structure. This model postulates that (1) the lipid and integral proteins are disposed in a kind of mosaic arrangement, and (2) biological membranes are quasi-fluid structures in which both the lipids and the integral proteins are able to perform translational movements within the overall bilayer.

To understand the molecular organization of the membrane, it is necessary to remember that not only the lipids but also many of the intrinsic proteins and glycoproteins are amphipathic molecules. (The term *amphipathy*, coined by Hartley in 1936, refers to the presence of hydrophilic and hydrophobic groups within the same molecule.) In the fluid mosaic model represented in Figure 3–1 the integral proteins of the membrane are intercalated to a greater or lesser extent into a rather continuous lipid bilayer. This arrangement is based on the amphipathic character of these proteins, whose polar regions protrude from the membrane surface while their nonpolar regions are embedded in the hydrophobic interior of the membrane.

This molecular disposition may explain why certain enzymes and antigenic glycoproteins have their active sites exposed on the outer

**fluid mosaic model—** currently accepted model of plasma membrane structure, postulating integral proteins embedded in the lipid bilayer but free to move laterally, and possibly projecting out of the bilayer at one surface or the other surface of the cell. It is well recognized that a protein of appropriate size or a cluster of protein subunits may pass across the entire membrane. It is possible that such transmembrane proteins could be in contact with the aqueous solvent on both sides of the bilayer.

Much supportive evidence for the fluid mosaic model comes from freeze-fractured samples of erythrocyte and other cell membranes. The red cell ghosts show a large number of particles, about 8 nm in diameter, that have been interpreted as representing proteins embedded within the plane of cleavage which passes through the middle of the lipid bilayer (Fig. 3–7). In a single erythrocyte there are some 500,000 such particles, of which more are attached to the protoplasmic (inner) half of the membrane than to the external half. It is now thought that these particles correspond to dimers or tetramers of the polypeptide of band 3 (Table 3–2), and that they represent channels for the passage of anions.

## Membrane Fluidity: Biological Importance

membrane fluidity—lateral freedom of movement of proteins and lipids within the bilayer

The concept of membrane fluidity refers to the fact that both lipids and proteins may have considerable freedom of lateral movement within the bilayer. However, vectorial movement across the membrane is severely constrained (meaning that a lipid or protein in the outer half of the bilayer cannot pass into the inner half).

Membrane fluidity is essentially a property of the lipids. Normally these are fluid at body temperature and the main consideration is the degree of saturation of the hydrocarbon chains. Unsaturated fatty acids (i.e., those containing double and triple bonds) have a lower melting point than saturated ones, and in most biological membranes there are sufficient unsaturated lipids that the melting point of the lipid bilayer remains below physiological temperature.

Several physical techniques can be used to study membrane fluidity, but here we shall confine our discussion to some of the biological techniques in use. One of the simplest methods involves binding gold or carbon particles to the cell surface and observing the movement of those particles under the light microscope. More information has come from studies in which different ligands, such as antibodies and plant lectins, interact with cell surface receptors. If these ligands are labeled with fluorescent dyes their movement can be followed by fluorescence microscopy.

If lymphocytes are treated with fluorescent antibodies to certain membrane antigens, it is possible to observe the *capping phenomenon*. The antigens visibly displace in the membrane, forming patches (Fig. 3–8). Then they aggregate at one pole of the cell, producing a kind of cap that is highly fluorescent. At this point the membrane may invaginate into vesicles (by a process called pinocytosis; Chapter 5) which are internalized into the cytoplasm. This capping process can be inhibited by lowering the temperature so that the lipid bilayer solidifies.

A classic experiment involving these phenomena is that of Frye and Edidin, in which two different cultured cells having different surface antigens are fused. *Cell fusion* is achieved by the use of an inactivated parainfluenza virus called Sendai (after a city in Japan).

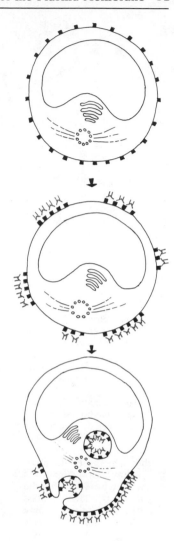

**Figure 3-8** Diagram representing the phenomenon of capping in lymphocytes. **Top,** the normal distribution of antigens on the cell surface. **Middle,** the antigens are now clustered into patches after being cross-linked with a bivalent antibody. **Bottom,** a cap is formed by the active transport of the patches toward the pole that contains the centrosomal-Golgi area. Observe that the membrane is being internalized by endocytosis. Antibodies are in blue. (From S. De Petris and M. C. Raff.)

*Sendai virus* facilitates fusion of the cells' plasma membranes, merging the cytoplasms and thus producing a *heterokaryon* with two nuclei (Fig. 3–9). If the two cells are originally labeled with fluorescent antibodies of different colors [such as fluorescein (green) and rhodamine (red)], it is possible at the outset of fusion to recognize the parts of the plasma membrane corresponding to each cell. However, intermixing occurs as the antigens are dispersed, and the two colors become less and less detectable. After 40 minutes the intermixing is complete and the two antigens can no longer be distinguished. Again, in this case, temperatures below 20°C impair the intermixing by causing solidification of the lipid bilayer.

## The Mobile Hypothesis of Receptors

Membrane receptors are macromolecules that have the dual function of recognizing a chemical signal and initiating a biological response. Several cellular regulatory agents (e.g., peptide hormones,

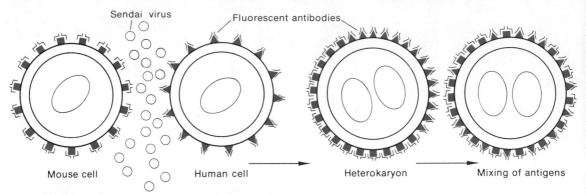

**Figure 3-9**　Diagram representing the cell fusion experiment of Frye and Edidin. A mouse and a human cell, having different surface antigens that are labeled with fluorescent antibodies, are fused by Sendai virus. A heterokaryon is produced in which half of the cell has one type of antigen and the other half has the other type. After 40 minutes, mixing of the antigens has occurred because of their movement within the plane of the membrane.

ligand-receptor interaction—recognition/binding process occurring between cellular regulatory agents (ligands) and specific membrane receptor sites

adenylate cyclase—membrane-bound enzyme that catalyzes the production of cAMP from ATP

mobile hypothesis—proposal that lateral diffusion of receptor molecules within the lipid bilayer allows coupling with adenylate cyclase in a flexible and controllable fashion

neurotransmitters) are ligands that may act as chemical signals which are recognized by such receptors. At the receptor site there is a specific type of binding that involves a *ligand-receptor interaction*. Several kinds of receptors produce a response by interacting within the membrane with the enzyme *adenylate cyclase*, which as we have seen produces cyclic AMP (cAMP) from ATP. Several receptors that are specific for different ligands may act on a single type of adenylate cyclase. The important point is that the effect of the ligands is not additive; in other words, the system behaves as though several receptors were coupled to one molecule of enzyme. To explain this phenomenon, Cuatrecasas and others have postulated the *mobile hypothesis*, which is based on the fluidity of the membrane. According to this hypothesis, the lateral diffusion of receptors and enzyme within the plane of the lipid bilayer allows several receptors to couple with a single adenylate cyclase.

This hypothesis has received experimental confirmation in fusion studies similar to those performed by Frye and Edidin. Cells having only the β-adrenergic receptor in their membrane were fused with others containing only adenylate cyclase (Fig. 3–10). The resulting hybrid cells contained both macromolecules, enabling coupling between the receptor and adenylate cyclase to occur. In these hybrid cells, under the influence of the ligand (in this case, isoproterenol), there was production of cAMP; the original unhybridized cells were unreactive to the ligand.

## 3–3 CELL PERMEABILITY

Permeability is fundamental to the functioning of the living cell and to the maintenance of satisfactory intracellular physiological conditions. This function determines which materials can enter the cell, many of which may be necessary to maintain its vital processes and the synthesis of living substances. It also regulates the excretion of waste material and water.

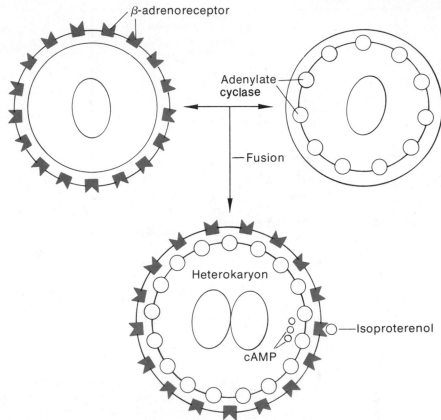

**Figure 3-10** Diagram representing the experiment of Schramm et al. (1977) in which a cell containing the β-adrenergic receptor was fused with another carrying adenylate cyclase. The resulting heterokaryon gave rise to a cell having the components that can couple. In this case stimulation of the receptor by isoproterenol results in the production of cAMP.

The presence of a membrane establishes a net difference between the *intracellular* fluid and the extracellular fluid in which the cell is bathed. This may be fresh or salt water in the case of unicellular organisms growing in ponds or the sea, but in multicellular organisms the internal fluid—blood, lymph, or *interstitial* fluid—is in contact with the outer surface of the cell membrane.

## Passive Permeability: Concentration Gradient

Permeability may be *passive* if it obeys only physical laws, as in the case of diffusion, or *active* if it involves the expenditure of energy. It is common knowledge that if a concentrated solution of a soluble substance (e.g., sugar) is placed in water, there will be a net movement of the solute along the *concentration gradient* (i.e., from the region of high to that of low concentration). However, if a lipoprotein membrane (such as the plasma membrane) is interposed, the diffusion process is greatly modified and the membrane acts as a barrier to the passage of water-soluble molecules.

passive transport—membrane transport that obeys only physical laws

active transport—membrane transport that requires the expenditure of energy

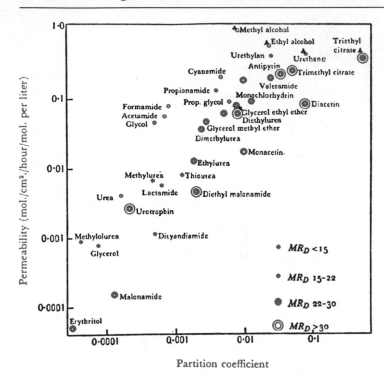

**Figure 3-11**  Rate of penetration (permeability) in cells of *Chara ceratophylla* in relation to molecular volume (measured by molecular refraction, $MR_D$), and to the partition coefficient of the different molecules between oil and water. (After R. Collander and H. Bärlund, *Acta Bot. Fenn.*, *11*:1, 1933.)

At the end of the 19th century, Overton demonstrated that lipid-soluble substances penetrate more easily into the cell. Collander and Bärlund, in their classic experiments on the cells of the plant *Chara*, demonstrated that the rate of entrance of molecules was dependent on their solubility in lipids and their size (Fig. 3–11). The more soluble they are, the more rapidly they penetrate, and given equal solubility in lipids, smaller molecules penetrate faster. The permeability (P) of molecules across membranes can be expressed as

$$P = \frac{KD}{t} \tag{1}$$

**partition coefficient—equilibrium ratio of a solute concentration in the oil phase to its concentration in the aqueous phase**

where K is the partition coefficient; D is the diffusion coefficient, which is a function of molecular weight; and t is the thickness of the membrane. The partition coefficient can be measured by mixing the solute with an oil-water mixture, waiting until the phases separate, and then determining the concentration of solute in each phase. The partition coefficient is the ratio of solute in the oil phase to solute in the water phase. Equation (1) shows that molecules of equal size will penetrate faster if their solubility in lipids is higher; in the case of molecules with equal lipid solubility, the smaller ones will penetrate the membrane faster.

**membrane electrical potential—potential energy change (voltage) across the cell membrane, due to unequal ion distribution**

## Passive Permeability: Electrical Gradient

All cells have a different ionic concentration than their extracellular medium, and an *electrical potential* exists across the membrane.

TABLE 3-3  IONIC CONCENTRATION†
AND STEADY POTENTIAL IN MUSCLE*

| | Interstitial Fluid | | Intracellular Fluid |
|---|---|---|---|
| Cations | Na+ | 145 | 12 |
| | K+ | 4 | 155 |
| Anions | Cl− | 120 | 3.8 |
| | HCO₃− | 27 | 8 |
| | A− and others | 7 | 155 |
| Potential | | 0 | −90 mV |

*Modified from J. W. Woodbury, The cell membrane: Ionic and potential gradients and active transport. In *Neurophysiology*. (Ruch, T. C., Patton, H. D., Woodbury, J. W., and Towe, A. L., eds.) Saunders, Philadelphia, 1961.
†Ionic concentration in mEq.

These two properties are closely related, since the electrical potential depends on the unequal distribution of the ions on both sides of the membrane. Using fine microelectrodes with a tip of 1 $\mu$m or less, investigators are able to penetrate through the membrane into a cell and to detect this potential (also called the *resting* or *steady* potential), which is always negative inside. The values of the membrane potential vary in different tissues between −20 and −100 millivolts (mV).

If we now consider the passive diffusion of ions across a membrane, we shall see that this process is fairly complex. Since ions are charged particles, their diffusion depends not only on the concentration gradient but also on the electrical gradient. Table 3–3 shows that cellular interstitial fluid has a high concentration of Na+ and Cl−, whereas the intracellular fluid has a high concentration of K+ and larger organic anions (A−). In 1911, Donnan predicted that if a cell is put in a solution of KCl, K+ will be driven into the cell by both the concentration and electrical gradients, whereas Cl− will be driven inside by the concentration gradient but repelled by the electrical gradient. At equilibrium the concentrations of K+ and Cl− in and out are reciprocal:

$$\frac{[K^+_{in}]}{[K^+_{out}]} = \frac{[Cl^-_{out}]}{[Cl^-_{in}]} \tag{2}$$

The relationship between the concentration gradient and the resting membrane potential is given by the Nernst equation:

$$E = RT \ln \frac{C_1}{C_2} \tag{3}$$

where E is given in millivolts, R is the universal gas constant, and T is the absolute temperature.

Donnan equilibrium—
relationship between intra-
and extracellular
concentrations of K$^+$ and Cl$^-$
ions, expressed in terms of
membrane potential

From (2) and (3) the Donnan equilibrium for KCl can now be expressed as follows:

$$E = RT \ln \frac{[K^+_{in}]}{[K^+_{out}]} = RT \ln \frac{[Cl^-_{out}]}{[Cl^-_{in}]} \qquad (4)$$

According to (4), any increase in the membrane potential will cause an increase in the ion asymmetry across the membrane, and vice versa. While the first measurements of membrane potentials and ion concentrations seemed to confirm this type of *passive* or *diffusion equilibrium,* more precise determinations in different cell types demonstrated that this was not the case. As discussed in the next section, this discrepancy can be explained by the active transport of ions.

## Active Transport: The Sodium Pump

In addition to the diffusion or passive movement of neutral molecules and ions across membranes, cell permeability includes a series of mechanisms that require energy. These mechanisms are collectively described as *active transport processes.* Adenosine triphosphate (ATP), which is produced mainly by oxidative phosphorylation in mitochondria, is generally used as the energy source. For this reason, active transport is generally related to or coupled with cell respiration.

When an ion is transported against an electrochemical gradient, extra consumption of oxygen is required. It has been calculated that 10 percent of the resting metabolism of a frog muscle is used for the transport of sodium ions. This consumption may increase to 50 percent in some experimental conditions in which the muscle is stimulated.

The resting membrane potential itself is maintained by active transport. This may be demonstrated in plant and animal cells that have been metabolically blocked by oxygen deprivation or specific poisons. In such cases leakage of K$^+$ occurs and the potential across the membrane may decrease to zero. Clearly, the active transport of ions is fundamental to the maintenance of cellular osmotic equilibrium. By regulating the specific concentrations of anions, cations, and other special ions needed for its metabolism, the cell keeps its osmotic pressure constant.

Potassium ions, which are concentrated inside the cell, must enter against a concentration gradient. This can be achieved by a pumping mechanism requiring energy. Sodium ions, which are continually exiting from the cell along with water, must also be transported by an active process. This process is sometimes called the *sodium pump.*

sodium pump—mechanism
that transports K$^+$ into and
Na$^+$ out of the cell against
their respective concentration
gradients, using energy
provided by ATP

The diagram in Figure 3–12 summarizes the relationship existing between the transfer of K$^+$ and Na$^+$ by passive and active mechanisms and the resulting steady state potential. The passive (downhill) ionic fluxes are distinguished from the active (uphill) ionic fluxes. Notice that the active pumping out of Na$^+$ is the main mechanism for maintaining a negative potential of −50 mV inside the membrane. The diagram illustrates that the distribution of ions

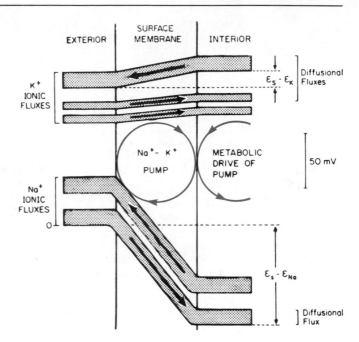

**Figure 3-12** Active and passive Na$^+$ and K$^+$ fluxes through the membrane in the steady state. The ordinate is the electrochemical potential of the ion ($\epsilon_S - \epsilon_K$ for K$^+$, $\epsilon_S - \epsilon_{Na}$ for Na$^+$). The abscissa is the distance in the vicinity of the membrane. The width of the band indicates the size of that particular one-way flux. Passive efflux of Na$^+$ is negligible and is not shown. (After J. C. Eccles, *Physiology of Nerve Cells*, Johns Hopkins, Baltimore, 1957.)

across the membrane depends on the summation of two distinct processes: (1) simple electrochemical diffusion forces which tend to establish a Donnan equilibrium (i.e., passive transport), and (2) energy-dependent ion transport processes (i.e., active transport).

## Na$^+$K$^+$ ATPase and Sodium Transport

We mentioned that Na$^+$, together with water, is eliminated from the cell by an active transport mechanism often called the sodium pump. This mechanism, discovered by Hodkin and Keynes in 1955, was soon associated by Skou with the enzyme *Na$^+$K$^+$ATPase*. This enzyme is able to couple the hydrolysis of ATP with the removal of Na$^+$ from the cytoplasm against an unfavorable electrochemical gradient.

Na$^+$K$^+$ATPase is tightly bound to the cell membrane, from which it may be released by the use of detergents. Its molecular weight has been estimated to be about 670,000, and it probably contains several polypeptides. A red cell contains about 5000 of these enzyme molecules, each of which may extrude 20 Na$^+$ ions per second.

Figure 3–13 shows an idealized diagram of Na$^+$K$^+$ATPase within the red cell membrane. Note that the hydrolysis of one ATP provides the energy for the linked transport of two K$^+$ ions toward the inside and three Na$^+$ ions toward the outside of the cell. This diagram shows the vectorial characteristics of the enzyme, which is sensitive to ATP on the inside of the membrane but not on the outside. This ATPase is stimulated by a mixture of both Na$^+$ and K$^+$. Figure 3–13 also shows that the Na$^+$ and K$^+$ sites are independent and are competitively inhibited by K$^+$ and Na$^+$, respectively. Na$^+$K$^+$ATPase has

Na$^+$K$^+$ATPase—membrane-bound enzyme that couples ATP hydrolysis with the transport of Na$^+$ and K$^+$

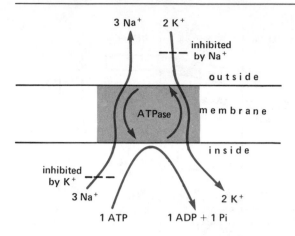

**Figure 3-13**  Diagram of the Na$^+$K$^+$ATPase in the cell membrane. Observe that for each molecule of ATP hydrolyzed at the inner part of the membrane 3 Na$^+$ ions are transported outside and 2 K$^+$ ions are transported inside. (See description in the text.)

been found to be concentrated in the membranes of nerve, brain, and kidney cells, and is particularly rich in the electric organ of the eel and the salt gland of certain marine birds in which the transport of ions is extremely active.

In the Na$^+$K$^+$ATPase system, the transport of Na$^+$ outward is compensated by that of K$^+$ inward and electrical neutrality is achieved. However, there are other cases in which the pump becomes electrogenic, since the exit of Na$^+$ is not compensated by the entrance of K$^+$. The *electrogenic pump* generates a potential which may provide the driving force to transport other solutes. Several substances (glucose, amino acids) may enter the cell by means of the Na$^+$ pump.

**electrogenic pump— mechanism that contributes to the electrical potential across the cell membrane, providing the driving force to transport certain solutes**

## Transport Proteins, Carrier and Fixed-Pore Mechanisms

The transport of many different molecules across the membrane shows a high degree of specificity. That is, the permeability of a molecule is related to its chemical structure. While one molecule may readily enter the cell, another of the same size but a slightly different molecular structure may be completely excluded.

A clear example of this selectivity is provided by the transport of two isomers, glucose and galactose, into a bacterial cell. Although the only difference between them is in the position of an —OH group at carbon 4, these two sugars penetrate the membrane by different transport mechanisms. This type of selectivity is attributed to *transport proteins*, also called *carriers* or *permeases*. It is thought that a permease functions in a manner somewhat similar to an enzyme or a receptor in having a binding site able to recognize the molecule to be transported. Permeases accelerate the transport process, provide for special selectivity, and are recycled, meaning that they remain unchanged after having assisted in the entry or exit of a molecule. Some permeases can transport only if there is a favorable concentration gradient, while others can do so against an unfavorable one. The first type of transport, being driven by a passive mech-

**transport protein— membrane-bound protein that provides for special selectivity among and accelerates the transport process for molecules entering or leaving the cell; also carrier or permease**

anism, is sometimes called *facilitated diffusion*. When the permease operates against the gradient, it is another case of active transport.

The details of the molecular mechanisms by which substances are selectively transported across the plasma membrane are largely unknown. Two general alternative hypotheses have been postulated: a carrier mechanism and a fixed-pore mechanism. A *carrier mechanism* implies that the molecule binds to the transport protein (carrier) at the outer surface of the cell, and that this complex rotates and translocates the molecule into the cytoplasm (Fig. 3–14). Rotation and translocation of the carrier are relatively unlikely, however, in view of what is known about the organization of the cell membrane. Rotation is thermodynamically difficult, and the translocation of a macromolecule from one half of the bilayer to the other is even less likely to occur. The *fixed-pore mechanism* seems more probable because it requires less expenditure of energy. In this mechanism the carrier is represented by integral proteins that traverse the membrane and which, once bound to the molecule to be transported, undergo conformational changes. A fixed pore or channel is generally thought to consist of several protein subunits (oligomers) having a hydrophilic lining in the middle (Fig. 3–14).

An interesting example of a fixed-pore mechanism of transport is provided by the penetration of anions (e.g., chloride, bicarbonate) into red blood cells. The use of special chemical probes that enter such channels and are then fixed covalently has made it possible to identify polypeptide 3 as the site involved in transport. As shown in Figure 3–2, this protein spans the membrane as a dimer (or tetramer). The proposed model is a proteinaceous channel across the membrane, having near the outer surface an anion binding site with three positive charges. There is also a hydrophobic barrier to limit the free diffusion of anions (Fig. 3–15). It is assumed that the segment of the channel that carries the binding site can exist in two conformations, one facing outward and the other inward. In this way it can act as a gate, swinging between the two positions and permitting access to anions originating either inside or outside the cell.

The vectorial function of $Na^+K^+ATPase$ can also be explained along the lines described for a carrier or fixed-pore mechanism. As diagrammed in Figure 3–13, the enzyme has binding sites for $Na^+$ and $K^+$ and a carrier mechanism coupled to the hydrolysis of ATP that generates the energy needed to move the ions against the concentration gradient.

A carrier mechanism has also been postulated to mediate the entrance of glucose into intestinal cells. This mechanism is activated by $Na^+$. The carrier in this model should have a binding site for $Na^+$ and another for glucose, and should exist in two conformational states, one with low affinity for glucose and one with high affinity. A carrier model can be postulated in which the glucose site is initially in a high affinity state and binds the glucose molecule at the surface of the cell. In a second step the carrier is translocated, and the glucose site changes to a low affinity, releasing the molecule into the cell. The two affinities are modulated by the $Na^+$ concentration inside and outside the cell (Fig. 3–16).

facilitated diffusion—passive transport mediated by a permease

carrier mechanism—postulated model in which the molecule to be transported binds to the carrier at the outer cell surface and is translocated into the cytoplasm by rotation of the molecule/carrier complex

fixed-pore mechanism—favored model in which the molecule to be transported binds to an integral carrier protein(s) and is transferred to the cytoplasm by a conformational change induced in the carrier

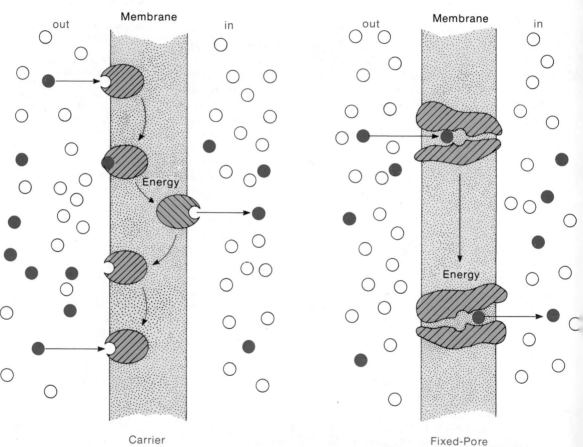

**Figure 3-14**   Diagram representing the carrier and the fixed-pore mechanisms of selective transport. (See description in the text.)

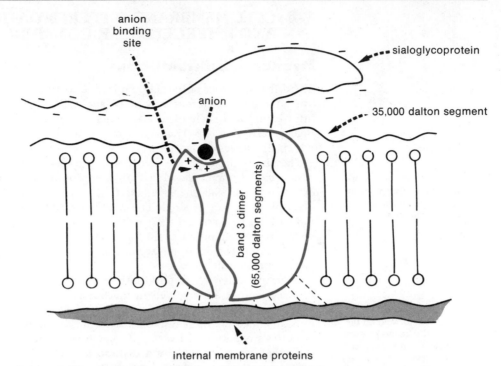

**Figure 3-15**   Schematic model of the disposition of the band 3 protein and of the anion transport site in the erythrocyte membrane. (From A. Rothstein, Z. I. Cabantchik, and P. Knauf, *Fed. Proc., 35*:3, 1976.)

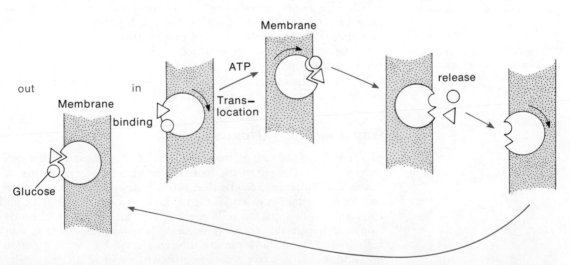

**Figure 3-16**   Coupling of the transport of sodium and glucose by a hypothetical carrier mechanism. (See description in text.)

## 3-4 CELL MEMBRANE DIFFERENTIATIONS AND INTERCELLULAR COMMUNICATIONS

### Membrane Differentiations

The differentiations of the cell membrane correspond to regions that are adapted to different functions, such as absorption, secretion, fluid transport, mechanical attachment, or interaction with neighboring cells. Figure 3–17 is a diagram of an idealized columnar cell which illustrates the various types of differentiations.

microvillus—plasma membrane differentiation that greatly increases the effective absorption surface of the cell

The outer surface of the membrane is projected into slender processes called *microvilli*. Microvilli are prominent in the intestinal epithelium, for example, in which they greatly increase the effective absorption surface. A single cell may have 3000 microvilli that are 0.6 to 0.8 $\mu$m long and 0.1 $\mu$m thick. The core of the microvillus contains a group of fine actin filaments that are attached to the tip by $\alpha$-actinin. The outer surface of this process is covered by a fuzzy coat of glycoproteins.

desmosome—mechanical attachment formed by the thickened plasma membranes of two adjacent cells, containing dense (probably strengthening) material in the intercellular gap

The three classes of differentiations described in the remainder of this section are all involved in mechanical attachment with other cells. The *desmosomes* (*macula adherens*), found in several cell types, appear under the light microscope as darkly stained bodies. These structures are formed by the plasma membranes of two adjacent cells that are separated by a distance of 30 to 50 nm. Numerous *tonofilaments* loop around through the desmosome and cytoplasm of both cells. Within the intercellular gap is a dense material that contains acid mucopolysaccharides and proteins, and presumably assists in strengthening the tissue of which the cells are a part.

intermediary junction—structure similar to a desmosome, in which sections of the adjacent membranes come into contact at certain points

The *intermediary junction* (*zonula adherens* or terminal bar) appears somewhat similar to the desmosome in the electron microscope. The membrane is thickened and the adjacent material is dense, but filaments are generally not present. Within the intermediary junction parts of the two adjacent membranes come into close contact at certain points, forming a kind of network that is especially visible in freeze-etched preparations.

tight junction—structure formed when adjacent cell membranes adhere to each other, creating a barrier to diffusion

In the *tight junction* (*zonula occludens*) the adjacent cell membranes adhere, and consequently there is no intercellular space for a variable distance. Cell contacts of this sort seem to be designed to create a barrier or "seal" to diffusion.

### Gap Junctions (Nexus)

The concept of the cell as the basic unit of all living matter should not lead us to disregard the fact that multicellular organisms are made of populations of cells that interact among themselves. Such *cellular interaction* is essential for the coordination of activities, and the propagation between cells of signals for growth and differentiation is indispensable for development. It is now known that most cells in an organized tissue are interconnected by *junctional channels* and that they share a common pool of many small metabolites and ions that pass freely from one cell to another.

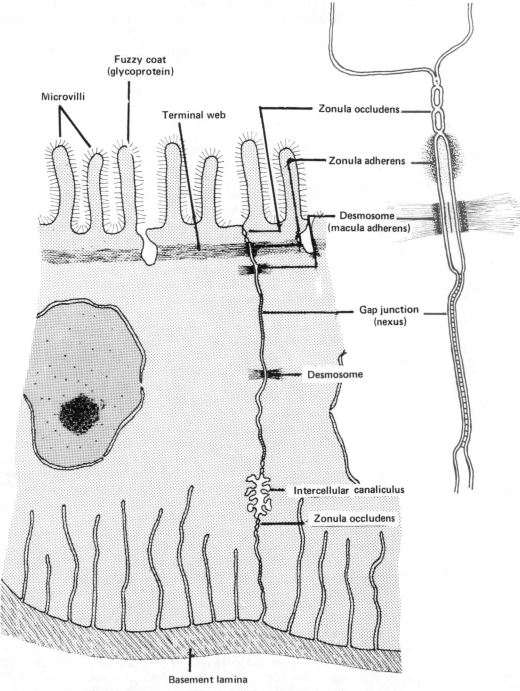

**Figure 3-17** Diagram of an idealized columnar epithelial cell showing the main differentiations of the cell membrane. **To the right,** at a higher magnification, the series of differentiations found between two epithelial cells are indicated. (See description in text.)

One of the manifestations of cellular interaction is electrical coupling between cells. By introducing intracellular microelectrodes into adjacent cells of a tissue, investigators have been able to demonstrate that there are intercellular communications in many animal cells. In this case the cells are electrically coupled and have regions of low resistance in the membrane through which there is a rather free flow of electrical current carried by ions. The other parts of the cell membrane, which are not coupled, show a much higher resistance. This type of coupling, called *junctional communication*, is found extensively in embryonic cells. In adult tissue it is usually found in epithelia, cardiac cells, and liver cells. Skeletal muscle and most neurons do not show electrical coupling.

gap junction—array of plaque-like connections between the plasma membranes of adjacent cells

It is now known that electrical coupling is only one of the forms of intercellular communication that is represented structurally by the *gap junction* (*nexus*) (Fig. 3–17). In a thin section the gap junction appears as a plaque-like contact in which the plasma membranes of adjacent cells are separated by a space of only 2 to 4 nm. This gap can be filled with electron-opaque stains. In tangential sections, gap junctions show a hexagonal array of 8- to 9-nm particles (Fig. 3–18). The electron-opaque material is able to penetrate in between the particles, thus delineating their polygonal arrangement, and also into the central region of each particle. This central region has a diameter of 1.5 to 2 nm and probably corresponds to the location of the channel.

Using freeze-fracturing it is possible to split the junctions and to define their internal structure in more detail. Isolated gap junction membranes have the same polygonal lattice structure as the intact junction; Figure 3–19 shows one such membrane isolated from liver and stained with a negative staining technique. Each unit has a ring structure made up of six identical protein subunits surrounding a hydrophilic channel. Figure 3–19, *B* shows a computer projection map of the units in which the six subunits are clearly revealed. The unit of the gap junction has been named the *connexon*, and it appears to span the bilayer of each of the two connected cell membranes as well as the gap in between (Fig. 3–20). The development of gap junctions has been observed in various cell types. The process involves the appearance of plaques between cells, with an accompanying reduction in the intercellular space, the appearance of larger particles, and finally the arrangement of the particles in polygonal arrays. When two cells are brought into direct contact, junctional communications may be formed in a matter of seconds; new protein synthesis is not required. Thus it is thought that the 8-nm protein particles are already present on the cell surface, and that when cells come into contact these particles diffuse and interact with similar particles on the other membrane. From a physiological viewpoint we can envision the gap junction as resulting from the apposition of two channels, one present in each membrane. We have mentioned that such channels not only bridge each membrane but also project into the intercellular gap, where they are joined to create the connexon. The essential features of this structure are represented in Figure 3–20.

connexon—single unit of a gap junction, consisting of hexagonally arranged particles surrounding a central channel

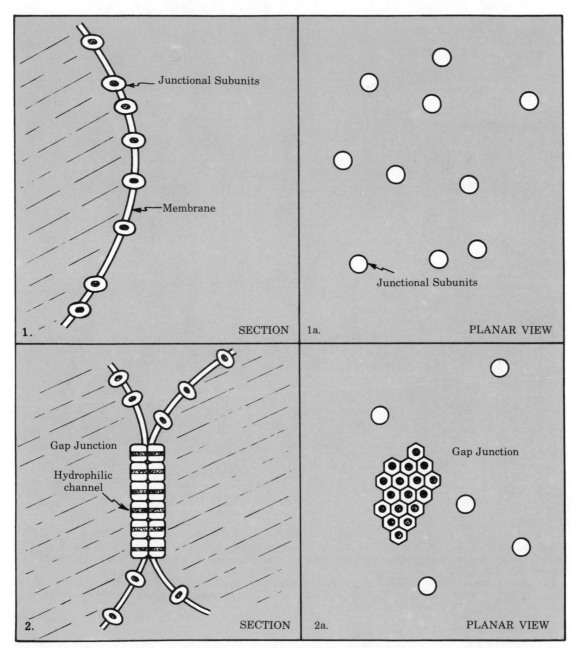

**Figure 3-18**  Diagram of the formation of gap junctions in transverse section and planar view. Larger precursor particles appear first (**1** and **1a**), the space between cells is reduced, and the particles are confluent with the plaques that make the channels (**2** and **2a**). In a planar view the gap junctions show a hexagonal array. (Courtesy of J. D. Pitts.)

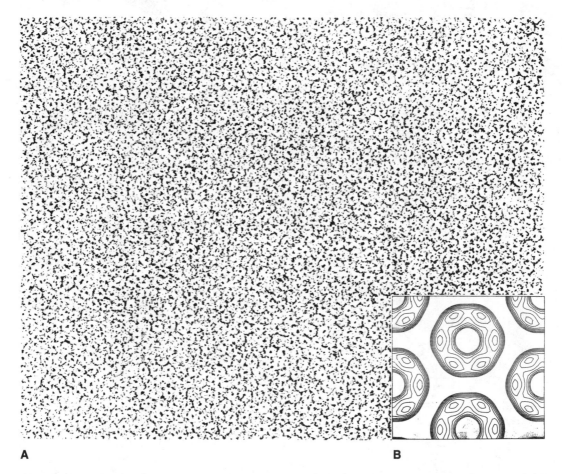

**A**                                                                                      **B**

**Figure 3-19   A,** Surface view of a gap functional membrane isolated from rat liver and negatively stained with uranyl acetate. Observe that each morphological unit has a ring-shaped appearance with a dot of stain in its center. Magnification 425,000×.

**B,** Projection map obtained by computer processing of phases and amplitudes from an area of **A.** The unit cell dimensions are 8.5 × 8.5 nm. Each ring-shaped unit is made of six identical protein subunits surrounding a hydrophilic channel. (Courtesy of G. Zampighi.)

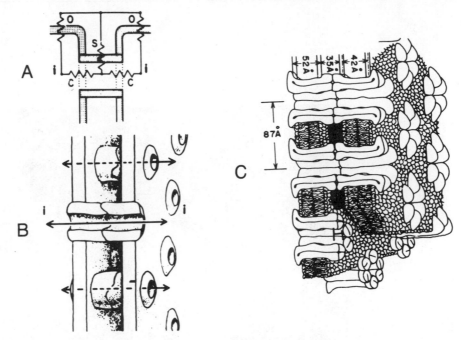

**Figure 3-20** **A,** Diagram of the gap junction channel as defined by electrical measurements and intracellular tracer diffusion. (O) nonjunctional membrane; (C) junctional membrane; (S) junctional insulation. **B,** Model of the gap junction in which two protein channels, one in each membrane, are in continuity. (Courtesy of W. R. Loewenstein.) **C,** Structure of the functional unit as inferred from x-ray diffraction and electron microscopy. The channel has a bore of 2 nm and is formed by two hexamers (six subunits) traversing the lipid bilayer (Redrawn from Makoski et al., 1977; courtesy of W. R. Loewenstein.)

The gap junctions represent a region of high permeability that is separated from the intercellular fluid by a "junctional seal" of low permeability. These junctions not only provide for the electrical coupling that is produced by the passage of ions, but also facilitate the passage of metabolites and even larger molecules. Using various fluorescent peptides it has been possible to observe the intercellular passage of molecules weighing up to 1900 daltons, corresponding to a channel diameter of about 1.4 nm.

## Coupling Implies Metabolic Cooperation

In addition to permitting the free passage of electric current, junctional channels allow the passage of ions and small molecules such as nucleotides, sugars, vitamins, and other metabolites. This exchange makes possible the phenomenon of metabolic cooperation between cells. Figure 3-21 shows the transfer of nucleotides between cells in a culture. A few cells prelabeled with a nucleotide were added to a culture of unlabeled cells; it can be seen that the nucleotide is transferred only between those cells that have direct contacts. Other experiments have shown that macromolecules such as enzymes and RNA are not transferred.

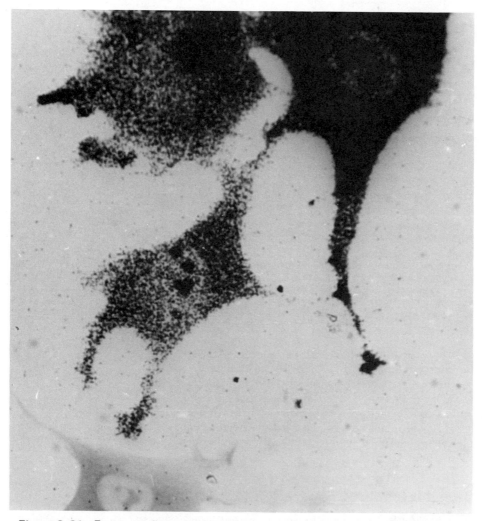

**Figure 3-21**   Experiment illustrating metabolic cooperation. Cultured BHK cells labeled with $^3$H-uridine were added to similar but unlabeled cells. (In this autoradiograph the highly labeled cell on the top right was added to the culture.) It may be observed that the label was transferred to the two cells that make direct contact but not to another which is not in contact (bottom). $\times 1160$. (Courtesy of J. D. Pitts.)

Another interesting finding is that certain cancer cells show no intercellular coupling and fail to communicate with normal cells. An experiment was performed in which such cancerous cells were fused with normal ones and the hybrid cells obtained showed noncancerous behavior. A factor, probably linked to one chromosome, appeared to correct for both the type of growth and the defect in channel communication. This experiment and others have suggested that growth and differentiation may be controlled by regulatory agents (probably nucleotides) which, in normal cells, are transferred from one cell to another. It is evident that the exchange of materials and information is essential for normal functioning.

## 3-5 CELL COAT AND CELL RECOGNITION

Animal cells are frequently covered by a *cell coat* or *glycocalyx*, which generally appears as a layer 10 to 20 nm thick in direct contact with the outer surface of the plasma membrane. The glycocalyx contains the oligosaccharide side chains of the glycolipids and glycoproteins which are exposed to the outer surface of the membrane. It can be considered a secretion product of the cell which is an integral part of the cell surface and is renewed continuously. All the components of the cell undergo active turnover, and some of them are shed into the surrounding medium. We shall see that this is particularly evident in cultured cells, which leave a kind of "footprint" on the glass to which they adhere.

cell coat—continually renewed, outermost layer of an animal cell, in direct contact with the plasma membrane; also glycocalyx

### Functions of the Cell Coat

The cell coat has many other functions in addition to that of protecting the plasma membrane. For example, it may operate as a *filter* in certain blood capillaries (e.g., in the kidney) and in connective tissue. It may provide the cell with a special *microenvironment* having a particular electrical charge, pH, and ion concentration. Certain *enzymes* may be concentrated in the glycocalyx; for example, the fuzzy coat that covers the intestinal microvilli contains all the enzymes used in the terminal digestion of carbohydrates.

*Molecular recognition* is one of the main functions of the cell coat. The number and position of the various oligosaccharides it contains constitute a molecular code or special fingerprint for each cell type. The antigenic properties of a given cell are related to its molecular recognition properties; for example, blood groups are antigens having specific terminal oligosaccharides. Other important antigens are those of *histocompatibility*, which permit the recognition of cells from the same organism and the rejection of alien cells (e.g., a graft). Molecular recognition is particularly significant in the nervous system, where a neuron can establish contacts with several other neurons to form circuits of immense complexity.

molecular recognition— ability of normal cells to recognize similar cells by a signal(s) at the cell surface

histocompatibility—self-recognition of cells or tissues from the same organism

The classic experiment of Wilson (1908) illustrates in a dramatic way the phenomenon of cell-to-cell recognition. If living sponges of different species and colors are forced through a fine silk mesh, they disintegrate into motile cells. If these cells are all mixed together, after some time the cells of the same color will reaggregate among themselves. It has also been discovered that when similar cells attach to each other, they form aggregates that are characteristic for a given cell population (e.g., retinal cell, kidney, or bone). Thus the reaggregation is not necessarily species-specific: if cells of chick and mouse embryos are mixed, they reaggregate according to the type of cell population rather than the species. There is considerable experimental evidence that these cases of cell-to-cell recognition are dependent on the carbohydrate composition and distribution in the cell coat.

## Contact Inhibition and Cancer Cells

contact inhibition—tendency of normal cultured cells to limit their own movement, cell division, and growth

Normal cells growing in tissue culture show *contact inhibition* of movement and growth. When they make contact with neighboring cells, junctions are formed and there is a slowing down of movement. The rate of cell division also decreases and there is a gradual inhibition of cell growth. Contact inhibition is apparently dependent on a signal that propagates only by cell contact and is not effective at a distance.

Cancer cells behave quite differently. They continue to move, and the mitotic rate is not inhibited; the cells tend to pile up, giving rise to multilayers (Fig. 3–22). These effects, referred to as *loss of contact inhibition,* can be studied in normal cultured cells that have been *transformed* into cancerous cells by certain oncogenic (i.e., cancer-causing) viruses.

transformation—conversion of normal cultured cells into cancerous cells

Cancer cells also undergo many changes in their cell surface properties. For example, gap junctions tend to disappear and the concentration of mucopolysaccharides in the cell coat increases. Cancer cells are less well regulated than normal cells, and enzymes may leak into the medium. They may also carry new antigens not present in the normal cell. Some of the many changes in the cell surface that occur in cancer cells are shown in Figure 3–23.

## Fibronectin and the Behavior of Cells in a Tissue

fibronectin—high molecular weight glycoprotein component of the cell coat, thought to aid in determining cellular distribution in embryonic and adult tissues

One of the major chemical components of the cell coat is *fibronectin,* a high molecular weight glycoprotein which can be isolated from normal cultured fibroblasts by mild procedures, such as washing with a solution of urea. Together with other glycoproteins, fibronectin is found in the "footprints" that a moving cultured cell leaves on the substratum with which it comes in contact (e.g., a glass surface). Fibronectin occurs widely in connective tissues, among others, and it is thought to have a role in determining the distribution of cells within both embryonic and adult tissues. The presence of fibronectin increases cell adhesion to the substratum and to other cells and influences the morphology of the cell as well as inducing locomotion and migration.

Fibronectin was originally called LETS (*l*arge *e*xternal *t*ransformation-*s*ensitive) protein because it is absent or drastically reduced

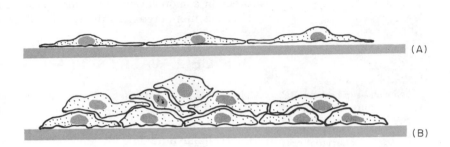

(A)

(B)

**Figure 3-22** Schematic representation of the growth pattern of normal **(A)** and transformed **(B)** cells cultured on a solid substratum. Normal cells are inhibited by contact and stop multiplying, forming a monolayer. Transformed cells continue to multiply, forming multilayers. (From N. R. Ringertz and R. E. Savage, *Cell Hybrids*. New York, Academic, 1976.)

**Figure 3-23** Highly schematic diagram of the surface changes that a cell may undergo after neoplastic transformation by an oncogenic virus. (Courtesy of G. L. Nicolson, 1977, ed.) From Nicolson, G. L., et al., Modifications in transformed and malignant tumor cells. In *International Cell Biology*. (Brinkley, B. R. and Porter, K. R., eds.) Rockefeller University Press, New York, 1977, p. 138.

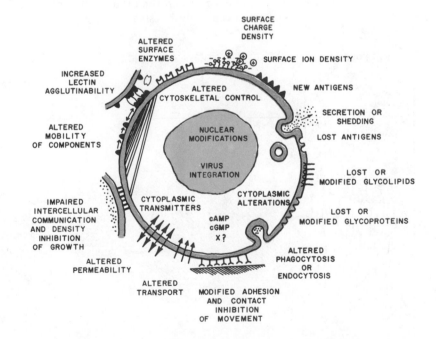

in cultured cells undergoing cancerous transformation. It is conceivable that the reduction of fibronectin, together with the uncoupling phenomenon mentioned above, could have adverse effects on the social behavior of transformed cells. It has been observed that the addition of fibronectin to cultured transformed cells produces changes in their behavior, causing them to acquire a more normal appearance. For example, they attach more readily to the substrate, adopt a more flattened morphology, and tend to align with one another in a monolayer. The cytoskeleton of a transformed cell also becomes more organized under the influence of fibronectin.

## 3-6  CELL WALL OF PLANT CELLS

cell wall—rigid exoskeletal structure enclosing and protecting the contents of most plant and some bacterial cells

The cell wall constitutes a kind of *exoskeleton* that provides protection and mechanical support for the plant cell. One of its functions is the maintenance of a balance between the osmotic pressure of the intracellular fluid and the tendency of water to penetrate the cell. When plant cells are placed in a solution that has an osmotic pressure similar to that of the intracellular fluid, the cytoplasm remains adherent to the cell wall. When the solution of the medium is more concentrated than that of the cell, it loses water, and the cytoplasm retracts from the cell wall. When the solution of the medium is less concentrated than that of the intracellular fluid, the cell swells and eventually bursts.

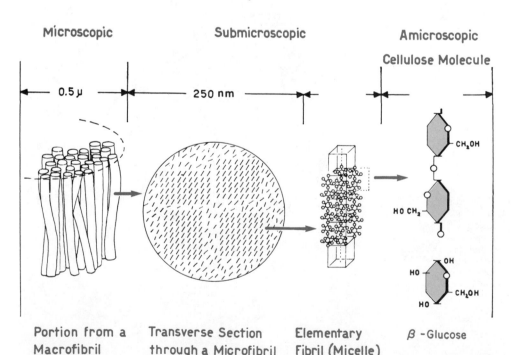

Microscopic      Submicroscopic      Amicroscopic

Cellulose Molecule

|← 0.5 μ →|   |← 250 nm →|

Portion from a        Transverse Section      Elementary        β -Glucose
Macrofibril           through a Microfibril   Fibril (Micelle)

**Figure 3-24**  Structural elements of cellulose at different levels of organization. (From K. Mühlethaler, Plant cell walls. In *The Cell*, Vol. 2. Brachet, J., and Mirsky, A. E., eds.). Academic, New York, 1961.)

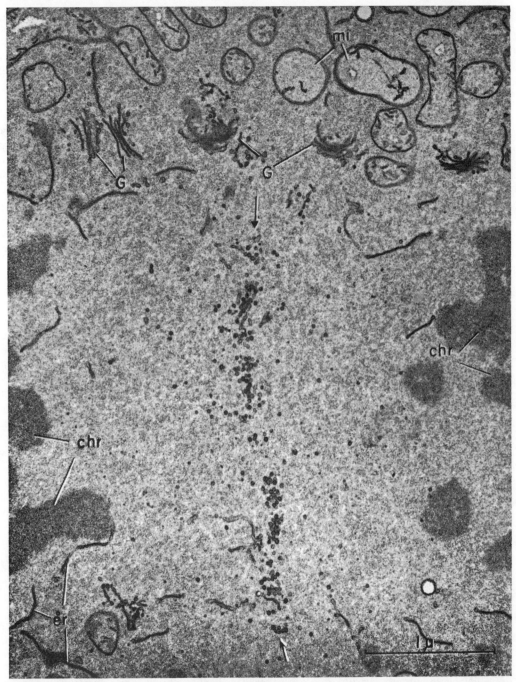

**Figure 3-25** Electron micrograph of root cells of *Zea mays* at telophase. This region corresponds to the cell plate. Note at the top the marginal mitochondria *(mi)* and the Golgi complex *(G)* (dictyosomes). Between the arrows the vesicles are aligned to form the first evidence of a cell plate. *chr,* telophase chromosomes in the two daughter cells; *er,* endoplasmic reticulum. ×45,000. (Courtesy of W. Gordon Whaley and H. H. Mollenhauer.)

The cell wall can be compared in some respects to a piece of plastic reinforced by glass fibers. The wall consists of a *microfibrillar* network lying in a gel-like *matrix* of interlinked molecules. The microfibrils are mostly *cellulose* (the most abundant biological product on earth), consisting of straight polysaccharide chains made of glucose units (Fig. 3–24). The wall also contains other polysaccharides such as pectin, hemicellulose, and lignin.

Most plant cells contain bridges of cytoplasmic material that establish communication between adjacent cells. These bridges, called *plasmodesmata*, pass through the thickness of the cell wall and membrane. Many of these bridges contain tubules that establish continuities between the endoplasmic reticulum of both cells. The plasmodesmata permit the free circulation of fluid, which is essential to the maintenance of plant cell tonicity, and probably also allow the exchange of solutes and macromolecules. According to these concepts, cell walls do not represent complete partitions between cells, but instead constitute a vast perforated structural framework for the organism.

Cell differentiation in plants results mainly from the special synthesis and assembly of the cell wall. The beginning of cell wall formation occurs during and after cell division. The *cell plate*, which is formed from vesicles of the Golgi complex that become aligned at the equatorial plane (Fig. 3–25), forms the intercellular layer or *middle lamellae* of the more mature cell wall. After cell division each daughter cell deposits other layers that constitute the *primary cell wall*. When the cell has enlarged to its mature size, the *secondary wall* appears. This consists of material that is added to the inner surface of the primary wall, either as a homogeneous thickening, as in phloem, or as localized thickenings, as in xylem vessels. The primary cell wall is composed of pectin, hemicellulose, and a loose network of cellulose microfibrils, whereas the secondary wall consists mainly of cellulose and hemicellulose.

**plasmodesmata—cytoplasmic bridges between adjacent plant cells**

# SUMMARY

### 3-1 MOLECULAR ORGANIZATION OF THE PLASMA MEMBRANE

Erythrocyte membranes isolated by hemolysis (ghosts) represent a convenient and fairly typical example for study. Proteins are the main component of the erythrocyte membrane and may be classified as integral or peripheral according to their degree of association with it; among the important proteins in this particular membrane are spectrin, actin, polypeptide 3, and glycophorin. Every protein is distributed asymmetrically with respect to the lipid bilayer, whether it is embedded completely or protrudes on one or both sides. The bilayer of the membrane is interrupted only by the proteins it contains, and within the bilayer the lipids are also distributed asymmetrically.

### 3-2 MOLECULAR MODELS OF THE PLASMA MEMBRANE

In 1935 Danielli and Davson proposed a molecular model of the cell membrane as a kind of sandwich, with a lipid bilayer in the middle and protein attached to both surfaces. This model was supported by the study of artificial monolayer and bilayer systems, with which it has been possible to study many biophysical phenomena such as permeability mechanisms and protein-lipid interactions. On the basis of details later observed by electron microscopy, Robertson (1959) postulated the unit membrane model, which was similar in its essentials to that of Danielli and Davson.

These models have been modified in the light of recent knowledge, and the currently accepted model is the fluid mosaic. This model postulates that (1) the lipid and integral proteins are disposed in a mosaic arrangement within the bilayer, and (2) the membrane is a quasi-fluid structure in which both lipids and integral proteins are able to move laterally within the overall membrane. Much support for this model has come from electron microscopic studies.

Membrane fluidity is a property of the lipids. Several biological techniques can be used to study membrane fluidity. In a classic experiment designed by Frye and Edidin, Sendai virus-initiated cell fusion was used to study the mobility of surface antigens on a heterokaryon. It was demonstrated that two cells labeled with different colored fluorescent antibodies and then fused will eventually appear to be of a uniform color.

Membrane receptors are macromolecules that recognize a chemical signal and initiate a biological response. At the receptor site there is a specific ligand-receptor interaction, although several receptors that are specific for different ligands may act on a single type of adenylate cyclase. The mobile hypothesis has been postulated to explain the finding that the effect of the ligands is not additive. This hypothesis has been confirmed by fusion studies similar to those of Frye and Edidin.

## 3–3  CELL PERMEABILITY

The cell membrane regulates the movement of materials in and out of the cell and establishes a net difference between the intracellular and extracellular fluid. Permeability across the membrane may be passive if it obeys only physical laws, or active if it involves the expenditure of energy. It was demonstrated early in this century that more liposoluble molecules penetrate the membrane more rapidly, and given equal lipid solubility, smaller molecules penetrate faster.

All cells have a different ionic concentration than their extracellular medium, and an electrical potential exists across the membrane, varying between $-20$ and $-100$ mV in different tissues. Passive diffusion of ions depends on both the concentration and electrical gradients. Cellular interstitial fluid has a high concentration of $Na^+$ and $Cl^-$; intracellular fluid has a high concentration of $K^+$ and larger organic anions.

Active transport mechanisms move ions and molecules in or out of the cell against an unfavorable gradient, using ATP as an energy source. Active transport is essential to the maintenance of the resting membrane potential and cellular osmotic equilibrium.

Sodium and water are eliminated from the cell by an active transport mechanism called the sodium pump. This mechanism is facilitated by the enzyme $Na^+K^+ATPase$, which is bound to the cell membrane.

The hydrolysis of one ATP provides energy for the linked transport of two $K^+$ ions toward the inside and three $Na^+$ ions toward the outside of the cell. In some cases the sodium pump becomes electrogenic, generating a potential which may provide the force necessary to transport other solutes.

Transport of molecules across the membrane is highly specific. This specificity is attributed to transport proteins, also called carriers or permeases. These substances are thought to be similar to enzymes or receptors in having a binding site able to recognize the molecule to be transported. Permeases may be driven either by a passive mechanism (facilitated diffusion) or an active one (active transport). Carrier and fixed-pore mechanisms have been postulated to describe the means by which materials are selectively transported across the plasma membrane. The fixed-pore mechanism seems more probable because it requires less energy; it involves conformational changes in the molecules constituting the fixed pore. An example of the fixed-pore mechanism is the penetration of anions into red blood cells.

## 3–4  CELL MEMBRANE DIFFERENTIATIONS AND INTERCELLULAR COMMUNICATIONS

The differentiations of the cell membrane correspond to regions that are adapted to different functions. Microvilli are slender processes that increase the effective absorption surface of the cell. Desmosomes and intermediary junctions are involved in mechanical attachments between cells; desmosomes are thought to have a strengthening function. Tight junctions are also mechanical attachments; they seem to be designed to create a barrier to diffusion.

The gap junction is a type of intercellular contact that facilitates communication between cells. A gap junction consists of a polygonal array of units called connexons; each connexon is made up of six identical protein subunits surrounding a hydrophilic channel. Gap junctions form very rapidly, probably from the apposition of channels in adjacent cell membranes. These channels bridge both membranes and the intercellular gap to allow for electrical coupling as well as the passage of metabolites and larger molecules.

Experimental evidence indicates that ions, nucleotides, sugars, vitamins, and certain other molecules may be exchanged between cells, although enzymes and RNA molecules, for example, may not. Cancer cells show no intercellular coupling and fail to communicate with normal cells. It is evident that normal cell development and function are dependent on the exchange of materials and information.

## 3-5  CELL COAT AND CELL RECOGNITION

Animal cells are frequently covered by a cell coat (glycocalyx) which is in direct contact with the outer surface of the plasma membrane and is a secretion product of the cell. The cell coat may have many functions, including (1) protection, (2) filtering, (3) the provision of a special microenvironment, (4) the localizing of certain enzymes, and (5) molecular recognition. Molecular recognition is thought to be a function of the particular oligosaccharide composition and arrangement in each cell coat type. When similar cells attach to each other, they form aggregates that are characteristic for a given cell population; aggregation is not necessarily species-specific.

Normal cells growing in tissue culture show contact inhibition of movement and growth. With cancer cells, however, this is not the case. Cancer cells also undergo many changes in their cell surface properties, such as loss of gap junctions, increasing concentrations of mucopolysaccharides, and the addition of new antigens not present in the normal cell.

One of the major chemical components of the cell coat is fibronectin, a protein that occurs widely in connective tissues, increases cell adhesion to the substratum and to other cells, influences the morphology of the cell, and is thought to induce cell locomotion and migration during tissue development. Fibronectin is absent or greatly reduced in cells that have undergone cancerous transformation, and the addition of fibronectin to cultured transformed cells causes them to acquire a more normal appearance.

## 3-6  CELL WALL OF PLANT CELLS

The cell wall provides protection and support for the plant cell, as well as maintaining an osmotic balance. The wall consists of a microfibrillar network lying in a gel-like matrix of interlinked molecules. Most plant cells contain bridges of cytoplasmic material, called plasmodesmata, that establish continuity between adjacent cells, permit the circulation of fluid, and probably also allow the exchange of solutes and certain molecules.

# STUDY QUESTIONS

- What proposed structures for the plasma membrane preceded the fluid mosaic model, and on what evidence were they based?
- Define the fluid mosaic model. To what properties of the membrane do *fluid* and *mosaic* refer?
- What is the mobile hypothesis of receptors?
- Explain the difference between passive and active transport.
- What is a Donnan equilibrium?
- How does the sodium pump work, and how is $Na^+K^+ATPase$ involved in its functioning?
- What are the carrier and fixed-pore mechanisms?
- Describe the structure of a gap junction. How is it assembled?
- What is the relationship between the cell coat and the process of molecular recognition?
- Compare normal and cancerous cells with regard to the following properties: intercellular coupling; contact inhibition; cell surface properties; role of fibronectin.

# SUGGESTED READINGS

**Abercrombie, M.** (1979) Contact inhibition and malignancy. *Nature, 281*:259.

**Albersheim, P.** (1975) The walls of growing plant cells. *Sci. Am., 232*:80.

**Edelman, M.** (1976) Surface modulation in cell recognition and cell growth. *Science, 192*:218.

**Keynes, R. D.** (1979) Ion channels in the nerve cell membrane. *Sci. Am., 240*:126.

**Lodish, H. F., and Rothman, J. E.** (1979) The assembly of cell membranes. *Sci. Am., 240*:48.

**Loewenstein, W. R.** (1979) Junctional intercellular communication and the control of growth. *Biochim. Biophys. Acta, 560*:1.

**Luftig, R. B., Wehrli, E., and McMillan, P. N.** (1977) The unit membrane image: A re-evaluation. *Life Sci., 21*:285.

**Nicolson, G. L.** (1979) Cancer metastasis. *Sci. Am., 240*:66.

**Pearlstein, E., and Garcia-Pardo, A.** (1980) Fibronectin: a review. *Mol. Cell. Biochem., 29*:103.

**Quinn, P.J., and Chapman, D.** (1980) The dynamics of membrane structure. CRC Rev. Biochem. *8*:1.

**Raff, M. C.** (1976) Cell-surface immunology. *Sci. Am., 234*:30.

**Roland, J. C., and Vian, B.** (1979) Wall of the growing plant cell. *Int. Rev. Cytol., 61*:129.

**Singer, S. J., and Nicolson, G. L.** (1972) The fluid mosaic model of the structure of cell membranes. *Science, 175*:720.

**Staehelin, L. A., and Hull, B. E.** (1978) Junctions between living cells. *Sci. Am., 238*:140.

**Unwin, P. N. T., and Zampighi, G.** (1980) Structure of the junction between communicating cells. *Nature, 283*:545.

# Cytoskeleton and Cell Contractile Systems

In Section 1–5 we introduced the features of the ultrastructure of the cytoplasm and defined the matrix or ground cytoplasm as the true internal milieu of the cell. The existence of an organized fibrous array of macromolecules in the cytoplasmic matrix was postulated many years ago, particularly by Koltzoff (1929), who concluded that "each cell is a system of liquid components and rigid skeletons which generate [its] shape. . . ." The name *cytoskeleton*, coined at approximately that time, has since been abandoned because many structures observed in the cytoplasm with the light microscope were considered to be fixation artifacts.

In recent years the electron microscope has revealed that the cytoplasmic matrix of most eukaryotic cells contains a cytoskeletal fabric made up of *microtubules* and various types of *microfilaments* arranged in a highly structured, three-dimensional lattice (Fig. 4–1). Considerable progress has been made in the isolation of the proteins that constitute these structures, and much has been learned about the mechanism of their assembly. In the first sections of this chapter we shall describe the microtubules and microtubular organelles that are involved in cell division and in certain forms of cell motility. Then we will discuss the microfilaments and the cytoskeletal lattice that are involved in other types of cell motility; and finally, we will look briefly at the myofibril of skeletal muscle as an example of highly elaborate microfilament organization.

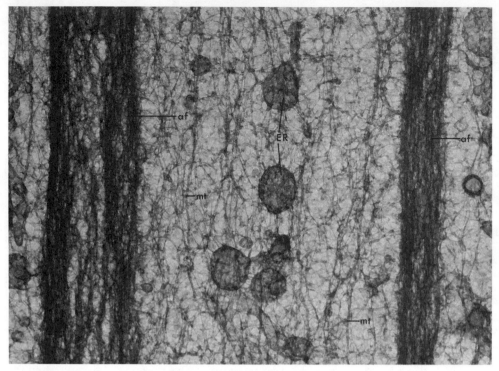

**Figure 4-1**   Electron micrograph of a thin section of a cultured cell made with the high-voltage electron microscope at $10^6$ volts. Two bundles of actin filaments (*af*) run vertically. Vesicles of the endoplasmic reticulum *(ER)* and a few microtubules *(mt)* are also visible. Observe the lattice of microtrabeculae that pervades the matrix. ×40,000. (Courtesy of K. R. Porter. From K. R. Porter, Introduction: Motility in cells. In *Cell Motility*, Vol. 1. (Goldman, R. D., et al., eds.). Cold Spring Harbor Laboratory, New York, 1976, p. 1.)

# 4-1  MICROTUBULES

*microtubule—hollow cylindrical structure made up of longitudinal fibrils and thought to assist in intracellular movements; also component of the cytoskeleton and numerous organelles*

Microtubules are structures universally present in the cytoplasm of eukaryotic cells. They were first observed in axon protoplasm by De Robertis and Franchi in 1953. Since most microtubules are rather labile and do not respond well to many fixatives, intensive studies began only after 1963 with the introduction of glutaraldehyde fixation in electron microscopy.

Cytoplasmic microtubules are uniform in size—about 25 nm in outer diameter and several micrometers in length—and remarkably straight. In cross-section they show a dense wall about 6 nm thick and a light center. The wall of the microtubule consists of individual linear or spiraling filamentous structures of about 5 nm in diameter. In a cross-section there are usually 13 of these filaments, arranged with a center-to-center spacing of 4.5 nm (Fig. 1–20). We shall see that although all microtubules seem to have nearly the same physiological characteristics, they differ in other respects, such as in their resistance to various treatments.

## Microtubule Assembly from Tubulin Dimers

Microtubules are composed of protein subunits that are rather similar, even though they are found in a variety of cell types. The term *tubulin* is used to designate the principal protein of cilia, flagella, and cytoplasmic microtubules. Tubulin is a dimer of 110,000 to 120,000 daltons. In most cases its two monomers are of different kinds but are similar in molecular weight. The 8 nm spacing along the longitudinal axis of microtubules (Fig. 1–20) probably reflects the pairing of the two types of tubulin monomers. One dimer of tubulin binds to a molecule of $^3$H-colchicine, and this property is used to assay for this protein.

The assembly of microtubules from the tubulin dimers is a specifically oriented and programmed process; as with the production of enzymes, the quantity of polymerized tubulin varies according to the need for it in the cell. Normally microtubules are in equilibrium with free tubulin; phosphorylation of the tubulin monomers by a cyclic AMP-dependent kinase favors the polymerization. The assembly of tubulin is polarized, meaning that the dimers assemble at one end of the microtubule while they disassemble at the other end. When colchicine binds to tubulin the assembly process is inhibited but disassembly continues, so that eventually the microtubules are completely depolymerized (Fig. 1–20).

Microtubules contain other proteins that are collectively called *microtubule-associated proteins* (MAPs), several of which are in a high molecular weight range (~280,000 daltons) and are involved in microtubular assembly. Highly purified tubulin does not polymerize into microtubules.

In the cell there are sites of orientation, such as centrioles and the basal bodies of cilia, from which the process of polymerization is directed in some way. Cytoplasmic microtubules often extend radially from the nucleus and appear as straight or curved filaments that seem to terminate near the cell surface (Fig. 4–2). These filaments disappear when treated with *Colcemid* (a derivative of colchicine) or when cooled, and they reappear if the conditions are reversed. This has revealed that microtubules arise near the nucleus, from one or two focal points corresponding to the centrosomal (i.e., centriole-containing) region or *centrosphere*. Many such observations have suggested that the centrosphere is the main microtubule-organizing center. In cells about to enter mitosis the cytoplasmic microtubules disappear and are replaced by those integrating the spindle and asters. In cells transformed by a virus or by certain chemicals, the microtubules are disorganized and sometimes disappear entirely.

## Functions of Cytoplasmic Microtubules

**MECHANICAL FUNCTION.**  The shape of some cell processes or protuberances has been correlated with the orientation and distribution of microtubules. They are considered to be the framework which determines cell shape and the distribution of its content.

tubulin—principal protein in many types of microtubules, occurring as a dimer whose two subunits differ slightly from each other

colchicine—drug that prevents microtubule polymerization and facilitates depolymerization of existing microtubules by binding to tubulin subunits; its derivative Colcemid has similar effects

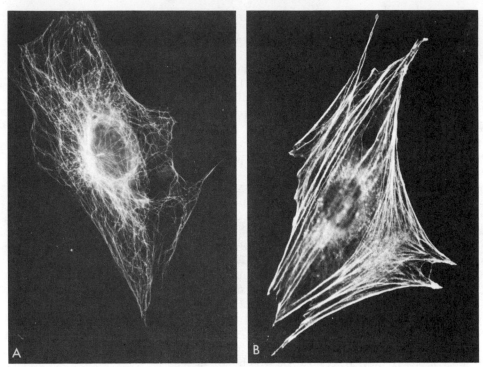

**Figure 4-2**   Cultured 3T3 cells stained by immunofluorescence with antibodies against tubulin **(A)** and actin **(B)**. ×600. (Courtesy of M. Osborn and K. Weber, 1976.)

Their integrity is necessary to maintain the characteristic shape of many cells and the rigidity of elongated structures. A clear example of such structures would be the axons and dendrites of neurons.

**MORPHOGENESIS.**   Related to their mechanical function is the role that microtubules play in the shaping of the cell during *cell differentiation*. The morphogenetic changes that occur during spermiogenesis provide an interesting example. The elongation that takes place in the nucleus of the spermatid is accompanied by the production of an orderly array of microtubules that are wrapped around it.

**CELLULAR POLARITY AND MOTILITY.**   The determination of the intrinsic polarity of certain cells is also related to the mechanical function of microtubules. Treatment of various culture cells with Colcemid generally results in a change of motion; although certain forms of motion persist, the directional gliding of the cell, for example, is replaced by a random movement.

**CIRCULATION AND TRANSPORT.**   Microtubules may also be involved in the transport of macromolecules within the cell; to this end, they probably form channels in the cytoplasm. One example of an association between microtubules and the transport of particulate material occurs in the melanocyte, in which the melanin granules have been observed moving between channels created by the microtubules in the cytoplasmic matrix.

**SENSORY TRANSDUCTION.**    Regularly arranged bundles of microtubules are common in sensory receptors, and a possible function in energy transduction has been postulated.

# 4–2 MICROTUBULAR ORGANELLES

Several cell organelles derive from special assemblies of microtubules. Some of them, such as asters and the spindle, are transitory organelles that appear and disappear in relation to the mitotic or meiotic cycles. Others, such as cilia, flagella, basal bodies, and centrioles, are more permanent.

## The Mitotic Apparatus

The function of the mitotic apparatus will be considered in Chapter 9 in the discussion of cell division. Here we shall simply describe its microtubular organization.

The *aster* appears as a group of radiating microtubules that converge toward a centriole, around which there is often a clear zone called the *microcentrum* or *centrosome*. Not much is known about the function of the asters, although in animal cells the spindle is usually formed between them during mitosis.

The *spindle* is also formed by microtubules and can be studied to best advantage in the electron microscope. [However, since both the aster and spindle microtubules have an optical property called *positive birefringence*, it is possible to observe them in living cells by the use of polarization and interference microscopy (Fig. 4–3).] Two types of mitotic apparatus are recognized, depending on the presence of asters (*astral spindle)* or their absence (*anastral spindle).* In plant cells there are neither centrioles nor asters (Fig. 4–4), and microtubule formation is not related to the poles but rather to the *kinetochores,* which are special regions of the centromeres of chromosomes (Chapter 9). Studies have revealed that the centrioles, spindle, and kinetochores all contain tubulin, suggesting that these structures contain intermediates for the assembly of tubulin into microtubules.

The spindle fibers were formerly divided into three types: (1) chromosomal, joining the chromosomes to the poles; (2) continuous, extending from pole to pole; and (3) interzonal, observed between the daughter chromosomes at anaphase and telophase (Fig. 4–4). With the electron microscope it has been demonstrated that there are (1) *kinetochoric* microtubules, which originate from this region of the chromosomes; (2) *polar* microtubules, which originate from the poles; and (3) *free* microtubules. Quantitative studies have shown that the so-called continuous fibers are not in fact continuous, since microtubules generally do not extend from pole to pole.

One of the most intriguing aspects of the spindle is the facility

aster—bundle of microtubule fibers radiating out from each cell pole during the metaphase stage of mitosis

spindle—microtubular structure spanning the cell from pole to pole during metaphase; involved in the correct partitioning of chromosomes to daughter cells

birefringence—property of some substances to affect the polarization of transmitted light in such a way that these substances appear prominent under a polarization microscope

kinetochore—proteinaceous region of the centromere to which spindle microtubules attach during cell division

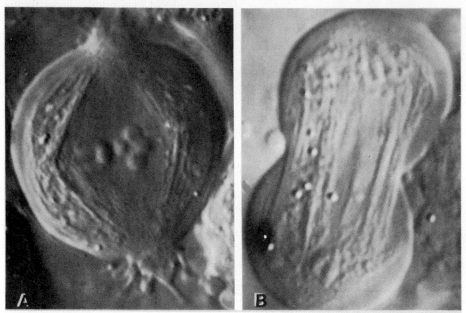

**Figure 4-3**   Metaphase **(A)** and anaphase **(B)** of meiosis in the insect *Nephrotoma soturalis* observed with Nomarski interference microscope. (Courtesy of F. Muckenthaler.)

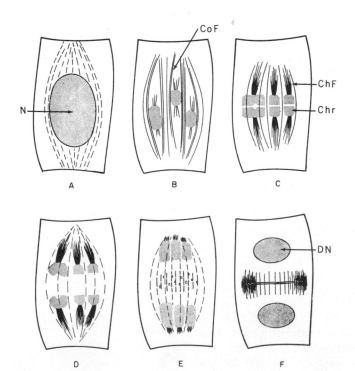

**Figure 4-4**   Diagram of mitosis in a plant cell showing the changes in birefringence of the spindle fibers. Note the absence of centrioles and asters. Abbreviations: *CoF,* continuous fibers; *ChF,* chromosome fibers; *Chr,* chromosome; *N,* nucleus. (Courtesy of S. Inoué.)

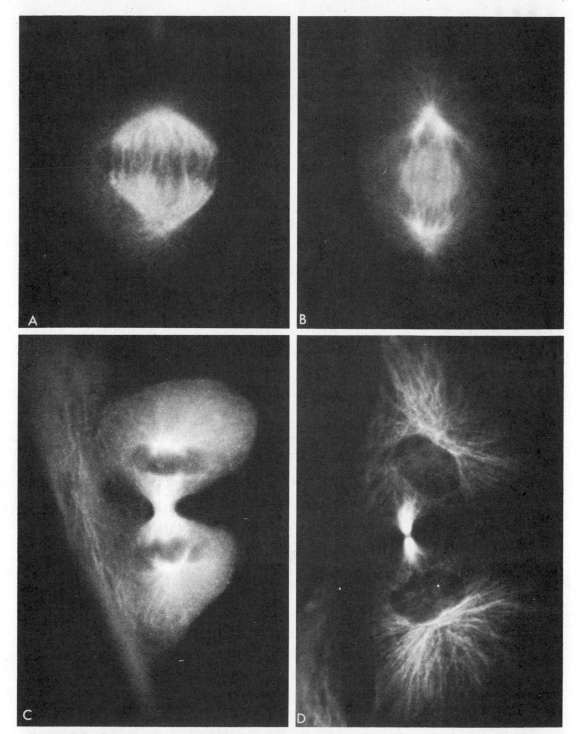

**Figure 4-5**  Cultured cells in mitosis stained by immunofluorescence with anti-tubulin antibodies. **A,** *Metaphase.* The stain is concentrated in the spindle fibers. **B,** *Anaphase.* Spindle fibers are apparent, but there is staining at the asters and the midzone of the spindle. **C,** *Telophase.* The asters and the midzone of the spindle are more heavily stained. **D,** *After telophase.* The tubulin fibers radiate from the asters, and the cytoplasmic bridge *(midbody)* between the two daughter cells is heavily stained. (Courtesy of B. R. Brinkley.)

with which it assembles and disassembles, both in vivo and in response to various treatments. We have seen that a cell normally contains a pool of tubulin monomers that are in a kind of dynamic equilibrium with the polymerized microtubules. As shown in the sequence of Figure 4–5, during mitosis there is also a cycle of assembly-disassembly between the cytoplasmic microtubules and those incorporated in the spindle. The intracellular concentration of $Ca^{2+}$ is thought to have an important role in this cycle. The function of $Ca^{2+}$ has been recently related to a $Ca^{2+}$ binding protein of low molecular weight called *calmodulin*.

## Cilia and Flagella in Cells and Tissues

cilium—short microtubular organelle projecting into the extracellular medium; used for cell locomotion or to create currents in the surrounding fluid

flagellum—long microtubular organelle projecting into the extracellular medium; used for cell locomotion

The organelles responsible for cell motility are adapted to liquid media and consist of minute, specially differentiated appendices that vary in number and size. They are called *flagella* if they are few and long, and *cilia* if they are numerous and short. There is an entire class of protozoa that is characterized by the presence of flagella; and the spermatozoa of metazoans move as isolated cells by means of flagella. There are also many classes of protozoa in which each cell has hundreds or thousands of cilia which permit the organism to progress rapidly through the liquid medium. Ciliated epithelial sheets covering large areas of the external surface may determine the motion of an organism, although such sheets are more often found lining cavities or internal tubes, such as the air passages of the respiratory system or the various parts of the reproductive tract. In these passages all the cilia move in unison in such a way as to produce currents in the surrounding fluid. In some cases, the currents serve to eliminate solid particles in suspension (e.g., the eggs of amphibians and mammals are driven along the oviduct with the aid of ciliary currents).

The essential components of the ciliary apparatus are (1) the *cilium*, which is the slender cylindrical process that projects from the surface of the cell, and (2) the *basal body* from which the cilium originates and about which we shall have more to say later. In some cells there is a third component consisting of fine fibrils or *ciliary rootlets*, which arise from the basal body and converge into a conical bundle that ends near the nucleus.

axoneme—axial microtubular component of cilia and flagella; thought to be their essential motile element

*Axoneme* is the term applied to the axial microtubular structure of cilia and flagella. This structure is thought to be the essential motile element. The axoneme of a cilium or flagellum may range from a few microns to 1 or 2 millimeters in length, but its outside diameter is about 0.21 $\mu$m. It is generally surrounded by an *outer ciliary membrane* which is continuous with the plasma membrane. All the components of the axoneme are embedded within the *ciliary matrix* (Fig. 4–6).

9 + 2 configuration—characteristic pattern of microtubules found in cilia and flagella, consisting of nine pairs of peripheral subfibers surrounding a central pair

Figure 4–7 shows a general diagram of the axoneme, illustrating the fundamental 9 + 2 microtubular pattern and the relationship of the essential structural components. A plane perpendicular to an imaginary line joining the two central tubules divides the axoneme into a right and a left symmetrical half. It is generally thought that

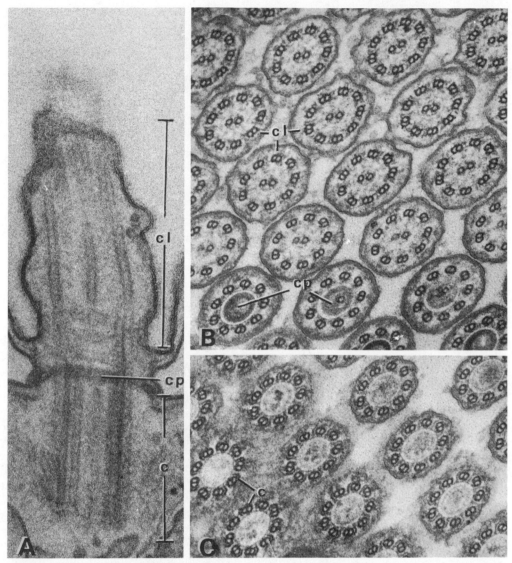

**Figure 4-6**  Electron micrographs of cilia in longitudinal and cross-sections. **A,** cilium of *Paramecium aurelia* showing the centriole or basal body *(c),* the ciliary plate *(cp),* and the cilium *(cl)* proper. **B** and **C,** cross-sections through cilia of *Euplotes eurystomes.* **B,** the section passes through the cilium proper showing the typical structure and the ciliary plate *(cp).* **C,** the section passes through the centriole or basal body *(c).* Notice the absence of central tubules and triple number of peripheral tubules. **A,** ×110,000; **B** and **C,** ×72,000. (Courtesy of J. André and E. Fauret-Fremiet.)

the plane of the ciliary beat is perpendicular to this plane of symmetry. Close examination of Figure 4–6 shows that the nine pairs of peripheral microtubules are ellipsoidal in cross-section, whereas the central ones are circular. Furthermore, the two microtubules of each peripheral pair can be distinguished by several morphological features. One tubule, designated subfiber A, lies closer to the central axis than the other (i.e., subfiber B). The microtubule of subfiber A

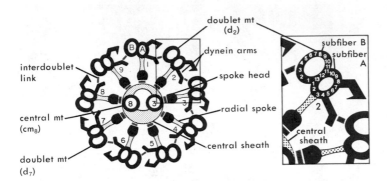

**Figure 4-7** Diagram of a cross-section of a cilium showing the axoneme with the "9 + 2" microtubular structure. The axoneme is viewed from base to tip, with the arms directed clockwise. The inset shows doublet 2 ($d_2$) at high magnification with indication of the tubulin subunits. (See description in text.) Modified from F. D. Warner, Macromolecular organization of eukaryotic cilia and flagella. In *Advances in Cell and Molecular Biology*, Vol. 2. (DuPraw, E. J., ed.) Academic, New York, 1972, p. 193.

is smaller but complete, whereas that of subfiber B is larger and incomplete, since it lacks the wall adjacent to A; A has 13 tubulin subunits, whereas B has only 11. Subfiber A also has processes, called the *dynein arms,* that are oriented in the same direction in all microtubules. As shown in Figure 4–7, the doublets are linked by *interdoublet* or *nexin links* (a name derived from the fact that the protein nexin has been isolated from them). There are also radial bridges or spokes between the A subfiber and the sheath containing the central microtubules. Observations of these spokes in straight and bent or tilted regions of the axis have led to the hypothesis that they may have an active role in local axial bending.

Ciliary movement can be studied on the free surface of an epithelial cell, where synchronized waves of contraction occur as long as continuity with the cytoplasm is maintained. The basic process underlying this movement appears to be the sliding of the microtubular doublets over each other, with the associated making and breaking of cross-bridges between adjacent doublets, represented by the dynein arms. The sliding of doublets during the bending of the cilium (or flagellum) has actually been observed and measured (Fig. 4–8).

dynein—microtubule subfiber protein having ATPase activity and thought to be essential for movement

The dynein arms contain dynein, a high molecular weight ATPase that is activated by $Mg^{2+}$ and $Ca^{2+}$. Most of the experimental work on ciliary motion has been aimed at demonstrating the involvement of ATP. When ATP is made available to cilia and flagella under certain experimental conditions, these organelles begin rhythmic activity that may persist for a few minutes or even hours. Interaction between tubulin and dynein seems to be essential in this type of cell motility. When we study muscle myofilaments later in this chapter, we shall see that the actin-myosin system is also involved in contraction.

A recent approach to the study of ciliary motion is the use of mutant cells in which various deficiencies in axoneme structure have been observed. From the medical point of view, this research is of great significance with regard to a human syndrome *(Kartagener's syndrome)* characterized by nonmotile sperm, chronic bronchitis, and sinusitis. It has been demonstrated that this syndrome results from a mutation involving the dynein arms.

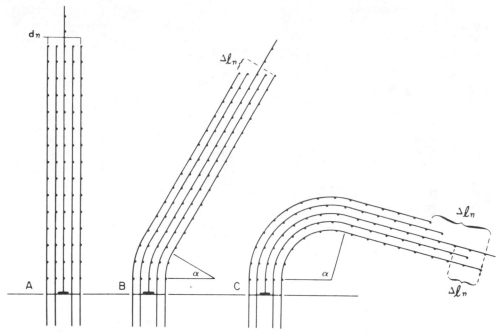

**Figure 4-8**  Diagram of the sliding model of the axoneme. The amount of displacement ($\Delta_{ln}$) is a function of the distance separating the microtubules ($d_n$) and the bend angle ($\alpha$). The central microtubules are represented by the longer lines. (From M. A. Sleigh, The physiology and biochemistry of cilia and flagella. In *Frontiers of Biology: Handbook of Molecular Cytology*. Vol. 15. (Lima-De-Faría, A., ed.) North-Holland, Amsterdam, 1970, p. 1243.)

## Basal Bodies and Centrioles

Since the classic studies of Henneguy and Lenhossek in 1897, it has been suggested that *basal bodies* (or *kinetosomes*) of cilia and flagella are homologous with the *centrioles* found in mitotic spindles (Fig. 4–9). It was discovered that in some cells a centriole engaged in mitosis could carry a cilium at the same time. This homology was fully confirmed by electron microscopy.

Centrioles are cylinders that measure, on the average, 0.2 µm × 0.5 µm; at times they may be as long as 2 µm. This cylinder is open at both ends unless it carries a cilium. In the latter case, it is separated from the cilium by a *ciliary plate* (Fig. 4–6).

The wall of the centriole has nine groups of microtubules arranged in a circle. Each group is a triplet, formed of three tubules (rather than two, as in cilia) that are skewed toward the center (Fig. 4–9). The tubules are designated subfiber A, B, and C from the center toward the periphery. Subfibers A and B both cross the ciliary plate and are continuous with the corresponding subfibers in the axoneme; subfiber C terminates near the ciliary plate. There are no central microtubules in the centrioles and no special arms.

The centriole grows from the distal end, and in the case of kinetosomes it is from this end that the cilium is formed. The *procentrioles*, which are formed at right angles to the centriole, are located

centriole—hollow cylindrical organelle involved in the organization of the spindle; its wall contains nine microtubule triplets

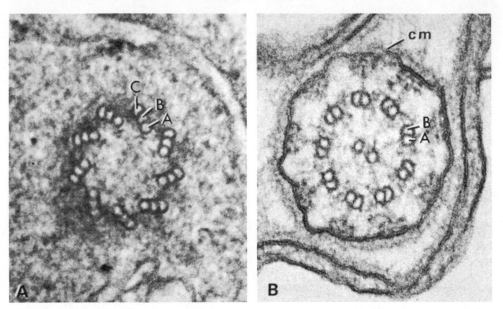

**Figure 4-9   A,** transverse section through a centriole of the chick embryo. Observe the three tubules (A, B, C) present in each of nine groups at the periphery of the centriole. Also note the absence of central tubules and the density of the centriolar wall. **B,** transverse section through a cilium (compare with the centriole structure); *cm,* ciliary membrane. ×150,000. (Courtesy of J. Andrè.)

near the proximal end. There are now several indications that centrioles may be generated by two different mechanisms. Centrioles destined to form mitotic spindles arise directly from the wall of the preexisting centriole. The *daughter centrioles* appear first as annular structures (procentrioles) which lengthen into cylinders. The groups of three tubules originate from single and double groups that first appear at the base of the procentriole. When they are half grown, the daughter centrioles are released into the cytoplasm to complete their maturation.

The other mechanism of centriole formation produces those destined to become basal bodies, as in a ciliated epithelium. These structures are assembled progressively from a precursor fibrogranular material located in the apical cytoplasm. The newly formed kinetosomes become aligned in rows beneath the plasma membrane and each one may then produce satellites from its side, a root from its base, and a cilium from its apex. Development of the cilium begins with the appearance of a vesicle that becomes attached to the distal end of the kinetosome. The growing ciliary shaft pushes out against the vesicular wall, which forms a temporary ciliary sheath until the permanent one is formed.

**basal body—structurally identical to the centriole but serving as a cilium or flagellum attachment site; also called a kinetosome**

## 4–3   MICROFILAMENTS

**microfilament—smallest of the cytoplasmic filamentous structures, thought to have some cytoskeletal and contractile function and to participate in cellular motion**

In recent years the active or motile function of the cytoskeleton has been attributed to some of the *microfilaments* that are observed by electron microscopy as thin structures between 6 and 10 nm in diameter. Various studies have suggested that such filaments represent contractile systems, and this concept has been reinforced by evidence that the contractile proteins of muscle (i.e., actin and myosin) are also widely distributed in other cells.

As shown in Figure 4–1, just beneath the plasma membrane there are bundles of filaments that are made of actin and are in continuity with a network of similar filaments that pervades the cell matrix or cytosol. The channels or spaces created in among this *trabecular lattice* are of the order of 50 to 100 nm, and in the living cell may provide for the rapid diffusion of fluids and metabolites throughout the cytosol. The filaments of the lattice are in contact with vesicles of the endoplasmic reticulum, with microtubules, and with polysomes, all of which seem to be contained or supported within the lattice. The assembly and disassembly of the lattice may also allow small movements of the contained organelles.

While the depolymerizing effect of colchicine has been of critical importance in the study of microtubules, *cytochalasin B* has had an equally decisive influence in the study of microfilaments. This drug has been found to impair numerous cell activities in which some types of microfilaments are involved (e.g., smooth muscle contraction, cell migration, endo- and exocytosis, and other processes) by disrupting the regular arrangement of microfilaments associated with some of these functions. Not all microfilaments, however, are sensitive to cytochalasin B; it has been generally concluded that the cytochalasin B-sensitive microfilaments are the contractile machinery of nonmuscle cells.

cytochalasin B—drug that disrupts some types of microfilaments without affecting microtubules

Before we discuss the structure and function of microfilaments in detail, we will mention that in addition to microtubules and microfilaments a variety of cells contain a third cytoskeletal component consisting of filaments of an intermediate size (7 to 11 nm). These have been observed in several kinds of cells (e.g., fibroblasts and muscle cells), and do not seem to be affected by either Colcemid or cytochalasin. This *intermediate system* of microfilaments appears to be related to the anchorage and maintenance of cell shape.

## Actin, Myosin, and Other Contractile Proteins

The finding that the proteins *actin* and *myosin* are present in nonmuscle cells has suggested a molecular mechanism for cellular movements. It is now known that all eukaryotic cells contain high concentrations of actin and low concentrations of myosin. In addition to these two force-generating proteins there are also others, such as *tropomyosin* and *troponin C*, that (as in muscle) have a regulatory function; there are *actin-binding* proteins that can cross-link actin filaments; and there is $\alpha$-actinin, which may participate in actin filament-membrane interactions.

Actin accounts for a large proportion of the cytoplasmic protein in many cells. It is present mainly in its globular form *(G-actin)*, which has a molecular weight of 42,000, and it may polymerize quickly to form the microfilaments of fibrous actin *(F-actin)*. Cyto-

actin—force-generating cytoplasmic and muscle protein, present in both globular (G-actin) and fibrous (F-actin) forms

plasmic actin is very similar in structure to muscle actin and forms identical 6-nm wide microfilaments consisting of a double helical array of globular actin molecules (see Fig. 4–16). Actin filaments can be identified in the electron microscope by virtue of the fact that they become "decorated" with myosin to form arrow-like complexes.

Compared with actin, cytoplasmic myosin is rather heterogeneous in size and composition. In most cases it has a molecular weight of 460,000 daltons, similar to that of muscle myosin, but contains different subunits. Despite these differences, all myosins bind reversibly to actin filaments.

cytoplasmic myosin— heterogeneous protein that binds reversibly to actin filaments

## Motility in Nonmuscle Cells: Cyclosis and Ameboid Motion

The molecular basis of contraction in nonmuscular systems is still hypothetical, but the scheme that is evolving is similar in some ways to that found in muscle in being based on the interaction of actin and myosin filaments, with the consequent production of a shearing force. The nonmuscular mechanism differs, however, in that the proteins are distributed more randomly and there is a much smaller concentration of myosin. These factors may help to account for the much slower contraction in these primitive contractile systems. Many motile events in nonmuscle cells—cell spreading and shape changes, locomotion, the formation of pseudopodia, and many others—require a tight coupling between the cell surface and the contractile apparatus. This coupling could be achieved by α-actinin molecules that bind the actin microfilaments to the plasma membrane.

In contrast to muscle cells, in which actin filaments are stable, most nonmuscle cells contain transitory filaments that may appear or disappear within one minute. Thus the cell must have some mechanism that controls the assembly and disassembly of the actin monomers and the formation of stress fibers.

We shall now consider two types of cell motility in which microfilaments and actin-myosin interactions are involved. *Cytoplasmic streaming*, or *cyclosis*, is easily observed in plant cells, in which the cytoplasm is generally confined to a layer next to the cellulose wall and to fine trabeculae crossing the large central vacuole (Fig. 1–1). It is possible to see continuous currents that displace chloroplasts and other cytoplasmic granules.

cyclosis—generation of cytoplasmic currents by the action of microfilaments

The classic experimental work of cyclosis has utilized the huge cylindrical cells of *Nitella*, which have a thin protoplasmic layer surrounding a central vacuole. This protoplasmic layer is divided into a cortical region of structured cytoplasm in which are embedded nonmotile chloroplasts and a layer of isolated cytoplasm (the *endoplasm*) where cyclosis takes place. The whole system is surrounded by the plasma membrane, under which is a single layer of microtubules. Bundles of actin filaments are situated in the cortical region just beneath the chloroplasts and in a direction parallel to that of cyclosis. It is supposed that the actual motive force may be provided by a mechanism of actin-myosin interaction that takes place in the region between the stationary ectoplasm and the moving endoplasm. It has been found that cytoplasmic streaming stops when the cortex of *Nitella* is treated with cytochalasin B.

*Ameboid motion* occurs mainly in certain protozoa and in animal cells. In this form of motion the cell changes shape actively, sending forth cytoplasmic projections called *pseudopodia* into which the protoplasm flows. Although this form of locomotion can be observed easily in amebae, it also occurs in numerous other types of cells such as leukocytes (Fig. 4–10).

ameboid motion—type of locomotion in which the cell changes shape actively, sending cytoplasmic pseudopodia in the direction of movement

In a culture consisting of a single layer of cells, a type of ameboid motion sometimes called *ruffling* is observed. The ruffle is a kind of surface projection that makes intermittent contact with the glass surface. By an alternating attachment and detachment of the ruffle, the cell is able to displace itself over the support surface and actively move into the zone of migration of the culture.

*Adhesion* to a solid support is also fundamental to the movement of amebae. An ameba that floats freely in the liquid medium can emit pseudopodia but cannot progress; only when it adheres to a solid surface can it commence this type of locomotion (Fig. 4–11).

It is generally assumed that microfilaments are involved in ameboid motion. Actin filaments and thicker myosin filaments have been found in amebae, and there seems no doubt that the actin-myosin interaction provides the actual motive force. However, the views are divided regarding the most likely site of contraction.

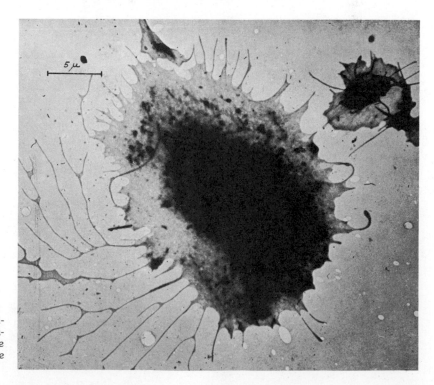

**Figure 4-10** Electron micrograph of a polymorphonuclear leukocyte showing fine filopodia. ×3000. (From E. De Robertis.)

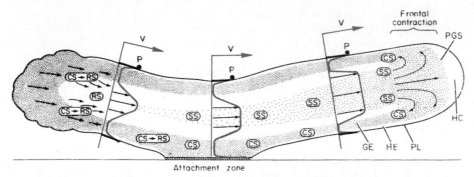

**Figure 4-11** A schematic diagram corresponding to a frontal contraction model for ameboid motion. *SS*, stabilized regions; *CS*, contracted and *RS*, relaxed regions; *V*, velocity profile; *HC*, hyaline cap; *PGS*, plasma gel sheet; *HE*, hyaline ectoplasm; *GE*, granular ectoplasm; *PL*, plasmalemma. (From D. L. Taylor et al., 1973.)

Whereas some investigators regard the posterior region as more active, others give more importance to the advancing end of the ameba. The diagram in Figure 4–11 emphasizes the latter viewpoint. An ameba is shown in a vertical section with the zone of surface attachment indicated. From the velocity profiles (V) it is evident that the streaming of the axial endoplasm increases toward the front, where the main region of contraction is indicated. On the other hand, contraction is propagated in a direction opposite that of the streaming.

# 4-4 MOLECULAR BIOLOGY OF MUSCLE

Cell contractility reaches its highest development in the various types of muscular tissues. The structural organization of muscle is adapted to unidirectional shortening during contraction. For this reason most muscle cells are elongated and spindle-shaped. The cytoplasmic matrix is considerably differentiated, and the major part of the cytoplasm is occupied by contractile fibrils. These *myofibrils* are large structures made up of *myofilaments* disposed in parallel. In smooth muscle cells the myofibrils are homogeneous and birefringent. In cardiac and skeletal muscle cells the myofibrils are striated and have dark, birefringent *(anisotropic)* zones alternating with clear *(isotropic)* zones (Fig. 4–12). In this section we will be examining the molecular structure of the myofibril and the way in which this structure is related to its function.

myofibril—long cylindrical contractile fibril of striated muscle tissue

sarcomere—structural unit of the myofibril, extending between adjacent Z lines and containing identifiable bands or zones

## Contractile Structures: Myofibril and Sarcomere

Myofibrils are long cylindrical structures about 1 $\mu$m in diameter. They show regular transverse striations resulting from the repetition of a fundamental unit, the *sarcomere*. The boundary between sarcomeres is the *Z line* (or *Z disc*) which divides the less dense zones or *I bands* into two adjacent sarcomeres. (The *I* stands for isotropic,

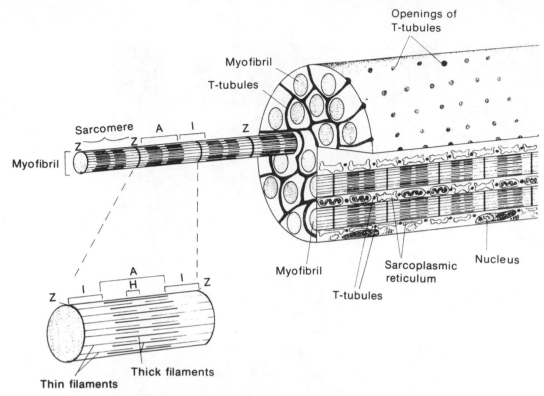

**Figure 4-12** Diagram of skeletal muscle fiber, showing the myofibrils and, between them, the sarcoplasmic reticulum and the mitochondria. A sarcomere with the thick and thin filaments marked is also illustrated. (From Ch. J. Flickinberg et al. *Medical Cell Biology.* Saunders, Philadelphia, 1979.)

meaning that it does not bend polarized light.) In the center of the sarcomere is the denser *A band* (anisotropic, or birefringent), which is subdivided into two semidiscs by the *H disc* (Hensen's disc). In the middle of the H disc the M line can be observed. Figures 4–13 and 4–14, respectively, provide an electron micrographic and a schematic illustration of these features. To give some idea of the relative dimensions, in a relaxed mammalian muscle cell the A band is about 1.5 $\mu$m long and the I band about 0.8 $\mu$m long.

The striated appearance of the myofibril results from the disposition of myofilaments along the axis of the sarcomere. Morphologically speaking, there are two kinds of myofilaments: *thick myofilaments*, about 1.5 $\mu$m long and 10 nm thick, and thin myofilaments, about 1.0 $\mu$m long and 5 nm thick. As shown in Figure 4–19, these two types of filaments are disposed in register and overlap to an extent that depends on the degree of contraction of the sarcomere. In a relaxed state, the I band contains only thin filaments; the H band contains only thick filaments; and within the A band the thick and thin filaments overlap. The regular disposition of these two types of filaments can best be observed in a cross-section through the A band (Fig. 4–14). In vertebrate muscle each thick filament is surrounded by six thin ones, and each thin filament is

myofilament—one of the individual thick or thin filaments arranged in register to make up the myofibril

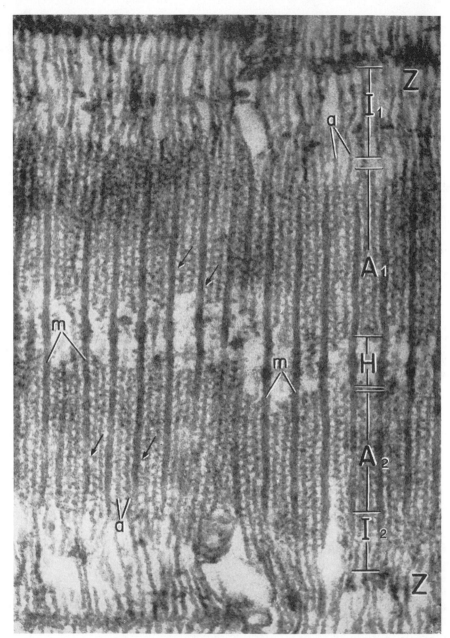

**Figure 4-13** Electron micrograph of two sarcomeres in adjacent myofibrils: $Z$, Z line; $A_1$ and $A_2$, anisotropic half bands; $H$, Hensen's band; $I_1$ and $I_2$, isotropic half bands; $m$, thick, and $a$, thin filaments. The cross-bridges between both types of filaments can be seen clearly. Some of them are indicated by arrows. $\times 175,000$. (Courtesy of H. Huxley.)

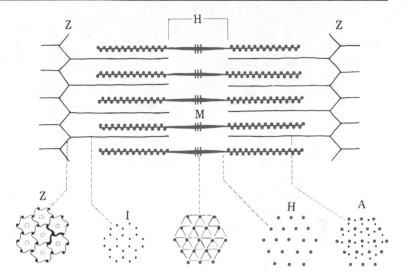

**Figure 4-14** Diagram representing the structure of one sarcomere in striated muscle in longitudinal and transverse sections. Observe that the I band has only thin filaments, the H band only thick ones, and the A band both thick and thin filaments. Special structures are observed at the Z and M lines.

surrounded by three thick ones. A cross-section through the H band shows only thick myofilaments; a cross-section through the I band shows only thin myofilaments.

Another interesting detail revealed by the electron microscope is that the two sets of filaments are linked together by a system of cross-bridges (Fig. 4–13). These arise from the thick filaments at intervals of about 7 nm. Each bridge is situated along the axis with an angular difference of 60 degrees. This means that the bridges describe a helix about every 43 nm, and as a result of this arrangement one thick filament joins the six adjacent thin ones. In the Z line there is a square lattice that has been compared to a woven basket, made of Z filaments (probably containing α-actinin) that join the thin filaments (Fig. 4–14).

In a living muscle fiber there are characteristic changes that take place, upon contraction, in the structures just described. These changes can be observed with phase-contrast and interference microscopy. One striking observation is that the A band remains constant in a wide range of muscle lengths, whereas the I band shortens in accordance with the contraction. The shortening of the I band results from the progressive sliding of the thin myofilaments into the arrays of thick filaments (Fig. 4–15). As contraction progresses the thin filaments penetrate into the H band and may even overlap, thereby producing a more dense band in the center of the sarcomere *(inversion of the banding)*. Finally, the thick filaments make contact and are crumpled against the Z lines. These events are involved in the *sliding filament mechanism* of contraction, which will be discussed further below.

## Molecular Organization of the Contractile System

The molecular components involved in muscle contraction include the following: (1) the force-generating proteins *myosin* and *actin;*

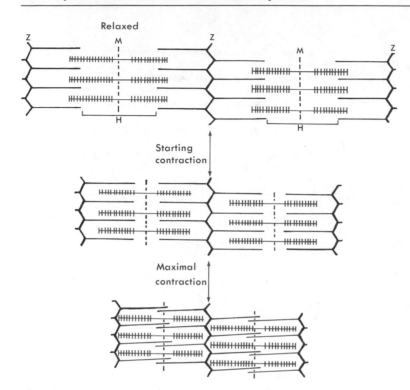

Relaxed

Starting
contraction

Maximal
contraction

**Figure 4-15**   Diagram showing the sliding model of contraction. In the relaxed condition the H bands are wide and contain only thick filaments. With the beginning of contraction, the thin filaments slide toward the center of the sarcomere. With maximal contraction, the thin filaments penetrate into the H band and produce inversion of the banding.

**muscle myosin—contractile protein molecule found in the thick myofilament, consisting of a helical rod and a globular head region**

(2) regulatory proteins such as *tropomyosin* and the various *troponins;* and (3) structural proteins such as *α-actinin* (present in the Z line), the *M band proteins,* and the *C protein.*

Myosin accounts for about half of the total protein of the myofibril. It is a huge molecule of approximately 500,000 daltons, having two polypeptide chains of 200,000 daltons each and four smaller subunits of 20,000 daltons each. As shown in Figure 4–16, the two heavy chains coiled around each other form a helix about 140 nm long and 2 nm in diameter. The molecule is folded into two globular regions at one end. Proteolytic enzymes act on the molecule to produce identifiable sections: a rod-shaped *light meromyosin* (LMM) and a *heavy meromyosin* (HMM); the latter can be subdivided further into a helical rod $S_2$ and a globular head $S_1$. HMM-$S_1$ is the most important part of myosin since it contains ATPase and the actin binding site.

The thick myofilament is made by the lateral association of several myosin molecules. The center of the myofilament is smooth in the region of the H band, while in each semidisc A it is rough and shows the *myosin cross-bridges* formed by the HMM-$S_1$ globular ends. Observe in Figure 4–17 (bottom) that within the thick filament the myosin molecules are polarized in opposite directions in the respective semidiscs A on either side of the H band.

Actin represents about 25 percent of the protein of the myofibril. The monomer units polymerize into two helical strands in which most of the actin (by virtue of having polymerized) is F- rather than G-actin. These actin helixes form the major structural feature of the

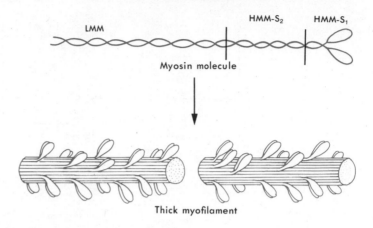

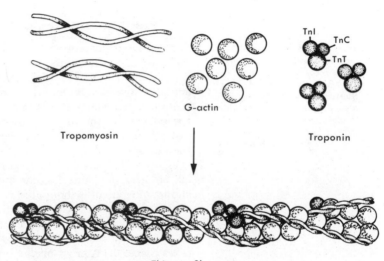

**Figure 4-16** Diagram indicating the molecular structure of the thick myofilaments **(top)** and thin myofilaments **(bottom)** in the myofibril. The myosin molecule is made of two polypeptide chains 140 nm long, having at one end the $S_1$ head (LMM, light meromyosin, HMM-$S_2$, heavy meromyosin-$S_2$, HMM-$S_1$, heavy meromyosin-$S_1$). Observe that the thick filament results from the assembly of myosin molecules. The thin filament is the result of the association of G-actin monomers, tropomyosin and the troponins (TnT, troponin T; TnC, troponin C; and TnI, troponin I). Note that each tropomyosin molecule extends over seven actin molecules.

thin myofilament, which also contains tropomyosin and the troponins. *Tropomyosin* is an elongated molecule 40 nm long made of two parallel polypeptide chains in an α-helix. As shown in Figure 4–16, this molecule extends over seven actin monomers and lies in the groove of the actin double helix. *Troponin* is a complex of three proteins that binds to the ends of tropomyosin and thus occurs along the thin myofilament at intervals of 40 nm. Its three components are *troponin C*, which binds $Ca^{2+}$; *troponin T*, which has a high affinity for tropomyosin; and *troponin I*, which inhibits the myosin ATPase.

Various labeling techniques have revealed the presence of the structural proteins: α-actinin at the Z line, M proteins in the M band, and C protein in the A band. For our purposes it is sufficient simply to recognize that these proteins are among the components of the myofilament structure.

tropomyosin—regulatory protein complexed with F-actin in the thin myofilament

troponin—regulatory protein bound to the actin-tropomyosin complex and having $Ca^{2+}$ binding properties

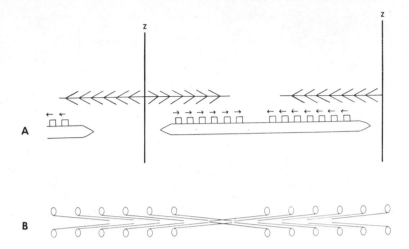

**Figure 4-17** **A,** diagram illustrating the polarity of cross-bridges in the myosin and actin myofilaments. The sliding forces tend to move the actin filaments toward the center of the sarcomere. (From Huxley, H. E.) **B,** arrangement of myosin molecules within the thick filaments. Each molecule has a tadpole shape with a globular head and a tail. The axis of the filament is formed by the assembly of the tails. Observe that in each half of the myofilament the molecules are polarized. (From H. E. Huxley, *Proc. Roy. Inst. Gr. Br.,* 44:274, 1970.)

## The Mechanism of Contraction

**sliding filament mechanism—currently favored hypothesis for muscle contraction, in which the thin and thick filaments are displaced with respect to each other by the repetitive interaction of the myosin cross-bridges with the actin filaments**

Contraction is currently explained by a sliding mechanism in which the thin actin filaments are displaced with respect to the thick myosin filaments. This movement depends on highly specific interactions between actin and myosin molecules and requires that these molecules be polarized within the sarcomere. Figure 4–17 illustrates the polarity of the myosin cross-bridges with respect to the actin myofilaments, emphasizing the polarity differences in each half of the sarcomere.

Figure 4–18, **A** shows that, in the resting state, the cross-bridges are not attached, and thus the muscle is plastic and extensible. The sliding movement is thought to result from the repetitive interaction of the cross-bridges with the actin filaments. (The diagram depicts each cross-bridge as the active end of the myosin molecule; that is, the $S_1$ globular unit.) The following sequence of events probably takes place: (1) A cross-bridge binds to a specific site on the actin filament. (2) The cross-bridge undergoes a conformational change which displaces the point of attachment toward the center of the A band, thereby pulling the actin filament; at the same time a second cross-bridge becomes attached. (3) At the end of the cycle the first bridge returns to the starting configuration in preparation for a new cycle. According to this theory, the actin filaments in each half-sarcomere are pulled toward the center of the sarcomere by the myosin arms as they move back and forth. The energy for this interaction is provided by the splitting of ATP because ATPase is present in the cross-bridge. It is estimated that the splitting of one ATP accounts for a displacement of 5 to 10 nm for each myosin cross-bridge.

We mentioned above that when muscle is in a relaxed state the cross-bridges are not attached to thin filaments. The reverse case could explain the rigidity of the muscle (state of rigor); in this case a permanent attachment of the cross-bridges could be in effect. It is understandable that in order to produce a contraction-relaxation cycle there should be some regulation of cross-bridge attachment and detachment. We now know that this transition between rest and activity depends on the concentration of free calcium ions in the vicinity of the contractile machinery.

Control by $Ca^{2+}$ requires the presence of the regulatory proteins tropomyosin and troponin, which, as we have seen, form part of the structure of the thin myofilament. The mechanism is thought to occur as follows: (1) In the absence of $Ca^{2+}$, the regulatory proteins inhibit the interaction of actin and myosin. (2) When the $Ca^{2+}$ concentration increases above $10^{-6}$ M, the contraction is triggered and $Ca^{2+}$ binds to troponin C. (3) The influence of troponin C plus $Ca^{2+}$ is supposed to be transmitted by way of tropomyosin to the seven adjacent G-actin monomers. Figure 4–19 shows an end view of the model, based on electron microscopic and x-ray diffraction observations, that has been postulated to account for what we know so far about this interaction. The model proposes that in the relaxed state, tropomyosin is deep in the groove of F-actin, covering the actin binding site of $S_1$ myosin. When $Ca^{2+}$ binds to troponin C, tropomyosin moves away from the groove and exposes the actin binding site of $S_1$, making it possible for the actin-myosin interaction and the resulting contraction to take place.

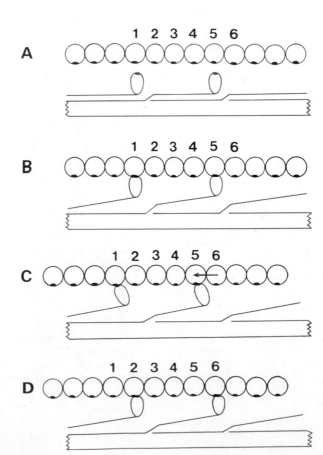

**Figure 4-18**  A very schematic interpretation of the mode of action of cross-bridges in the sliding of myofilaments. **A,** at relaxation there is no attachment of the $S_1$ cross-bridges to the actin filament. **B,** at the start of contraction interaction of $S_1$ and actin occurs. **C,** a conformational change in $S_1$ causes the actin myofilament to move toward the center of the A band. **D,** at the end of the stroke $S_1$ detaches, returns to the original configuration, and re-attaches to the following G-actin.

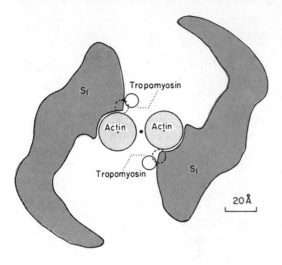

**Figure 4-19** A possible molecular model of regulation of contraction. The diagram shows an end-on view of the relative positions of actin and the $S_1$-myosin. Tropomyosin is shown with a line contour in the activated condition, in which there is interaction between actin and myosin. The dotted contour corresponds to the relaxed condition, in which tropomyosin is deep in the groove of F-actin and covers the actin binding site of $S_1$-myosin (Courtesy of H. Huxley.)

## Excitation-Contraction Coupling

*Excitation-contraction coupling* refers to the mechanism by which an electrical impulse is able to induce contraction in a muscle. Upon the arrival of a nerve impulse at a muscle cell, signal transmission occurs by a chemical process that involves the release of acetylcholine and its interaction with a receptor protein. This interaction results in the propagation of an *action potential* along the cell membrane of the muscle fiber, and this potential penetrates into the muscle and induces a contraction. The mechanism by which this occurs involves (1) the *T-system*, which conducts the excitation into the muscle fiber; (2) the release of $Ca^{2+}$ from the *sarcoplasmic reticulum*, which initiates the contraction; and (3) the re-uptake of $Ca^{2+}$ into the sarcoplasmic reticulum by way of a *$Ca^{2+}$-activated ATPase*, which produces muscle relaxation.

In Chapter 5 we will study the endoplasmic reticulum (ER) as part of the endomembrane system. The *sarcoplasmic reticulum* is one of the most interesting specializations of the ER. It was discovered by Veratti in 1902 as a reticulum present in the sarcoplasm (i.e., the cytoplasmic matrix of muscle cells) and extending between the myofibrils, but was neglected until 1953, when Porter and Bennett obtained the first electron micrographs of this structure.

The sarcoplasmic reticulum is a membrane-limited reticular system whose organization is regularly superimposed upon that of the sarcomeres. As shown in Figure 4–20, it has two main components: (1) a *longitudinal* component that represents the endoplasmic reticulum, and (2) a *transverse* component or *T-system* of vesicles and tubules that are disposed at the level of the Z lines and are connected to the muscle cell membrane or *sarcolemma*. The longitudinal reticulum is made up of interconnected wide tubules that cover the sarcomere at the level of the A and I bands and terminate at special cisternae found at the ends of the I bands near the Z line. The terminal cisternae of two consecutive sarcomeres flank another flat-

**excitation-contraction coupling—electrochemical mechanism by which a signal is propagated along the cell membrane of a muscle fiber and penetrates into the muscle to induce contraction**

**sarcoplasmic reticulum— specialized component of the endoplasmic reticulum that is found in muscle fibers**

**T-system—component of the sarcoplasmic reticulum consisting of vesicles and tubules connected to the muscle cell membrane**

**sarcolemma—muscle cell plasma membrane**

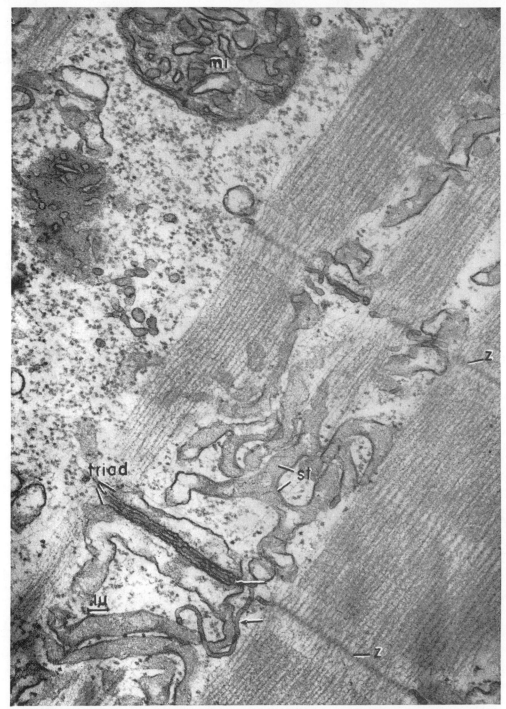

**Figure 4-20** Electron micrograph of striated muscle showing two myofibrils, one of which is tangentially cut and shows the arrangement of the sarcoplasmic reticulum. The two components of this system can be seen clearly. The transverse component is represented by the *triad* and especially by the central cisternae of the triad, which continues in special tubules (arrows). Notice the relationship of the triad to the Z line. The longitudinal component of the sarcoplasmic reticulum forms tubules *(st)* on the surface of the sarcomere; *mi,* mitochondrion. (Courtesy of K. R. Porter.)

tened vesicle belonging to the T-system, and the three vesicles together comprise a unit called the *triad*. The transverse components of the triad are continuous at certain points with the cell membrane, and this structure conducts the impulses from the points of surface contact to the deepest portions of the muscle fiber (Fig. 4–12). The presence of the T-system explains the physiological paradox that all myofibrils of a large fiber may contract synchronously once the activating action potential has passed over the surface.

Within the muscle fiber, $Ca^{2+}$ is stored mainly in the longitudinal components of the sarcoplasmic reticulum. It is now assumed that the release of $Ca^{2+}$ from these stores is caused by the arrival of the action potential through the T-system, opening $Ca^{2+}$ channels in the terminal cisternae of the triad and releasing the $Ca^{2+}$ from the sarcoplasmic reticulum. The increase in $Ca^{2+}$ concentration then causes the contraction by the mechanism described above. Muscle relaxation involves the opposite phenomenon. $Ca^{2+}$ is rapidly removed by a *$Ca^{2+}$-activated ATPase*, also called the *$Ca^{2+}$ pump*, that is present in the membrane of the sarcoplasmic reticulum. In summary, the series of events that take place after the arrival of the electrical signal includes the following: (1) The signal is received at the level of the Z line or the A-I junction by way of the T-system. (2) The coupling of the T-system with the terminal cisternae produces the release of $Ca^{2+}$ (this may occur in a millisecond). (3) $Ca^{2+}$ induces the contraction by virtue of its effect on the regulatory proteins troponin and tropomyosin, allowing actin and myosin to interact. During this step, ATP is used as an energy source. (4) $Ca^{2+}$ is quickly removed and restored to the sarcoplasmic reticulum by means of the $Ca^{2+}$ pump. $Ca^{2+}$ re-uptake results in muscle relaxation. This entire system is one of the best examples studied so far of a tight coupling between the processes furnishing energy and the actual machinery that utilizes it.

*$Ca^{2+}$ pump—$Ca^{2+}$-activated ATPase found in the membrane of the sarcoplasmic reticulum; removes $Ca^{2+}$ from the sarcomeres after muscle contraction, allowing relaxation*

# SUMMARY

## 4–1 MICROTUBULES

Microtubules are universally present in the cytoplasm of eukaryotic cells. They have a uniform size of about 25 nm in diameter and several micrometers in length. They have a dense wall, 6 nm thick, made up of 13 linear or spiraling filaments each about 5 nm in diameter. Despite their structural similarities, different kinds of microtubules may respond differently to various treatments.

Microtubules are constructed principally of dimers of the protein tubulin. The polymerization of these dimers is a specifically oriented process in that assembly occurs at one end of the microtubule while disassembly occurs at the other. The assembly process is favored by a cAMP-dependent kinase and the high molecular weight microtubule-associated proteins. In the cell there are sites of orientation from which polymerization is directed. Generally microtubules arise from the centrosomal region, near the cell nucleus.

Cytoplasmic microtubules are involved in many cellular functions of a mechanical and morphogenetic nature, as well as in cellular polarity and motility, circulation and transport, and sensory transduction.

## 4–2 MICROTUBULAR ORGANELLES

The mitotic apparatus consists of the asters and the spindle; in animal cells the latter generally appears between the former and is called an astral spindle. Plant cells do not contain asters and form anastral spindles. Spindle fibers are of three types: kinetochoric, polar, and free. During mitosis there is a cycle of assembly-disassembly between the cytoplasmic microtubules and those incorporated in the spindle.

The microtubular organelles responsible for cell motility are flagella, which are long and usually occur singly, and cilia, which are shorter and usually occur in groups. These organelles can either move a cell or set up fluid currents in the medium around it. The essential components of the ciliary apparatus are the cilium, the basal body, and (in some cases) the ciliary rootlets.

The axoneme is the axial microtubular structure of cilia and flagella, and it is thought to be the essential motile element. The axoneme has a characteristic 9 + 2 microtubular pattern. The microtubules of each peripheral pair are ellipsoidal in cross-section and can be distinguished by several morphological features. Subfiber A lies closer to the central axis and has 13 subunits in its outer wall (subfiber B has only 11); it also contains processes called dynein arms, which are considered to have an important mechanical role in ciliary/flagellar movement. The interaction of tubulin and dynein, fueled by ATP, is an integral part of this process.

Basal bodies and centrioles are essentially homologous structures. Each has a wall containing nine groups of microtubules; each group is a triplet skewed toward the center. Subfibers A and B of each triplet both cross the ciliary plate and are continuous with the corresponding axoneme fibers; subfiber C terminates at the plate. There is no central pair of microtubules. Centrioles are probably generated by two different mechanisms. Those destined to form mitotic spindles arise directly from the wall of the preexisting centriole; those destined to become basal bodies are assembled progressively from a precursor fibrogranular material located in the apical cytoplasm.

## 4-3  MICROFILAMENTS

The microfilaments found throughout the cell matrix are thought to represent contractile systems. The channels or spaces created in among this trabecular lattice may provide for the rapid diffusion of fluid and metabolites throughout the cytosol. Some types of cells also contain an intermediate system of microfilaments that appears to be related to the anchorage and maintenance of cell shape.

All eukaryotic cells contain actin, myosin, tropomyosin and troponin C, various actin-binding proteins, and $\alpha$-actinin; the interactions of these proteins are responsible for cell movement. Actin is present mainly in its globular form (G-actin) and may polymerize rapidly to form microfilaments of fibrous actin (F-actin). Cytoplasmic actin is very similar to muscle actin, although cytoplasmic myosin is not quite so similar to muscle myosin. Despite these differences, all myosins bind reversibly to actin filaments.

The molecular basis of contraction in nonmuscle systems is thought to be based, like that of muscle, on the interaction of actin and myosin filaments with the consequent production of a shearing force. The more random distribution of these proteins in nonmuscle cells accounts for the relative slowness of the contraction. Two examples of this type of cell motility are cyclosis (or cytoplasmic streaming) in plant cells and ameboid motion in certain protozoa and animal cells. Adhesion to a solid support is a necessary prerequisite to ameboid motion.

## 4-4  MOLECULAR BIOLOGY OF MUSCLE

The mechanism for cell contractility is highly developed in muscle tissues. The cytoplasmic matrix of muscle cells contains large numbers of myofibrils, which are made up of myofilaments arranged in parallel.

Myofibrils are long cylindrical structures that are subdivided transversely into fundamental units called sarcomeres. The various sections of the sarcomere have been identified as the Z line, I band, A band, H disc, and M line, from the boundary of the sarcomere toward its center. The striated appearance of the myofibril results from the disposition of thick and thin myofilaments along the axis of the sarcomere. These two types of filaments are arranged in register and overlap to an extent that depends on the degree of contraction of the sarcomere. In vertebrate muscle each thick filament is surrounded by six thin ones, and each thin filament is surrounded by three thick ones. The two sets of filaments are linked by a system of cross-bridges that arise from the thick filaments at 7 nm intervals. These cross-bridges have a central role in the sliding filament mechanism of contraction.

The molecular components involved in muscle contraction include force-generating proteins (actin and myosin), regulatory proteins (tropomyosin and troponin C), and various structural proteins. Myosin is a major constituent of the myofibril and occurs in the form of two heavy chains in a helical configuration with two globular regions (HMM-$S_1$) at one end. HMM-$S_1$ is the most important part of myosin since it contains ATPase and the actin binding site. The thick myofilament is made up of several of these myosin molecules associated laterally and arranged in a polarized fashion on either side of the H band. The thin myofilament consists largely of two helical strands of F-actin, with which tropomyosin and troponins are associated in a regular periodic configuration. Tropomyosin extends over seven actin monomers and lies in the groove of the helix; troponins bind to the ends of tropomyosin and thus appear at 40 nm intervals.

Contraction is currently explained by a sliding mechanism in which the thin actin filaments are displaced with respect to the thick myosin filaments due to the action of the cross-bridges between these structures. The cross-bridge binds to a specific site on the actin filament; the cross-bridge then undergoes a conformational change that pulls the actin filament toward the center of the A band; and at the end of the cycle the cross-bridge returns to its starting configuration and can repeat this process. The energy for this interaction is provided by ATP, and the transition between rest and activity is regulated by the concentration of free calcium ions in the vicinity of the contractile machinery. A model has been postulated in which tropomyosin interferes sterically with the ac-tin-myosin binding until it receives a signal from $Ca^{2+}$-activated troponin C.

Excitation-contraction coupling refers to the mechanism by which an electrical signal induces muscular contraction. The signal arrives at the muscle cell membrane as an action potential which is conducted into the cell by the T-system, a component of the sarcoplasmic reticulum. The propagation of the action potential through the T-system opens $Ca^{2+}$ channels in the triad, releasing $Ca^{2+}$ from storage in the sarcoplasmic reticulum. After $Ca^{2+}$ has been utilized in the contraction it is rapidly removed by the $Ca^{2+}$ pump, causing muscle relaxation.

# STUDY QUESTIONS

- How are microtubules assembled? Describe their structure.
- Compare the effects on a cell of Colcemid, colchicine, and cytochalasin B.
- What is birefringence, and what cellular components have this property?
- What are the structural and functional similarities between cilia and flagella, and between basal bodies and centrioles? What are the differences?
- Explain how microfilaments participate in cyclosis and ameboid motion.
- What are the relationships between myofibrils, myofilaments, sarcomeres, and the sarcoplasmic reticulum?
- What is the structural and molecular organization of the sarcomere?
- Explain the sliding mechanism of contraction.
- What is involved in the process of excitation-contraction coupling?
- What is the role of $Ca^{2+}$ in the various kinds of interactions described in this chapter?

# SUGGESTED READINGS

**Amos, L. A.** (1977) Tubulin assembly. *Nature, 270:*98.

**Cheung, W. Y.** (1980) Calmodulin plays a pivotal role in cellular regulation. *Science, 207:*19.

**Dustin, P.** (1980) Microtubules. *Sci. Am., 243:*66.

**Ebashi, S.** (1976) Excitation-contraction coupling. *Ann. Rev. Physiol., 38:*293.

**Lazarides, E.** (1980) Intermediate filaments as mechanical integrators of cellular space. *Nature, 283:*249.

**Lazarides, E., and Revel, J. P.** (1979) The molecular basis of cell movement. *Sci. Am., 240:*100.

**Murray, J. M., and Weber, A.** (1974) The cooperative action of muscle protein. *Sci. Am., 230:*59.

**Sleigh, M.** (1979) Contractility of the roots of flagella and cilia. *Nature, 277:*263.

**Spencer, M.** (1978) Sliding microtubules. *Nature, 273:*595.

**Squire, J.** (1977) Contractile filament organization mechanics. *Nature, 267:*75.

# chapter 5

# The Endomembrane System: Cell Secretion and Digestion

The existence of an endomembrane system that pervades the cytoplasm was discovered after the application of electron microscopy to whole-mounted cultured cells. The cytoplasm, which until that time had been thought to be structureless, was shown to contain a vast network of limiting membranes, tubules, and vesicles. This network was named the *endoplasmic reticulum* (Fig. 5-1). The use of thin sections made it possible to analyze the three-dimensional morphology of this endomembrane system—also called the *cytoplasmic vacuolar system*—and to recognize its three main components. (1) The *nuclear envelope* consists of two nonidentical membranes, one apposed to the nuclear chromatin and the other separated from the first by perinuclear cisternae and apposed to the cytoplasm. The two membranes come into contact at the nuclear pores. (2) The *endoplasmic reticulum* (ER) is generally the most developed portion of the endomembrane system. It consists of two parts distinguished by the presence or absence of ribosomes on the outer (i.e., cytoplasmic) surface. The portion that has ribosomes is referred to as the *rough* or *granular* endoplasmic reticulum (rER), and the portion

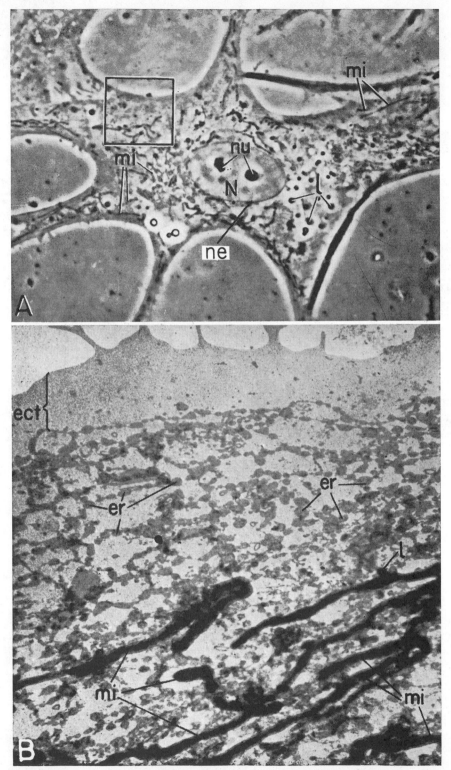

**Figure 5-1  A,** living cell of a tissue culture observed under the phase-contrast microscope: *l,* lipid; *mi,* mitochondria; *ne,* nuclear envelope; *nu,* nucleoli. The region indicated in the inset is similar to B. (Courtesy of D. W. Fawcett.) **B,** electron micrograph of the marginal region of a mouse fibrocyte in tissue culture: *er,* endoplasmic reticulum; *mi,* filamentous mitochondria; *l,* lipid. The peripheral region (*ect*) is homogeneous. ×7000. (Courtesy of K. R. Porter.)

that lacks ribosomes is the *smooth* or *agranular* endoplasmic reticulum (sER) (Fig. 5–2). (3) The *Golgi complex* is a specialized region of the ER that is involved mainly in some of the terminal processes of cell secretion.

The existence of this complex endomembrane system clearly sets the eukaryotic cell apart from the less organized primitive bacterium. The numerous membrane-bound compartments of the eukaryotic cell are responsible for vital cellular functions, among which are the separation and association of enzyme systems; the creation of diffusion barriers; the regulation of membrane potentials, ionic gradients, and different intracellular pH values; and various other manifestations of cellular heterogeneity.

The fact that the entire system of endomembranes represents a kind of barrier separating cytoplasmic compartments is clearly indicated in Figure 5–3. This diagram deliberately emphasizes the two faces of each membrane: the *cytoplasmic* (or protoplasmic) face and the *luminal* (or endoplasmic/extracellular) face.

The nuclear envelope will be considered in Chapter 7 along with the nucleus. This chapter will be devoted primarily to the endoplasmic reticulum and the Golgi complex, and we will see that among their numerous functions are the synthesis, processing, packaging, and secretion of proteins. Related to this system are other membrane-bound organelles that have important roles in cell metabolism. *Lysosomes* are packets of hydrolytic enzymes directly involved in the digestion of foods or other materials (including parts of the cell itself). *Peroxisomes* contain enzymes related to the metabolism of peroxides, and in plants are involved in photorespira-

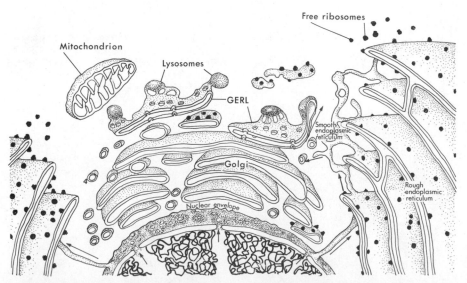

**Figure 5-2**   Three-dimensional diagram of the endomembrane system of the cell. The nucleus with its chromosomal fibrils shows interchromatin channels (arrows) leading to nuclear pores. Note the double-membrane organization of the nuclear envelope. Cisternae of rough endoplasmic reticulum are interconnected and have ribosomes attached to the outer surface. Some of these cisternae are extended by tubules of smooth endoplasmic reticulum. The Golgi apparatus shows the GERL region. The large arrows indicate the probable dynamic relationship of the portions of the endomembrane system.

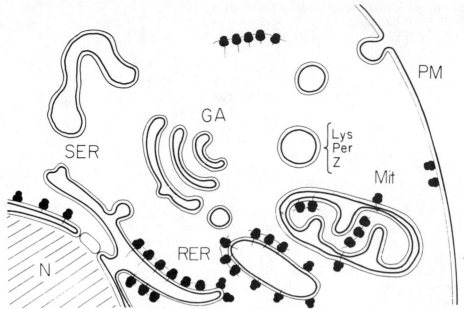

**Figure 5-3**   Diagram of cellular membranes and their relationship to compartments containing the ribosomes. In each membrane the luminal faces are shown with thick lines, while the cytoplasmic faces are depicted by thin lines. Ribosomes are always on the cytoplasmic (or matrix) side. *N*, nucleus; *GA*, Golgi apparatus; *Lys*, lysosome; *Mit*, mitochondria; *RER*, rough endoplasmic reticulum; *Per*, peroxisome; *PM*, plasma membrane; *SER*, smooth endoplasmic reticulum; *Z*, zymogen granule. (Courtesy of D. D. Sabatini and G. Kreibich. From D. D. Sabatini and G. Kreibich, Functional specialization of membrane-bound ribosomes in eukaryotic cells. In *The Enzymes of Biological Membranes,* Vol. 2. Martonosi, A., ed. Plenum, New York, 1976, p. 432.)

**TABLE 5-1   RELATIVE CYTOPLASMIC VOLUMES AND MEMBRANE SURFACE AREAS OF SECRETORY COMPARTMENTS IN RESTING GUINEA PIG PANCREATIC EXOCRINE CELLS***

| Compartment | Relative Cytoplasmic Volume Percent | Membrane Surface Area $\mu m^2/$ Cell |
|---|---|---|
| RER | ~20 | ~8000 |
| Golgi complex | ~8 | ~1300 |
| Condensing vacuoles | ~2 | ~150 |
| Secretory granules | ~20 | ~900 |
| Apical plasmalemma | | ~30 |
| Basolateral plasma-lemma | | ~600 |

*Data originally from Bolender, 1974; and Amsterdam and Jamieson, 1974. (From Jamieson, 1977.)

tion. Glyoxysomes, also present in plant cells, are involved in lipid metabolism.

# 5–1  THE ENDOPLASMIC RETICULUM (ER)

The development of the endoplasmic reticulum varies considerably in different cell types and is related to their functions. It is often small and relatively undeveloped in eggs and in embryonic or undifferentiated cells, but increases in size and complexity upon differentiation. In cells whose major activity does not involve the production of proteins for export, the ER tends to remain poorly differentiated or nonexistent, although the cell may contain a large number of ribosomes.

The rough ER is especially well developed in cells that are actively engaged in protein synthesis, such as the enzyme-producing cells. In the cells of the pancreas, for example, it is highly developed and consists of parallel stacks of flattened vesicles or *cisternae* that occupy large areas of the cytoplasm. Table 5–1 gives an idea of the large volume and surface area occupied by the ER in this cell.

Because the ribosomes are rich in RNA (Chapter 12), the regions of the cell containing rough ER are basophilic when viewed with the light microscope. These basophilic regions were formerly called ergastoplasm (Gr., *ergazomai*, to elaborate or transform), a name that implies their fundamental role in biosynthesis. The ribosomes of the ER are present as polysomes (see Chapter 12) held together by messenger RNA (Fig. 5–3) and are often arranged in "rosettes" or spirals on the surface of the ER membrane. The ribosomes are always attached to the membrane by their large (60S) subunit. This attachment is rather complex, involving both electrostatic interactions and the nascent polypeptide chain that originates from the ribosome and penetrates across the membrane into the cavity *(lumen)* of the ER. Recent studies have suggested that ribosome binding involves two transmembrane glycoproteins — the ribophorins I and II — which are present only in the rough ER.

## Biochemical Studies of the ER: Microsomes

In Chapter 1 we introduced the concept of the *microsomal* fraction of the cell; this fraction can be isolated by differential centrifugation after the nucleus and mitochondrial fractions, which sediment at lower centrifugal forces (Fig. 1–8). During cell homogenization, the cisternae and tubules of the ER break into smaller vesicles that reseal, maintaining the normal topological relationships between the membranes, the lumen, and the ribosomes (Fig. 5–4). The smooth and rough microsomes can be separated because they have different densities, and the ribosomes can also be isolated by various treatments (such as the use of high-salt solutions). If the membrane is then dissolved with detergents, the secretory proteins in the lumen can be separated by the use of mild detergents and sonication.

Microsomes constitute 15 to 20 percent of the total mass of a liver

endoplasmic reticulum (ER)—system of membrane-enclosed cytoplasmic channels involved in cellular transport processes; rough ER has ribosomes attached to its outer surface whereas smooth ER does not

cisterna—one of the flattened, membrane-enclosed sacs found in parallel stacks in, for example, the nuclear envelope or the ER

ribosome—cytoplasmic granule composed of RNA and protein, at which protein synthesis takes place

lumen—central channel of the ER

microsome—small closed vesicle of ER membrane obtained upon differential centrifugation

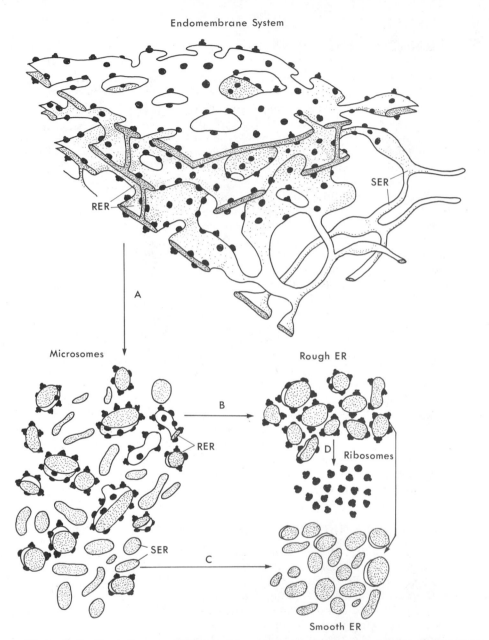

**Figure 5-4**   Three-dimensional diagram of the endomembrane system. *A*, isolation of microsomes by homogenization and differential centrifugation; *B* and *C*, separation of rough and smooth vesicles of the endoplasmic reticulum; *D*, separation of ribosomes from the RER. *RER*, rough endoplasmic reticulum; *SER*, smooth endoplasmic reticulum.

cell (Fig. 5–5) and contain 50 to 60 percent of all its RNA. The membranes are rich in lipids and, as in the case of the plasma membrane, contain a lipid bilayer with peripheral and integral proteins. Microsomes contain many enzymes that are utilized in the synthesis of triglycerides, phospholipids, and cholesterol. These products are not only incorporated into the reticular membrane itself, but are transferred to other organelles by way of specific carrier proteins which can shuttle through the cytosol.

In microsomes there are two electron transport systems which contain two flavoproteins (NADH-cytochrome C-reductase and NADH-cytochrome $b_5$-reductase) and two hemoproteins (cytochromes $b_5$ and P-450). Cytochrome P-450 functions as a terminal oxidase, and in liver cells it is used to detoxify or inactivate many substances. These *oxidative enzymes* are distributed mainly toward the outer or cytoplasmic face of the microsomal membrane.

At its luminal face, the ER has other enzymes that process secretory products. For example, there are *peptidases* that can remove part of a polypeptide chain, and *hydrolases* and *transferases* that can hydroxylate or glycosidate amino acid residues. *Glucose-6-phosphatase*, present mainly in the smooth ER, facilitates the degradation of glycogen and probably acts in a vectorial manner, sending glucose into the lumen of the ER (Fig. 5–6).

## Functions of the ER

As suggested above, the ER is not only different in different kinds of cells but has many functions within the same cell. This polymorphism has made an analysis of its multiple activities very difficult. The great complexity of this membrane system arises, as we have seen, from the fact that a large portion of it is associated with the ribosomes and thus has a fundamental role in the storage and processing of proteins destined for export from the cell. This aspect of the system's activity will be discussed in more detail in the next two sections of this chapter. Other functions of the endoplasmic reticulum are related to the fact that it divides the cytoplasm into two compartments, the cytosol and the area within the lumen.

**MECHANICAL SUPPORT.**  The endoplasmic reticulum, along with the system of microtubules and microfilaments described in the previous chapter, may participate in many of the mechanical functions of the cell by providing supplementary mechanical support for the colloidal areas of the cytoplasm.

**EXCHANGES.**  The enormous internal surface area of the endoplasmic reticulum gives an idea of the importance of the exchanges taking place between the matrix and the inner compartment. It is known that the vacuolar system has osmotic properties; after isolation, microsomes expand or shrink according to the osmotic pressure of the fluid. Diffusion and active transport may take place across the membranes of the vacuolar system. As in the plasma membrane, the presence of carriers and permeases that are involved in active transport across the membrane has been postulated.

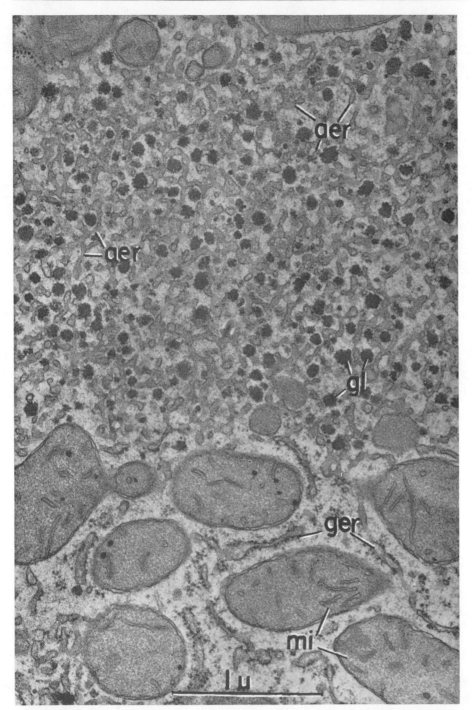

**Figure 5-5**   Electron micrograph of the cytoplasm of a liver cell. At the bottom, the rough or granular endoplasmic reticulum (*ger*) and mitochondria (*mi*); at the top, the smooth or agranular endoplasmic reticulum (*aer*) mixed with glycogen particles (*gl*). ×45,000. (Courtesy of G. E. Palade.)

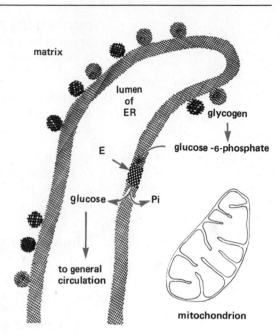

**matrix**

**lumen
of
ER**

**glycogen**

E

**glucose -6-phosphate**

**glucose**  **Pi**

**to general
circulation**

**mitochondrion**

**Figure 5-6**  Diagram of the intervention of the smooth endoplasmic reticulum in glucogenolysis with the consequent release of glucose. The enzyme (E), glucose-6-phosphatase, is present in the membrane and has a vectorial disposition by which it receives the glucose-6-phosphate from the matrix surface. The product—glucose—penetrates the lumen of the endoplasmic reticulum.

**IONIC GRADIENTS AND ELECTRICAL POTENTIALS.**  There is considerable reason to suppose that ionic gradients and electrical potentials exist across the intracellular membranes of the ER. This concept has been applied especially to the sarcoplasmic reticulum, the specialized form of the ER found in striated muscle fibers (Section 4–4).

**MEMBRANE FLUIDITY.**  The endoplasmic reticulum may act as a kind of circulatory system for the intracellular distribution of various substances. Membrane flow may also be responsible for carrying particles, molecules, and ions into and out of the cell. The continuities observed in some cases between the endoplasmic reticulum and the nuclear envelope suggest that membrane flow may also be active at such junctures, thus providing one of the several mechanisms for export of RNA and nucleoproteins from the nucleus to the cytoplasm.

We may consider the membrane of the endoplasmic reticulum at physiological temperature to be highly dynamic and fluid in the same sense as the plasma membrane. There is evidence from freeze-fracturing studies that the bound ribosomes are mobile on the membrane and that their mobility is controlled by its fluidity. Different studies have also suggested that the transfer of secretory proteins is accompanied by a flow of newly synthesized membrane proteins that are incorporated into the rough ER; these membrane proteins, however, have a slower rate of synthesis than the secretory ones.

The functions just mentioned, with the exception of protein synthesis, apply to both parts of the endoplasmic reticulum. The following activities occur primarily in the smooth portion.

**DETOXIFICATION.** The administration of large amounts of drugs to an animal results in the increased activity of enzymes related to detoxification, as well as other enzymes, and a considerable hypertrophy of the smooth ER.

**LIPID SYNTHESIS.** While the rough endoplasmic reticulum predominates in cells engaged in protein synthesis, the smooth type is abundant in those involved in lipid synthesis.

**GLYCOGENOLYSIS.** We mentioned above the association of glycogen deposits in the cytosol with the smooth ER (Fig. 5–5), and also that the withdrawal of glucose from glycogen could be mediated by way of glucose-6-phosphatase.

## Biogenesis of the ER: A Multi-Step Mechanism

The origin of the endoplasmic reticulum is not definitely known. Some electron microscopic observations of differentiating cells have suggested that it may develop by evagination from the nuclear envelope; at telophase, however, the nuclear envelope is re-formed by vesicles of the endoplasmic reticulum.

Several lines of experimental evidence suggest that the synthesis of membrane components proceeds along lines of structural and functional continuity. (There is temporal continuity as well, meaning that every cell receives from its parent cell a full set of membranes.) Current concepts of membrane biogenesis generally assume a multi-step mechanism involving the synthesis of a basic membrane of lipids and intrinsic proteins, followed by the addition, in a sequential manner, of other constituents such as enzymes, specific sugars, or other lipids. The process by which a membrane is modified chemically and structurally may be regarded as *membrane differentiation*.

The distribution of various membrane biosynthetic enzymes leads to the conclusion that most of the intracellular membranes are synthesized in the rough endoplasmic reticulum and then transformed, through a progressive series of changes, into the membranes of the smooth portion and the Golgi complex. The process of membrane biogenesis, however, cannot be understood without taking into consideration the fluidity of the membrane, which allows for the circulation, assortment, and segregation of its component parts.

## 5–2 THE GOLGI COMPLEX

In 1898, by means of a silver staining method, Camillo Golgi discovered a reticular structure in the cytoplasm of nerve cells (Fig. 5–7). More than fifty years later, the electron microscope provided a distinct image of this organelle and its structure was studied in detail. As with the ER, the isolation of the Golgi complex in purified fractions allowed investigators to perform biochemical studies and to discover a considerable amount about its function.

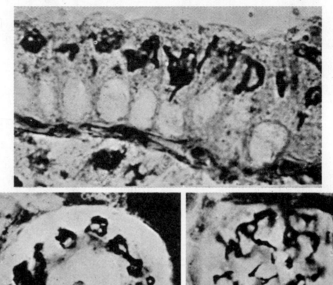

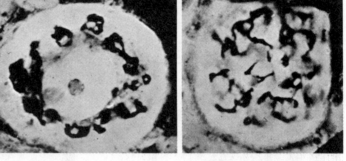

Figure 5-7  **Above,** Golgi complex (Golgi apparatus) in cells of the thyroid gland of the guinea pig, apical position. Osmic impregnation. **Below left,** ganglion cell, perinuclear Golgi apparatus. **Below right,** same structure as at left, optical section tangential with respect to the nucleus. Silver impregnation. (From E. De Robertis.)

The *Golgi complex* (or *apparatus*) is a differentiated portion of the endomembrane system that is located, in both spatial and temporal senses, between the ER on one side and secretory vesicles and the plasma membrane on the other. In this intermediary position it is involved in the transport and processing of many substances that are produced in the ER and eventually discharged outside the cell.

The Golgi complex is morphologically very similar in both plant and animal cells (Fig. 5–8). It consists of *dictyosome* units formed by stacks of flattened disc-shaped cisternae and associated secretory vesicles. In cells that have a polarized structure, the Golgi complex is generally a single large region that occupies a definite position between the nucleus and the pole of the cell at which release or secretion takes place. In other cells there are many dictyosomes that do not show any special polarity and are dispersed throughout the cytoplasm.

Although the location, size, and development of the Golgi complex vary from one cell to another, and also with the physiological state of the cell, this structure has morphological characteristics that permit its differentiation from the other parts of the endomembrane system. One such characteristic is that there are no attached ribosomes, and the whole complex appears to be surrounded by a zone from which ribosomes, glycogen, and mitochondria are excluded (the *zone of exclusion*).

In general, the Golgi complex consists of three membranous components that can be recognized in the electron microscope: (1) flattened sacs or cisternae; (2) clusters of tubules and vesicles of about

Golgi complex—cytoplasmic membrane system consisting of stacks of flattened sacs and other vesicles; involved in secretion processes

dictyosome—stack of cisternae making up the Golgi complex

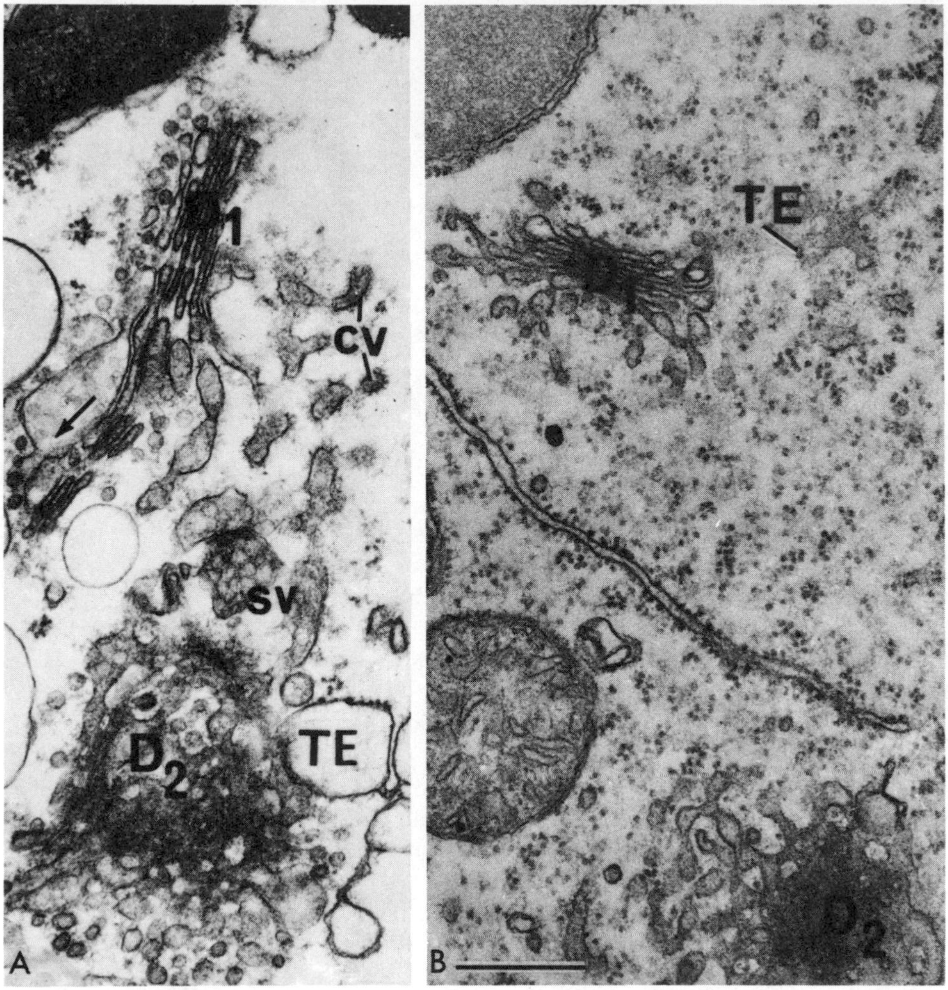

**Figure 5-8**  Electron micrographs of the Golgi apparatus in rat liver **(A)** and onion *(Allium cepa)* stem **(B)**. $D_1$, dictyosome in cross-section showing the stacked cisternae; $D_2$, dictyosome in tangential section showing a face view of cisternae with a central plate-like region and a fenestrated margin; *TE*, transition element of the endoplasmic reticulum adjacent to the Golgi; *cv*, coated vesicles; *sv*, secretory vesicles. An arrow shows a connection between the smooth endoplasmic reticulum and a secretory vesicle. ×35,000. (Courtesy of D. J. Morré. From D. J. Morré, Membrane differentiation and the control of secretion: A comparison of plant and animal Golgi apparatus. In *International Cell Biology*. Brinkley, B. R., and Porter, K. R., eds. Rockefeller University Press, New York, 1977, p. 293.)

forming (proximal) face—that side of the dictyosome closest to the nuclear envelope or the ER

maturing (distal) face—that side of the dictyosome from which secretory vesicles are released toward the plasma membrane

60 nm; and (3) larger vacuoles filled with an amorphous or granular material. The Golgi cisternae are arranged in parallel and are separated by a space of 20 to 30 nm (Fig. 5–8). Each stack of cisternae forming a dictyosome is a polarized structure having a *proximal* or *forming* face, generally convex and closer to the nuclear envelope or the endoplasmic reticulum, and a *distal* or *maturing* face of concave shape, which encloses a region containing large secretory vesicles. The forming face is characterized by the presence of small *transition vesicles* or tubules that converge upon the Golgi cisternae, forming

a kind of fenestrated plate. These transition vesicles are thought to form from the endoplasmic reticulum and to migrate to the Golgi where, by coalescence, they form new cisternae. A mechanism of membrane flow has been postulated in which new cisternae are formed at the proximal face and thus compensate for the loss of the distal face that occurs with the release of secretory vesicles.

Associated with the maturing face there is often a related region which is rich in acid phosphatase and has been called the GERL (Fig. 5–2). This acronym indicates that it has been interpreted as a structure near the Golgi, consisting of smooth endoplasmic reticulum, and involved in the production of lysosomes. More recent work relates the GERL to Golgi condensing vacuoles or presecretory granules.

GERL—acid phosphatase-rich region of the Golgi maturing face; thought to be involved in the production of lysosomes and/or presecretory granules

The dynamic polarization of the dictyosome structure is also expressed in terms of membrane differentiation. In fact, there is a progressive increase in the thickness of the membrane from the ER to the plasma membrane, and this change is apparent even among the various cisternae of the dictyosome. The more distal cisternae have thicker membranes, and there is an increase of glycoprotein staining in the lumen.

The transitions between the various parts of the Golgi complex are illustrated in Figure 5–9. In this micrograph it is possible to trace the synthesis and transport of lipoproteins, which appear as discrete dense granules of about 40 nm. These granules first appear within the tubules of the smooth ER and then enter the outer cisternae of the Golgi. Eventually they accumulate in the large vacuoles that are formed by dilatation of the edge of the sacs and, finally, they are detached as secretory vesicles. (Although not recognizable morphologically, the protein moiety of the lipoprotein is synthesized in the rough endoplasmic reticulum. The granules become visible in the smooth portion upon the addition of triglycerides.)

## Biochemical Studies of the Golgi Complex

Gentle methods of homogenization that preserve the stacks of cisternae have permitted the isolation of Golgi complexes by differential and gradient centrifugation (Fig. 5–10). This separation is favored by the fact that the Golgi complex has a lower specific density than the ER.

Biochemical studies carried out in parallel for the Golgi complex, the ER, and the plasma membrane of the same cell type have shown that, in general, the chemical composition of the Golgi complex is intermediary between that of the ER and the plasma membrane. This is reflected in the composition of the phospholipids and in the number of protein bands. The decreasing complexity in protein band patterns, from endoplasmic reticulum through Golgi complex to plasma membrane, is in some ways consistent with the view that membrane proteins are synthesized by ribosomes at the rough ER and then transported to other portions of the endomembrane system.

The most characteristic enzymes of the Golgi fraction are those

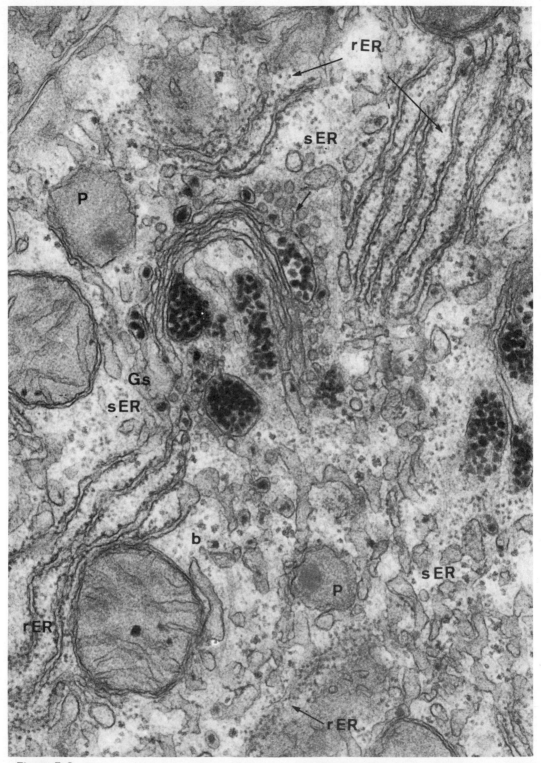

**Figure 5-9** Electron micrograph of a liver cell of an animal having a diet rich in fat. The synthesis and transport of the lipoprotein granules are visible. The rough endoplasmic reticulum (*rER*) appears as two stacks of lamellae converging toward a Golgi complex having Golgi sacs (*Gs*) on its convex or "forming face." There are portions of smooth endoplasmic reticulum (*sER*) connecting both parts of the vacuolar system. Mitochondria and peroxisomes (*P*) are observed. ×56,000. (Courtesy of A. Claude.)

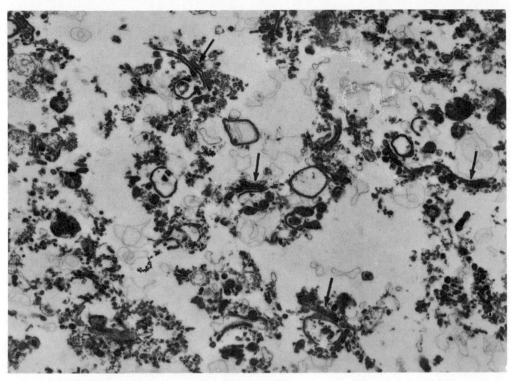

**Figure 5-10**   Isolated Golgi complexes from liver cells. The complexes that best show the stacks of cisternae are indicated by arrows. (Courtesy of D. J. Morré.)

that are able to transfer oligosaccharides to proteins to produce glycoproteins. These *glycosyltransferases* are of several types, and occur prominently along with acid phosphatase and other lysosomal and oxidative enzymes.

glycosyltransferase—characteristic type of Golgi enzyme that catalyzes glycoprotein formation

## Functions of the Golgi Complex

The major functions of the Golgi complex appear to be related to its intermediary position between the endoplasmic reticulum and the extracellular space. Through this system there is a continuous traffic of substances which may have been synthesized elsewhere but which are modified and transformed while being transported. Virtually every class of macromolecule is transported through the Golgi and secreted, and this implies a continuous and fast turnover of Golgi membranes. Dictyosomes in algae and liver cells, for example, are estimated to turn over every 20 to 40 minutes, and cisternae are released at the rate of one every few minutes.

We mentioned above that the glycosyltransferases are the most characteristic enzymes of the Golgi complex. The function of these enzymes is to glycosidate lipids and proteins, thus producing *glycosphingolipids* and *glycoproteins*. The transfer of the glycosyl residues is fueled by the use of one sugar nucleotide (e.g., UDP-galactose), and in the reaction the sugar is added to an amino acid or to another carbohydrate.

The experimental protocol for the study of glycoprotein biosynthesis involves the administration of radioactive precursors (e.g., $^3$H-galactose, $^3$H-fucose) followed at different time intervals by the fractionation of the cell and the isolation of different parts of the endomembrane system. The results obtained demonstrate that the protein backbone of the glycoprotein is synthesized by membrane-bound ribosomes and that the monosaccharides are added one by one as the protein moves through the channels of the ER and Golgi complex. The process of glycoprotein biosynthesis can be followed in the electron microscope using autoradiography (see next section). For example, 2 minutes after the injection of $^3$H-fucose, an intestinal cell is labeled almost exclusively in the Golgi region. Later on, at 4 hours, the label is concentrated at the apical membrane.

Figure 5–11 is an idealized diagram of the synthesis and secretion of glycoproteins and the intervention of the various segments of the

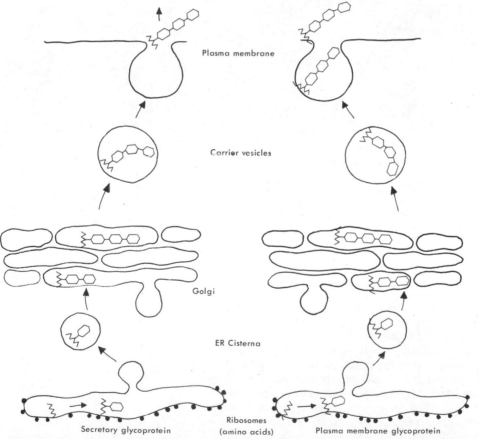

Plasma membrane

Carrier vesicles

Golgi

ER Cisterna

Secretory glycoprotein

Ribosomes
(amino acids)

Plasma membrane glycoprotein

**Figure 5-11**   Diagram comparing possible mechanisms for the elaboration of secretory glycoproteins **(left)** and plasma membrane glycoproteins **(right)**. Both proteins are first synthesized by ribosomes. The secretory proteins are released into the lumen, while the plasma membrane glycoproteins remain inserted in the wall of the ER and are transported by membrane flow. The various oligosaccharides are added in a sequential manner in the ER and the Golgi apparatus. The glycoproteins are transported from the Golgi by carrier vesicles and released by exocytosis at the plasma membrane. (Modified from Schachter, 1974. Courtesy of C. P. Leblond and G. Bennett. From C. P. Leblond and G. Bennett, Role of the Golgi apparatus in terminal glycosylation. In *International Cell Biology*. Brinkley, B. R., and Porter, K. R., eds. Rockefeller University Press, New York, 1977, p. 326.)

vacuolar system. The left side of the diagram depicts a secreted glycoprotein; the right side shows glycoproteins that are incorporated into the plasma membrane.

## 5–3 ROLES OF THE ER AND GOLGI COMPLEX IN CELL SECRETION

There is a complex interdependent relationship between the endoplasmic reticulum and the Golgi complex in the process of cell secretion. This is a process in which both plant and animal cells participate, and involves a continuous change that can be understood most clearly by studying the cell throughout the different stages of its activity. In some cells secretion is continuous; the product is discharged as soon as it is assembled. In these cells all the phases of the secretory process take place simultaneously. In other cells the secretory cycle is discontinuous: it is timed so that synthesis and intracellular transport are followed by the accumulation of the secretion product in special storage granules which are finally released to the extracellular space. In these cells the secretory cycle can be "observed" with the microscope: visible products accumulate in the cell and are ultimately eliminated.

### Stages in Cell Secretion

Six consecutive steps are generally recognized in the secretion of enzymes from the cells that produce them. Figure 5–12 is a diagram of this process which shows the involvement of the various parts of the endomembrane system.

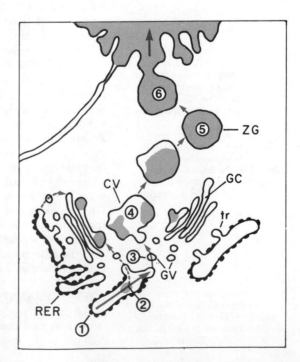

**Figure 5-12** Diagram of the processing of secretory proteins in a typical glandular cell. Steps 1 through 6 are described in the text. *RER*, rough endoplasmic reticulum; *tr*, transitional vesicles; *GC*, Golgi cisternae; *CV*, condensing vacuole; *ZG*, zymogen granules. (Courtesy of J. D. Jamieson and G. E. Palade. From J. D. Jamieson and G. E. Palade, Production of secretory proteins in animal cells. In *International Cell Biology*. Brinkley, B. R., and Porter, K. R., eds. Rockefeller University Press, New York, 1977, p. 308.)

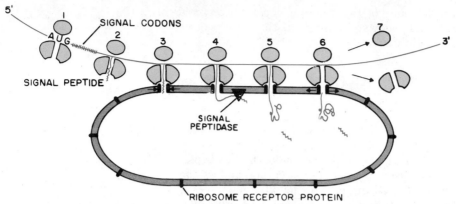

**Figure 5-13**   Schematic diagram illustrating the signal hypothesis. (Steps are described in the text.) (Courtesy of G. Blobel. From G. Blobel, Synthesis and segregation of secretory proteins, the signal hypothesis. In *International Cell Biology*. Brinkley, B. R., and Porter, K. R., eds. Rockefeller University Press, New York, 1977, p. 318.)

**STEPS 1 AND 2: RIBOSOMAL AND CISTERNAL STAGES.**   Step 1 involves the synthesis of proteins on the polysomes attached to the rough ER. It is now established that, in the case of those proteins to be secreted, the polypeptide chain grows through a "groove" or "tunnel" in the large ribosomal unit that is linked directly to a channel in the membrane of the ER. In this way, at some stage of its elongation, the protein emerges into the lumen of the endoplasmic reticulum.

An important, and not completely resolved, question concerns the mechanism by which the mRNA that selectively translates a protein for export is able to use bound rather than free ribosomes. In other words, what is the difference between the mRNAs that are translated by free and those that are translated by bound ribosomes?

The *signal hypothesis* of Blobel postulates that the mRNA for secretory proteins contains a set of special *signal codons* located immediately after the initiation codon AUG (Chapter 12). As shown in Figure 5–13, it is thought that synthesis starts on a free ribosome, and only when the *signal peptide* emerges from the large ribosomal subunit does the ribosome become attached to the membrane. It is known that the signal peptide is a sequence of 15 to 30 amino acids, most of which are hydrophobic, and that the function of the signal hydrophobic peptide is apparently to establish the initial association of the ribosome with the membrane. The signal hypothesis also postulates that there are special receptor proteins on the membrane which make the tunnel through which the polypeptide moves. Once inside the lumen of the ER, the signal peptide is removed by the action of a signal peptidase. The protein chain is then completed and the ribosome is removed from the membrane. (The ribosomal cycle is discussed in greater detail in Chapter 12.)

**STEP 3: TRANSPORT FROM THE ER TO THE GOLGI.**   In this stage the secretory proteins diffuse through the rough endoplasmic reticulum and enter the so-called *transitional elements*, which are located at the boundary between the rough endoplasmic reticulum

*signal hypothesis—proposal that the mRNA sequence of a secretory protein contains signal codons that direct its synthesis on ER-bound as opposed to free ribosomes*

*signal peptide—sequence of 15 to 30 hydrophobic amino acids whose function is to establish the initial ribosome-mRNA association with the ER membrane*

*transitional elements— vesicles found at the ER-Golgi boundary; thought to be involved in continuous dictyosome renewal*

and the Golgi complex. According to the studies of Palade and co-workers, these vesicles eventually fuse with the large condensing vacuoles of the Golgi that are present at the maturing face.

The intracellular transport of secretory protein has been followed cytochemically by the use of *radiolabeled precursors*, such as ³H-leucine, either injected into the organism or applied to tissue slices. Then, at specific times, the tissues are fixed and processed for *autoradiography*. In this method the tissue section is immersed in a photographic emulsion which leaves a thin deposit of silver halide grains on the surface of the section. The sample is kept in the dark for a certain time, during which the β-rays of the radioisotope are able to hit some of the halide grains in the photographic emulsion. These grains are then revealed after the emulsion is processed (Fig. 5–14).

One experimental procedure is carried out as follows. Slices of pancreas are exposed first to a leucine-free medium (i.e., a saline solution containing all amino acids except leucine), resulting in depletion of the leucine pool already available in the cell. After a short time, the tissue is submitted to a medium containing ³H-leucine for a few minutes (i.e., a *pulse*), and this procedure is then followed by one which involves washing and incubation with a solution containing excess unlabeled leucine (i.e., a *chase*). In the fixed tissue it is possible to observe that after a few minutes the isotope is concentrated in the endoplasmic reticulum (Fig. 5–15). A short time later, the newly synthesized protein passes into the Golgi complex. From this region it becomes progressively concentrated into *prozymogen granules* or *condensing vacuoles* surrounded by a membrane. After a longer time, the label is found principally in the *zymogen granules* (Fig. 5–16). Quantitative measurements of the grains have confirmed that transport follows the sequence rough endoplasmic

**autoradiography**—technique by which emissions from a radioactively labeled microscopic specimen can be visualized on photographic film by exposing an emulsion to the labeled material

**pulse**—submission of a tissue sample to medium containing a radioactively tagged precursor

**chase**—washing and incubation of a "pulsed" tissue sample in medium containing excess unlabeled precursor

**Figure 5–14** Diagrammatic representation of an electron microscope autoradiograph preparation. **Top,** *during exposure:* The silver halide crystals, embedded in a gelatin matrix, cover the section. A beta particle, from a tritium point source in the specimen, has hit a crystal (cross-hatched), causing the appearance of a latent image on the surface. **Bottom,** *during examination and after processing:* The exposed crystal has been developed into a filament of silver; the nonexposed crystals have been dissolved. The total thickness has decreased because the silver halide occupied approximately half the volume of the emulsion. (Courtesy of L. G. Caro.)

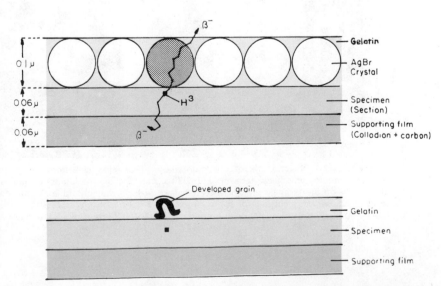

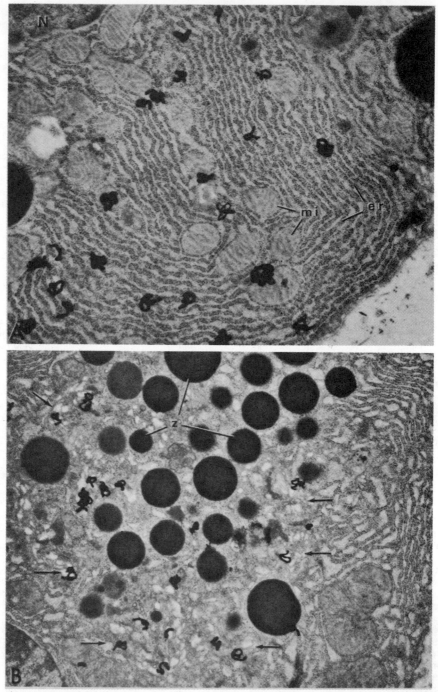

**Figure 5-15**  Electron microscopic autoradiograph of acinar cells from the pancreas of a guinea pig. **A,** three minutes after pulse labeling with ³H-leucine. The autoradiographed grains are located almost exclusively on the granular endoplasmic reticulum (*er*); *mi*, mitochondria; *N*, nucleus. ×17,000. **B,** the same as above, but incubated for seven minutes after pulse labeling. The label is now in the region of the Golgi complex (arrows); *z*, zymogen granules. ×17,000. (Courtesy of J. D. Jamieson and G. E. Palade.)

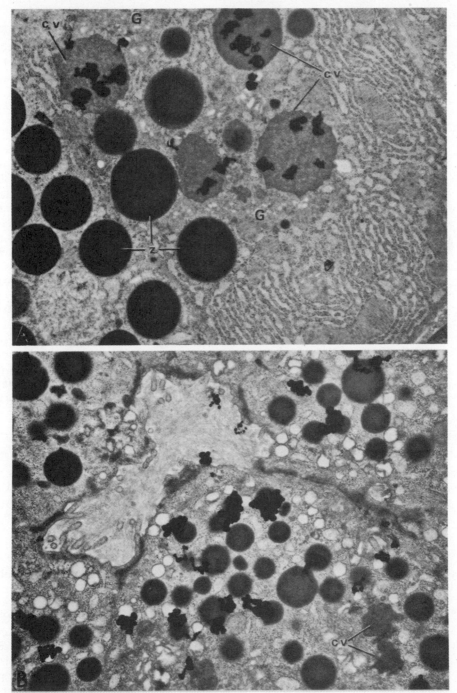

**Figure 5-16 A,** the same experiment as in Figure 5-15, but 37 minutes after pulse labeling. The label is now concentrated in the condensing vacuoles (*cv*) of the Golgi complex (*G*). The zymogen granules (*z*) are unlabeled. ×13,000. **B,** the same as above, but incubated for 117 minutes after pulse labeling. The autoradiographed grains are now localized primarily over the zymogen granules, while the condensing vacuoles (*cv*) are devoid of label. Some grains are in the lumen of the acinus, indicating the secretion: ×13,000. (Courtesy of J. D. Jamieson and G. E. Palade.)

reticulum → Golgi complex → immature granules (i.e., condensing vacuoles and prozymogen) → mature secretory (zymogen) granules. In the rat pancreas, for example, the total life span of a zymogen granule has been estimated at 52.4 minutes.

Since the intracellular transport is made against an apparent concentration gradient, it is of interest to determine its possible metabolic requirements. The transport of the newly synthesized polypeptide chain from the ribosome into the cisternae of the ER does not require additional energy and seems to be controlled mainly by the structural relationship between the large ribosomal subunit and the membrane of the reticulum. Transport from the ER to the condensing vacuole, however, is blocked by inhibitors of cell respiration or of oxidative phosphorylation. An important conclusion to be drawn from these studies is that at the periphery of the Golgi complex, in transitional elements between the endoplasmic reticulum and the small vesicles of the Golgi complex, there is a kind of *energy-dependent lock* in the transport, in which ATP is indispensable.

### STEP 4: CONCENTRATION IN THE CONDENSING VACUOLE.

condensing vacuole (prozymogen granule)— immature secretory granule

During this period the *condensing vacuoles* are converted into *zymogen granules* by progressive filling and concentration, until finally they acquire their characteristic electron-opaque appearance. This conversion does not require a supply of metabolic energy. It has been postulated that it may depend in part on the formation of osmotically inactive aggregates in the vacuoles, resulting in a passive flow of water from the vacuoles to the relatively hyperosmotic cytosol.

zymogen granule—mature, electron-opaque secretory granule that releases its contents into the extracellular medium when the appropriate stimulus is provided

### STEPS 5 AND 6: INTRACELLULAR STORAGE AND RELEASE BY EXOCYTOSIS.

Step 4 culminates with the storage of the secretory product in *secretory granules*, which are released when the appropriate stimulus (e.g., a hormone or neurotransmitter) acts on the cell. This storage provides a mechanism by which the cell can cope with a demand for secretory product that exceeds its rate of synthesis.

exocytosis—energy-requiring process by which secretory granules discharge their contents by fusing with the cell membrane

The discharge of the secretory granule takes place by a process called *exocytosis*, which involves movement of the granule toward the apical region of the cell followed by fusion of the granular and plasma membranes. When the membranes fuse, an orifice is produced through which the secretory product is discharged (Fig. 5–17). Biochemically, exocytosis requires the intracellular elevation of $Ca^{2+}$ and the production of ATP. Thus, exocytosis is the second energy-requiring step in the secretory process (see step 3, above). Presumably part of this energy requirement is related to the process of membrane fusion (disruption taking place during exocytosis) but it is possible that energy could be consumed in the propulsion of the granule to the site of secretion.

The membrane of the zymogen granule may be regarded as a vacuole that shuttles between the Golgi apparatus and the cell surface. Analysis of excess membrane removed from the apical region of the cell has suggested that patches of membranes are invaginated from the surface as small vesicles that move back into the Golgi

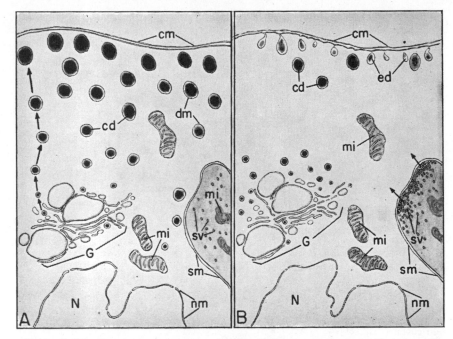

**Figure 5-17**  Diagrammatic interpretation of the mechanism of secretion in the chromaffin cell. **A,** cell in the resting stage, showing the storage of mature catechol droplets in the outer cytoplasm. Near the nucleus within the Golgi complex new secretion is being formed at a slow rate. At the right, a portion of a nerve terminal, showing the synaptic vesicles *(sv)* and mitochondria *(mi); cd,* catechol droplets; *cm,* cell membrane; *dm,* droplet membrane; *ed,* evacuated droplets; *G,* Golgi complex; *N,* nucleus; *nm,* nuclear membrane; *sm,* surface membrane. **B,** cell after strong electrical stimulation by way of the splanchnic nerve. Most of the catechol droplets have disappeared; the few that remain can be seen in different stages of excretion into the intercellular cleft. The Golgi complex is now forming new droplets at a higher rate. The nerve ending shows an increase of synaptic vesicles with accumulation at "active points" on the synaptic membrane. (From E. De Robertis and D. D. Sabatini.)

region, to be re-utilized in the packaging of more secretion products. Thus the process of exocytosis is coupled with that of endocytosis.

## Insulin Biosynthesis: A Good Example of Cell Secretion

It is well known that many secreted proteins are initially synthesized as biologically inactive precursors which are activated later. Activation consists essentially of the removal of a portion of the polypeptide chain and may occur at different sites. Various polypeptide hormones are produced as inactive *prohormones* which are then activated intracellularly by the *converting* (i.e., proteolytic) *enzymes* presumably present in the Golgi apparatus.

A beautiful example of molecular processing occurs during the course of insulin secretion. *Insulin* is a hormone (produced by the β-cells of the endocrine pancreas) which is essential in the metabolism of glucose. (Insulin deficiency results in the well-known disease *diabetes*.) The insulin molecule is a small protein (12,000 dal-

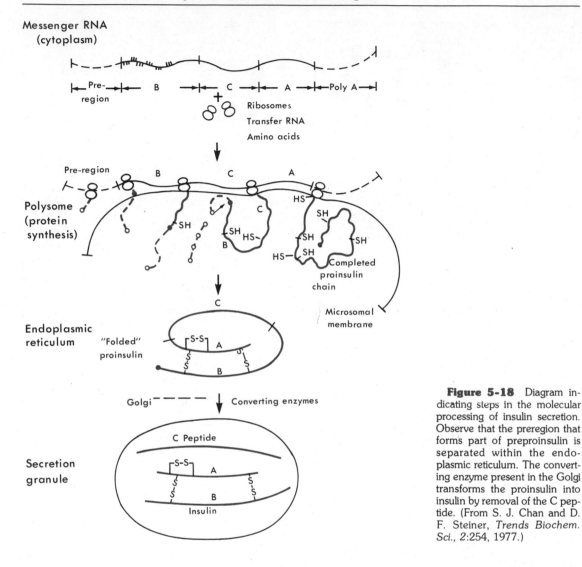

Messenger RNA
(cytoplasm)

Pre-region

Polysome
(protein
synthesis)

Endoplasmic
reticulum

"Folded"
proinsulin

Golgi — — — Converting enzymes

Secretion
granule

**Figure 5-18**  Diagram indicating steps in the molecular processing of insulin secretion. Observe that the preregion that forms part of preproinsulin is separated within the endoplasmic reticulum. The converting enzyme present in the Golgi transforms the proinsulin into insulin by removal of the C peptide. (From S. J. Chan and D. F. Steiner, *Trends Biochem. Sci.*, 2:254, 1977.)

tons) having an A chain of 21 amino acids and a B chain of 30 amino acids. The chains are linked by two S—S bonds. Figure 5–18 shows a diagram of the insulin messenger RNA. Observe that it has two regions *(cistrons)* coding for the A and B chains, as well as a C segment and a preregion or *signal sequence* that codes for the signal peptide. The complete translation of this mRNA on free ribosomes gives rise to *preproinsulin*. Normally, in the endoplasmic reticulum the signal peptide is removed by the *signal peptidase* and a chain of *proinsulin* is produced. Later the proinsulin chain is activated in the Golgi complex through the action of a converting enzyme that removes the C peptide. The active hormone and the C peptide are packaged within the zymogen granule for ultimate release.

**TABLE 5-2 LYSOSOMAL ENZYMES**

| Function | Enzyme | Function | Enzyme |
|---|---|---|---|
| Transferases transferring phosphorus groups | | Enzymes acting on peptide bonds (peptide hydrolases) | carboxypeptidase cathepsin A acid carboxypeptidase dipeptidase cathepsin $B_1$ and $B_2$ cathepsin C cathepsin D cathepsin E renin collagenase plasminogen activator |
| Nucleotidyltransferases | ribonuclease | | |
| Hydrolases acting on ester bonds | esterase lipase phospholipase $A_1$ and $A_2$ acid phosphatase (several types) phosphoprotein phosphatase phosphodiesterase (exonuclease) phospholipase C deoxyribonuclease II arysulfatase A and B | Enzymes acting on acid anhydride bonds in phosphoryl-containing anhydrides | pyrophosphatase arylphosphatase |
| | | Enzymes acting on P—N bonds | phosphoamidase |
| Hydrolases acting on glycosyl compounds | lysozyme (muramidase) neuroaminidase $\alpha$-glucosidase $\beta$-glucosidase $\alpha$-galactosidase $\beta$-glucuronidase hyaluronidase | | |

# 5-4 LYSOSOMES

The concept of the lysosome originated from the development of cell fractionation techniques, by which different subcellular components could be isolated. By 1949, a class of particles having centrifugal properties intermediate between those of mitochondria and microsomes was isolated by De Duve and found to have a high content of acid phosphatase and other hydrolytic enzymes. On account of their enzymatic properties, in 1955 these particles were named *lysosomes* (Gr., *lysis*, dissolution; *soma*, body). At present we know of about 50 lysosomal *hydrolases*, which are able to digest most biological substances. Table 5-2 lists only a few of them, together with their main functions. Lysosomes have been found in both animal and plant cells as well as in protozoa.

One important property of lysosomes is their stability in the living cell. This is due to the fact that the enzymes are enclosed within a membrane and the whole process of digestion is carried out within the organelle. Most lysosomal enzymes act in an acidic medium, which is maintained by a *proton pump* that accumulates $H^+$ inside the lysosome.

lysosome—one of a diverse group of single-membrane organelles containing many kinds of hydrolytic enzymes; involved in intracellular digestion

hydrolase—type of enzyme that digests biological substances into smaller molecules

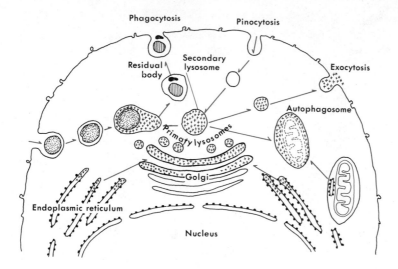

**Figure 5-19** Diagram representing the dynamic aspects of the lysosome system. Observe the relationships between the processes of phagocytosis, pinocytosis, exocytosis, and autophagy.

## Primary and Secondary Lysosomes

The most remarkable physical characteristic of lysosomes is their polymorphism of size and internal structure. According to the current interpretation, this polymorphism is the result of the association of primary lysosomes with the different materials that are *phagocytized* (literally, "eaten") by the cell. A summary of the various aspects of the lysosome system is presented in Figure 5–19.

At present four types of lysosomes are recognized, of which the first is the *primary lysosome;* the other three may be grouped together as *secondary lysosomes.*

primary lysosome—small storage vesicle formed at the GERL region

(1) The *primary lysosome* or *storage granule* is a small body whose contained enzymes are synthesized by the ribosomes and accumulated in the endoplasmic reticulum. From there they penetrate into the Golgi region. The GERL region of the Golgi complex maturing face has been implicated in primary lysosome formation.

heterophagosome—digestive vacuole resulting from the ingestion of foreign substances by the cell

(2) The *heterophagosome* or *digestive vacuole* results from the ingestion (by *phagocytosis* or *pinocytosis*) of foreign material by the cell. This body contains the engulfed material within a membrane, and the extent of the digestion that takes place depends on the amount and chemical nature of the material and the activity and specificity of the lysosomal enzymes. Under ideal conditions, digestion leads to products of low molecular weight that pass through the lysosomal membrane and are incorporated into the cell to be re-used in many metabolic pathways.

residual body—vacuole formed when the digestion of foreign substances is incomplete

(3) *Residual bodies* are formed if the digestion is incomplete. In some cells, such as the ameba and other protozoa, these residual bodies are eliminated by defecation (Fig. 5–19). In other cells they may remain for a long time and may have some role in the aging process. (For example, the pigment inclusions found in nerve cells of old animals may be a result of this type of process.)

autophagosome—lysosome specialized to digest parts of the cell that contains it

(4) The *autophagic vacuole, cytolysosome,* or *autophagosome* is a special case, found in normal cells, in which the lysosome contains

and digests a part of the cell itself (e.g., a mitochondrion or portions of the endoplasmic reticulum; Fig. 5–19).

## Endocytosis

The intracellular secretion of primary lysosomes is coupled to another system of extracellular origin that is generated by *endocytosis*, the collective name for several phenomena related to the activity of the plasma membrane. Endocytosis includes the processes of phagocytosis, pinocytosis, and micropinocytosis, by which solid or fluid materials are ingested in bulk by the cell (Fig. 5–19). (We have seen that *exocytosis* is the reverse process, by which membrane-enclosed products such as zymogen granules are released at the plasma membrane.)

*Phagocytosis* (Gr., *phagein*, to eat) occurs in a large number of protozoa and among certain cells of the metazoa. In protozoa, phagocytosis is closely linked to ameboid motion. An ameba ingests large particles, including microorganisms, by surrounding them with pseudopodia to form a food vacuole within which digestion takes place. In metazoa, rather than serving for cell nutrition, phagocytosis is generally a means of defense, by which particles that are foreign to the organism (e.g., bacteria, dust, and various colloids) are disposed of. In both protozoa and metazoa the process of phagocytosis involves two distinct stages. First the particle *adheres* (is *adsorbed*) to the cell surface, and then it actually penetrates into the cell. In some cases it has been possible to dissociate these two events; for example, at low temperatures bacteria may adhere to the surface of a leukocyte without being ingested.

*Pinocytosis* (Gr., *pinein*, to drink) is a mechanism by which proteins and other soluble materials are incorporated into the cell. This can be demonstrated easily using proteins labeled with fluorescent dyes. The presence of the protein seems to act as a stimulus to pinocytosis, and the uptake of protein is surprisingly high. Pinocytosis is induced not only by proteins but also by amino acids and certain ions, and is related to the cell coat that covers the plasma membrane.

*Micropinocytosis* is a related mechanism that can be observed only in the electron microscope. In endothelial cells lining capillaries, in smooth muscle, and in other cell types it is possible to observe vesicles of about 65 nm that result from the invagination of the plasma membrane. These vesicles are involved in the transport of fluids across the endothelium, a process which is assumed to occur in quantal amounts (i.e., in packets containing a certain amount of fluid). The sequence of events that occurs begins with the invagination of the membrane, followed by the formation of a closed vesicle by membrane fission, the movement of the vesicle across the endothelial cytoplasm, the fusion of the vesicle with the opposite plasma membrane, and the discharge of the vesicular content. Micropinocytosis is frequently associated with the formation of *coated vesicles* in which there are tiny and regularly spaced bristles on the cytoplasmic side under the membrane. Such vesicles are involved

endocytosis—collective term for several phenomena involving the surrounding and ingestion of various substances by the plasma membrane

phagocytosis ("eating")— ingestion of particulate material by the cell, either as a means of feeding or defense

pinocytosis ("drinking")— ingestion of proteins and other soluble materials by the cell

in the uptake of proteins by pinocytosis, but they may have other roles as well since they are often found in the Golgi region.

Phagocytosis and pinocytosis are essentially similar phenomena, as can be demonstrated by allowing an ameba to phagocytize some ciliated cells and then inducing pinocytosis. The number of channels formed is much smaller under these conditions. Furthermore, both phagocytosis and pinocytosis are active mechanisms in the sense that they require energy to operate. In cultured cells, the rate of endocytosis is increased by the addition of ATP and inhibited by respiratory and other metabolic poisons.

In endocytosis the attachment of the particles to the plasma membrane apparently leads to some kind of change beneath the attachment site, which may in turn be related to the actin microfilaments inserted in the cytoplasmic side of the membrane. These may be associated with membrane invagination and the formation of pseudopods. By endocytosis, large amounts of plasma membrane may be taken up into the cell. A macrophage (i.e., large phagosome) may incorporate almost twice its surface area in an hour, suggesting that there is a continuous recycling of the membrane back to the cell surface.

## Functions of Lysosomes

Lysosomes are involved in numerous cellular activities, several of which are mentioned above. The list can be elaborated upon and added to as follows.

**AUTOPHAGY.**   We have seen that many cellular components, such as mitochondria, are constantly being removed from the cell by *autophagy*. Cytoplasmic organelles become surrounded by membranes of the smooth endoplasmic reticulum, lysosomal enzymes are discharged into the autophagic vacuoles, and the organelles are digested. Autophagy is a general property of eukaryotic cells and is related to the normal renovation and turnover of cellular components. Autophagy may be greatly increased in certain conditions; for example, in protozoa deprived of nutrients or in the liver of a starving animal, many autophagic vacuoles appear. This is a mechanism by which parts of the cell are broken down to provide raw materials essential to survival.

autophagy—digestion of cellular material by the cell's own enzymes; part of the normal regeneration and turnover of eukaryotic cellular components

**SHEDDING OR REMODELING OF TISSUES.**   Many developmental processes involve extensive removal of whole cells and extracellular material. During the metamorphosis of amphibians, for example, considerable tissue remodeling is accomplished by lysosomal enzymes. The degeneration of the tadpole tail is caused by the action of certain proteolytic enzymes contained in the lysosomes. In addition, some tissues that undergo regression after a period of activity may do so by the action of lysosomes. For example, the human uterus just after giving birth weighs 2 kg, and in 9 days returns to its former size, weighing 50 g.

**HUMAN DISEASES.**   Lysosomes are involved in many diseases and syndromes. In rheumatoid arthritis, silicosis and asbestosis

Ref. 4

DeVita, V.T., Hellman, S., and
Rosenberg, S.A. _Cancer:
Principles and Practice of
Oncology_. Philadelphia:
Lippincott, 1982.

# RHODE ISLAND GROUP HEALTH ASSOCIATION
## EMERGENCY ROOM/HOSPITAL USAGE CARD

Patient's Name _____ I.D. # _____ Date: _____

Time: _____

Emergency Room

☐ Payment Authorized

Hospital _____

Complaint/Diagnosis _____

☐ Payment Denied

Transportation Authorized

☐ Ambulance

☐ Taxi

Hospital Admission

☐ Yes

Hospital _____

Complaint/Diagnosis _____

Expected LOS _____

Attending MD _____

Check if:

☐ Auto Accident

☐ Work-related

Provider Signature _____

(produced by the inhalation of silica or asbestos particles, respectively), and gout (in which crystals of urate accumulate in the joints), there is a release of lysosomal enzymes from the macrophages and an acute tissue inflammation that may lead to an increase in collagen synthesis. There are also several congenital diseases in which the main alteration is the accumulation within the cells of substances such as glycogen or various glycolipids. These are sometimes called *storage diseases*, and are caused by a mutation that affects one of the lysosomal enzymes involved in the catabolism of a certain substance.

**SEED GERMINATION IN PLANT CELLS.** Membrane-enclosed organelles containing digestive enzymes have been found in plant cells. During seed germination these enzymes, which are present in the seedling, are released and attack the stored material, making it available to the developing plant.

# 5-5 PEROXISOMES AND GLYOXYSOMES

With improved cell fractionation methods a second group of particles, in addition to the lysosomes, has been isolated from liver cells and other sources. These particles, which are rich in enzymes including peroxidase and catalase (among others), are called *peroxisomes*. Electron microscopic studies have suggested that these particles correspond morphologically to the *microbodies* found in kidney and liver cells. These microbodies or peroxisomes are limited by a single membrane; they contain a fine, granular substance that may condense in the center, forming an opaque core or nucleoid.

peroxisome (microbody)—
membrane-enclosed
organelle containing enzymes
for the metabolism of
hydrogen peroxide

In contrast to the nucleoid-containing peroxisomes found in liver and kidney cells are others that are smaller and lack a nucleoid. These *microperoxisomes* are found in all cells and are associated with the endoplasmic reticulum. They may be considered regions of ER in which catalase and other enzymes are compartmentalized.

Both types of peroxisomes are formed in the endoplasmic reticulum, and the enzymes they contain are synthesized by ribosomes bound to the rough ER. It is assumed that the peroxisome grows and is destroyed, probably by autophagy, after four or five days.

Isolated liver peroxisomes contain four enzymes related to the metabolism of $H_2O_2$. Three of them — urate oxidase, D-amino oxidase, and $\alpha$-hydroxylic acid oxidase — produce peroxide ($H_2O_2$), and catalase destroys it. Since $H_2O_2$ is toxic to the cell, catalase probably plays a protective role.

Green plants contain peroxisomes that carry out a process called photorespiration. In this process, glycolic acid, a two-carbon product of photosynthesis that is released from chloroplasts, is oxidized by glycolic acid oxidase, an enzyme present in peroxisomes. This oxidation, carried out by oxygen, produces hydrogen peroxide, which is then decomposed by catalase inside the peroxisome. Photorespiration is so named because light induces the synthesis of glycolic acid in chloroplasts. The entire process involves the intervention of both chloroplasts and peroxisomes.

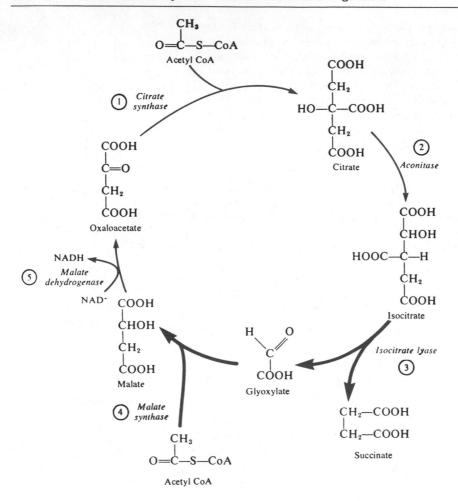

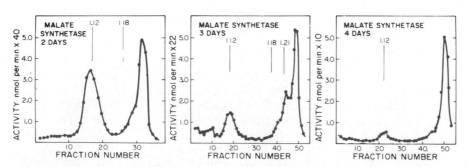

**Figure 5-20**  The glyoxylate cycle in castor bean endosperm and an experiment showing the localization of enzyme synthesis. **Top,** the glyoxylate cycle, showing the special steps (heavy arrows) that differentiate it from the Krebs cycle of mitochondria. **Bottom,** fractionation of the enzyme malate synthetase during germination of the castor bean endosperm. Germination was carried out for two, three, and four days, and the endosperm homogenate was centrifuged on a continuous gradient. Observe that at two days there are two peaks of enzyme activity corresponding to the endoplasmic reticulum and the glyoxysomes; with time, the first peak is reduced and the second becomes predominant. (From E. Gonzalez and H. Beevers, *Plant Physiol., 57:*406, 1976.)

*Glyoxysomes* are similar in some respects to peroxisomes, although they are found only in plant cells and are involved primarily in triglyceride metabolism. These organelles consist of an amorphous protein matrix surrounded by a limiting membrane. The enzymes of the glyoxysome are used to transform the fat stores of the plant seed into carbohydrates by way of the *glyoxylate cycle* (Fig. 5–20), which is a modification of the Krebs cycle (see Chapter 6). There is experimental evidence that these organelles are also synthesized in the endoplasmic reticulum.

glyoxysome—membrane-enclosed organelle found in plant cells and involved mainly in triglyceride metabolism

# SUMMARY

## 5-1 THE ENDOPLASMIC RETICULUM (ER)

The extent of development of the ER varies considerably in different cell types and is related to their functions. The rough ER is particularly well developed in cells that are actively engaged in protein synthesis, often consisting of parallel stacks of cisternae that occupy large areas of the cytoplasm. The ribosomes of the ER are present as polysomes and are always attached to the membrane by the 60S subunit.

The ER can be studied biochemically in microsomal form; microsomes are essentially small vesicles of ER membrane that maintain the normal topological relationships between the membranes, the lumen, and the ribosomes. The membrane contains a lipid bilayer with peripheral and integral proteins. Within the microsome are enzymes involved in the synthesis of biological macromolecules, the components of two electron transport systems, and enzymes that process secretory products.

The ER has many different functions within the same cell. Both rough and smooth ER are thought to have a role in mechanical support, intracellular exchanges, the maintenance of ionic gradients and electrical potentials, and the circulation of materials within the cell. The rough ER is involved primarily in protein synthesis; the smooth ER is involved in detoxification, lipid synthesis, and glycogenolysis.

The developmental origin of the ER is not definitely known, although experimental evidence supports the concept of structural and functional continuity. Current theories generally assume a multi-step mechanism in which most intracellular membranes are synthesized in the rER and then progressively transformed into those of the sER and the Golgi complex.

## 5-2 THE GOLGI COMPLEX

The Golgi complex is located between the ER and the secretory vesicles/plasma membrane. It is morphologically similar in plant and animal cells, consisting of dictyosome units formed by various cisternae and secretory vesicles. There are no ribosomes attached to the Golgi apparatus, and the complex is surrounded by a zone of exclusion. Each dictyosome is polarized, having a proximal face near the nuclear envelope or ER and a distal face which yields secretory vesicles. The GERL, a structure associated with the distal face, is thought to have some relationship to condensing vacuoles or presecretory granules.

Biochemical studies carried out in parallel for the Golgi complex, the ER, and the plasma membrane of the same cell type have indicated that the chemical composition of the Golgi is generally intermediary between that of the ER and the plasma membrane. The most characteristic enzymes of the Golgi fraction are the glycosyltransferases, along with acid phosphatase and other lysosomal and oxidative enzymes.

The major functions of the Golgi complex are related to its intermediary position between the ER and the plasma membrane; virtually all cell products are transported through the Golgi and secreted. A good example of this process is the synthesis and secretion of glycoproteins.

## 5-3 ROLES OF THE ER AND GOLGI COMPLEX IN CELL SECRETION

Secretion of cell products can be either continuous or discontinuous; six consecutive steps are generally recognized in this process. (1) Proteins are synthesized on polysomes attached to the rough ER. (2) The proteins emerge into the lumen of the ER. These first two steps are thought to be regulated to some extent by the signal codons, which determine whether synthesis will take place on free or ER-bound ribosomes. (3) The proteins are transported from the ER to the Golgi. At this stage it is possible to trace protein transport cytochemically by the use of radiolabeled precursors and a technique called autoradiography. (4) The proteins are concentrated into condensing vacuoles, which are converted into zymogen granules. (5) The products are stored for a certain length of time in storage granules, and finally (6) they are discharged by the process of exocytosis. Steps (3) and (6) require energy in the form of ATP.

Insulin is a very good example of a protein that is initially synthesized as a biologically inactive precursor which is activated later by a series of modifications taking place in the ER and Golgi complex.

## 5–4 LYSOSOMES

Lysosomes have a digestive function in both plant and animal cells. The hydrolases they contain are able to digest most biological substances, and this process takes place entirely within the organelle.

Lysosomes differ remarkably in size and internal structure. There is one type of primary lysosome: the storage granule; and there are three types of secondary lysosomes: the digestive vacuole, the residual body, and the autophagic vacuole.

Endocytosis includes the related processes of phagocytosis, pinocytosis, and micropinocytosis. These are active mechanisms in the sense that they require energy to operate.

In addition to their general involvement in storage, digestion, and protection, lysosomes play an important part in such phenomena as autophagy, shedding or remodeling of tissues, various human diseases, and seed germination in plant cells.

## 5–5 PEROXISOMES AND GLYOXYSOMES

Peroxisome particles are rich in enzymes including peroxidase and catalase. There are two types in animal cells: those corresponding to the microbodies found in kidney and liver cells, and a smaller type found in all cells and associated with the ER. Isolated liver peroxisomes contain enzymes related to the metabolism of $H_2O_2$. The peroxisomes of green plants carry out the process of photorespiration, which also requires the activity of chloroplasts. Glyoxysomes are found only in plant cells and are involved primarily in triglyceride metabolism by way of the glyoxylate cycle.

# STUDY QUESTIONS

- Protein A is synthesized for use within the cell, protein B is synthesized for continuous export, and protein C is secreted only under certain physiological conditions. Compare proteins A, B, and C with regard to site of synthesis and possible mechanisms of intracellular transport, processing, packaging, storage, and secretion.
- Describe the structure and function of microsomes.
- How are endomembrane components thought to be assembled in a cell that has just completed mitosis?
- Describe the location and structure of an idealized dictyosome. Where and what are the forming and maturing faces, the transition vesicles, and the GERL?
- How would you trace the intracellular transport of a protein using radiolabeled precursors and autoradiography?
- Which stages of intracellular protein transport require energy, and what physical and/or chemical properties govern those stages that do not?
- Describe the phenomena referred to collectively as "endocytosis."
- What are the functions of the different kinds of lysosomes?
- What are the functions of peroxisomes and glyoxysomes?

# SUGGESTED READINGS

Aterman, K. (1979) Development of the concept of lysosomes. *Histochem. J.*, 11:503.

Blobel, G. (1977) Synthesis and segregation of secretory proteins; the signal hypothesis. In *International Cell Biology*. (B. R. Brinkley and K. R. Porter, eds.), p. 318. Rockefeller University Press, New York.

Davis, B. D., and Tai, P.-C. (1980) The mechanism of protein secretion across membranes. *Nature*, 283:433.

**Dean, R. T.** (1977) Lysosomes and membrane recycling: A hypothesis. *Biochem. J., 168:*603.

**De Duve, C.** (1975) Exploring the cell with a centrifuge. *Science, 189:*186.

**Hand, A. R.** (1980) Cytochemical differentiation of the Golgi apparatus from GERL. *J. Histochem. Cytochem., 28:*82.

**Leighton, F., Coloma, L., and Koenig, C.** (1975) Structure, composition, physical properties and turnover of peroxisomes. *J. Cell Biol., 67:*281.

**Neutra, M., and Leblond, C. P.** (1969) The Golgi apparatus. *Sci. Am., 220:*100.

**Northcote, D. H.** (1971) The Golgi apparatus. *Endeavour, 30:*26.

**Palade, G. E.** (1975) Intracellular aspects of the process of protein secretion. *Science, 189:*347.

**Pearse, B.** (1980) Coated vesicles. *Trends Biochem. Sci. 5:*13.

# chapter 6

# Energy Transducing Organelles: Mitochondria and Chloroplasts

We mentioned in Chapter 1 that all cells and organisms can be grouped into two main classes that differ in the mechanism of extracting energy for their own metabolism. In the first class, called *autotrophs*, $CO_2$ and $H_2O$ are transformed by the process of *photosynthesis* into the elementary organic molecule *glucose*, from which more complex molecules are made (Fig. 1–9). The second class of cells, called *heterotrophs*, obtains energy from the different nutrients (carbohydrates, fats, and proteins) synthesized by autotrophic organisms. The energy contained in these organic molecules is released mainly by combustion with $O_2$ from the atmosphere in a process called *aerobic respiration*. The release of $H_2O$ and $CO_2$ by heterotrophic organisms completes this cycle of energy.

Energy transformation in cells takes place with the intervention of two main *transducing systems* (i.e., systems that produce energy transformations) represented by *mitochondria* and *chloroplasts*. These two organelles in some respects function in opposite direc-

## TABLE 6-1 DIFFERENCES BETWEEN PHOTOSYNTHESIS AND OXIDATIVE PHOSPHORYLATION

| Photosynthesis | Oxidative Phosphorylation |
|---|---|
| only in presence of light; thus periodic | independent of light; thus continuous |
| uses $H_2O$ and $CO_2$ | uses molecular $O_2$ |
| liberates $O_2$ | liberates $CO_2$ |
| hydrolyzes water | forms water |
| endergonic reaction | exergonic reaction |
| $CO_2 + H_2O + energy \rightarrow$ foodstuff | foodstuff $+ O_2 \rightarrow CO_2 + H_2O + energy$ |
| in chloroplasts | in mitochondria |

tions. Chloroplasts, present only in plant cells, are specially adapted to capture light energy and to transduce it into chemical energy, which is stored in covalent bonds between atoms in the different nutrients or *fuel molecules*. On the other hand, mitochondria are the "power plants" which, by oxidation, release the energy contained in the fuel molecules and make other forms of chemical energy. The main function of chloroplasts is *photosynthesis*, while that of mitochondria is *oxidative phosphorylation*. Table 6–1 shows some of the basic differences between these two transducing systems. Note that photosynthesis is an *endergonic* reaction, which means that it captures energy; oxidative phosphorylation is an *exergonic* reaction, meaning that it releases energy.

Mitochondria and chloroplasts are characteristic of eukaryotic cells but, as we have seen in Chapter 1, the two main functions they perform also take place in prokaryotes. The oxidation of organic material is carried out by enzymes present in the plasma membrane; and in blue-green algae, for example, the membrane is also able to perform photosynthesis.

# 6-1 THE MITOCHONDRION: STRUCTURE AND FUNCTION

## Structural Organization of Mitochondria

In 1894, Altmann first described mitochondria (calling them "bioblasts") and predicted their association with cellular oxidation. In 1913, Warburg observed that respiratory enzymes were associated with cytoplasmic particles. By 1950 the electron microscope allowed investigators to learn a great deal more about the structural organization of these organelles.

mitochondrion—membrane-enclosed organelle that generates chemical energy in the form of ATP for use in cellular metabolic processes

Mitochondria (Gr., *mito*-, thread + *chondrion*, granule) are present in the cytoplasm of protozoa and animal and plant cells, and are characterized by specific morphological, biochemical, and func-

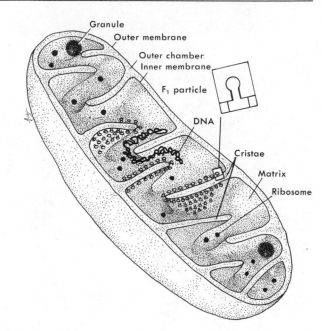

Labels on figure: Granule, Outer membrane, Outer chamber, Inner membrane, F₁ particle, DNA, Cristae, Matrix, Ribosome

**Figure 6-1** Three-dimensional diagram of a mitochondrion cut longitudinally. The main features are shown. Observe that the cristae are folds of the inner membrane and that on their matrix side they have the F₁ particles. The inset shows an F₁ particle with the head piece and stalk.

tional properties. They are generally cylindrical structures about 0.5 $\mu$m in diameter and several $\mu$m in length. A normal liver cell may contain between 1000 and 1600 mitochondria, while some oocytes may have as many as 300,000; green plant cells contain fewer mitochondria than many animal cells because some of their functions are taken over by chloroplasts. These organelles may be observed in the living cell, in which they display passive and active motion and show changes in volume that are related to their activity. They have a characteristic structural organization and lipoprotein composition, and contain a large number of enzymes and coenzymes that take part in cellular energy transformations.

As indicated in Figure 6–1, a mitochondrion consists of two membranes and two compartments. An outer limiting membrane, about 6 nm thick, surrounds the organelle. Within this membrane, separated from it by a space of about 6 to 8 nm, is an inner membrane that projects into the mitochondrial cavity complex infoldings called *mitochondrial crests* or *cristae*. This inner membrane, which is also about 6 nm thick, divides the mitochondrion into two compartments or chambers: (1) the outer chamber, contained between the two membranes and in the core of the cristae, and (2) the inner chamber, bound by the inner membrane. This inner chamber is filled with a relatively dense proteinaceous material usually called the *mitochondrial matrix*. The matrix is generally homogeneous, but in some cases it may contain a finely filamentous material or small, highly dense granules (see Fig. 6–2). These granules are now considered to be sites for binding $Mg^{2+}$ and $Ca^{2+}$. The cristae that project into the matrix are usually incomplete septa or ridges that do not divide the inner chamber into separate sections; thus, the matrix is continuous within the mitochondrion. The matrix also contains

crista—infolding of the inner mitochondrial membrane, containing components for respiratory metabolism and oxidative phosphorylation and providing a relatively large surface area on which these reactions can take place

mitochondrial matrix—dense proteinaceous material found in the inner chamber, containing $Mg^{2+}$ and $Ca^{2+}$ binding sites as well as the mitochondrial ribosomes and DNA

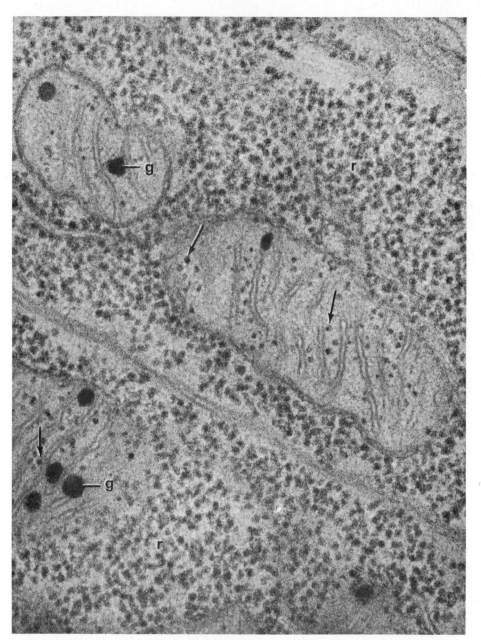

**Figure 6-2**   Electron micrograph of the intestinal epithelium showing a large accumulation of ribosomes *(r)* in the cytoplasm. In mitochondria, dense granules *(g)* and ribosomes (arrows) are visible. ×95,000. (Courtesy of G. E. Palade.)

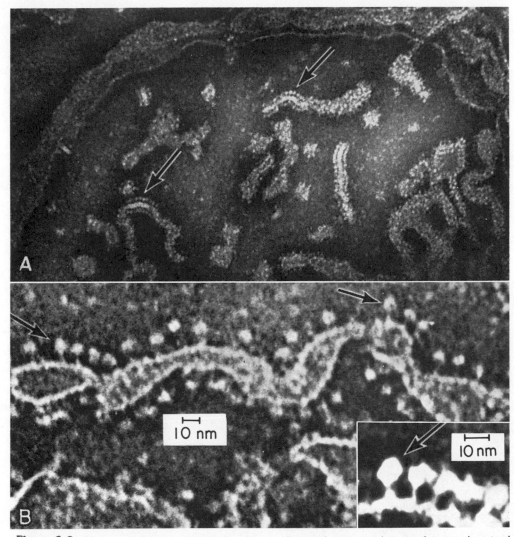

**Figure 6-3**   Electron micrograph of a mitochondrion swollen in a hypotonic solution and negatively stained with phosphotungstate. **A,** at low power; isolated crests can be observed in the middle of the swollen matrix. Arrows point to some of these crests. **B,** at higher magnification (×500,000), a mitochondrial crest showing the so-called "elementary particles" on the surface adjacent to the matrix. Inset at ×650,000, showing the elementary particles with a polygonal shape and the fine attachment to the crest. (Courtesy of H. Fernández-Moran.)

small ribosomes *(mitoribosomes)* and one or more molecules of circular DNA.

If a mitochondrion is allowed to swell and break in a hypotonic solution and is then immersed in a material that is opaque to electrons (e.g., phosphotungstate), a *negative stain* is obtained. (This means that the membranes and other structures are more transparent to the electron beam than is the surrounding medium.) In these conditions, as shown in Figure 6-3, the mitochondrial crests appear covered with mushroom-like particles of 8.5 nm that have

**oxidative phosphorylation—
ATP synthesis by
phosphorylation of ADP,
using the energy provided by
electron transfer during
aerobic respiration**

**mitoplast—mitochondrion
whose outer membrane has
been removed**

a stem linking them to the membrane. These so-called *F1 particles* are spaced at 10 nm intervals, and there are $10^4$ to $10^5$ of them per mitochondrion. We will see later that these particles represent a portion of a special ATPase involved in the coupling between oxidation and phosphorylation. The F1 particles are on the matrix side (M side) of the inner membrane, giving it a characteristic asymmetry that is related to the function of the ATPase.

In the same way that cell fractionation methods make it possible to separate mitochondria from other cell components, it is also possible to separate the mitochondrial membranes and compartments from each other. For example, the outer membrane may be stripped off, leaving a *mitoplast* that contains the inner membrane and the matrix. The mitoplast is able to perform most of the functions of the mitochondrion (Fig. 6-4).

More drastic treatments lead to the separation of the matrix and the inner membrane including the cristae, and have demonstrated that all mitochondrial enzymes are highly compartmentalized. For example, the matrix contains all the soluble enzymes that are involved in the Krebs cycle (see below) in addition to DNA and the machinery for protein synthesis. The inner membrane carries all the components related to the respiratory chain and to phosphorylation, as well as specific carrier proteins involved in the permeation of metabolites such as ADP, ATP, and phosphate.

The internal conformation of mitochondria may change dramatically between two extreme states. If the external ADP concentration is low or the respiratory chain is inhibited, mitochondria will be found in the *orthodox state*. This is the state usually observed in tissue sections (Fig. 6-5, **A**). In this case the matrix occupies most

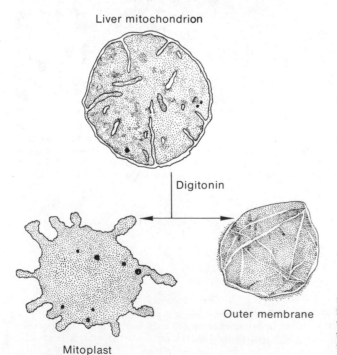

Liver mitochondrion

Digitonin

Outer membrane

Mitoplast

**Figure 6-4** Diagram showing an intact liver mitochondrion and its dissection by the action of digitonin. The mitoplast comprises the inner membrane with the finger-like unfolded mitochondrial crests and the matrix. The outer membrane revealed by negative staining shows a folded-bag appearance.

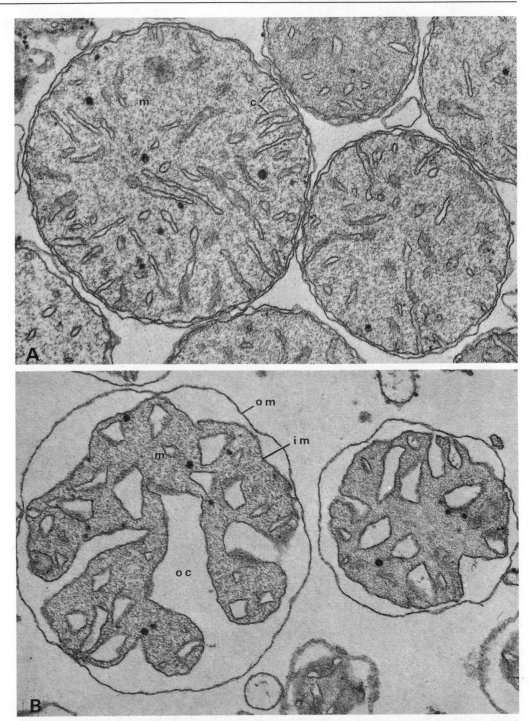

**Figure 6-5**  Electron micrographs of isolated mitochondria from rat liver in two extreme conformation states.  ×110,000. **A,** the *orthodox* conformation. The inner membrane is organized into crests *(c)*, and the matrix *(m)* fills the entire mitochondrion. **B,** the *condensed* conformation. Mitochondrial crests are not observed, and the outer chamber *(oc)* represents about 50 percent of the volume; *om,* outer membrane; *im,* inner membrane. (Courtesy of C. R. Hackenbrock.)

of the mitochondrion and the outer chamber is small. However, if ADP is added to the medium there is a sudden contraction of the inner compartment. This is the *condensed state*, which corresponds to an actively phosphorylating mitochondrion (Fig. 6–5, **B**). These changes in mitochondrial conformation are due to the fact that cell respiration and phosphorylation generate not only ATP but also $H_2O$, which accumulates in the outer chamber.

The fact that within a mitochondrion more than 70 enzymes, coenzymes, cofactors, and metals interact in an orderly fashion suggests the presence of an extremely complex molecular system. At this point we will discuss briefly the way in which chemical energy is transformed, and this will introduce the more thorough description of mitochondrial function that follows.

## Chemical Energy and ATP

**potential energy—energy stored in the covalent bonds between the atoms of a molecule**

The *chemical* or *potential* energy of foodstuffs is locked in the covalent bonds between the atoms of a molecule. For example, during the hydrolysis of typical chemical bonds, about 3000 calories per mole are liberated. In a mole ($\sim$ 180 g) of glucose, between the various atoms of C, H, and O, there are about 686,000 calories of potential energy which can be liberated by combustion, as in the following reaction:

$$C_6H_{12}O_6 + 6 O_2 \rightarrow 6 H_2O + 6 CO_2 + 686,000 \text{ calories} \qquad (1)$$

Within the living cell this enormous amount of energy is not released suddenly, as in combustion by a flame. It is made available in a stepwise and controlled manner, requiring a great number of enzymes that finally convert the fuel into $CO_2$ and $H_2O$.

The liberated energy may be used by the cell (1) to synthesize new molecules (i.e., proteins, carbohydrates, and lipids), by means of endergonic reactions, which can then be used to replace others or for the normal growth and metabolism of the cell; (2) to perform mechanical work such as cell division, cyclosis, or muscle contraction; (3) to carry out active transport against an osmotic or ionic gradient; (4) to maintain membrane potentials, as in nerve conduction and transmission, or to produce electrical discharges (e.g., in certain fish); (5) to perform cell secretion; or (6) to produce radiant energy (e.g., in bioluminescence). Only in the reactions of group (1) is the energy provided by nutrients transformed into chemical bond energy. In all the other types of reactions, chemical energy is transformed into other forms of energy.

**ATP—nucleotide containing high energy bonds; provides energy for many biochemical cellular processes by undergoing enzymatic hydrolysis**

The common link in all these transformations is the compound *adenosine triphosphate* (ATP). ATP is found in all cells. Its most significant chemical characteristic is that it has two terminal bonds with a potential energy much higher than that of all the other chemical bonds. Figure 2–2 shows that ATP is composed of the purine base adenine, a ribose moiety, and three molecules of phosphoric acid. (Adenine plus ribose form the nucleoside *adenosine*.) ATP and the closely related molecule ADP (adenosine diphosphate) are the most important compounds in energy transformation. If phosphate is represented by P, the simplified formula of ATP and its transformation into ADP or AMP is as follows:

$$\boxed{\text{Adenosine}}\text{———}\textcircled{P}\sim\textcircled{P}\sim\textcircled{P} \rightleftharpoons \boxed{\text{Adenosine}}$$
$$\text{———}\textcircled{P}\sim\textcircled{P} + \text{Pi} + 7300 \text{ calories} \tag{2}$$

or:

$$\boxed{\text{Adenosine}}\text{———}\textcircled{P}\sim\textcircled{P}\sim\textcircled{P} \rightleftharpoons \boxed{\text{Adenosine}}$$
$$\text{———}\textcircled{P} + \text{PPi} + 7300 \text{ calories} \tag{3}$$

Note that the release of any one of the two terminal phosphates of ATP yields about 7300 calories per mole, instead of the 3000 calories from common chemical bonds. The high energy $\sim$P bond enables the cell to accumulate a great quantity of energy in a very small space and keep it ready for use whenever it is needed.

Other nucleotides having high energy bonds, such as cytosine triphosphate (CTP), uridine triphosphate (UTP), and guanosine triphosphate (GTP), are involved in biosynthetic reactions. However, the energy source for these nucleotide triphosphates derives ultimately from ATP.

## The Krebs Cycle

The *Krebs cycle*, also called the *tricarboxylic acid cycle*, takes place in the mitochondrial matrix. It serves as the first step in a common pathway for the degradation of fuel molecules such as carbohydrates, fatty acids, and amino acids. In the cell cytoplasm these substances are first acted upon metabolically to produce acetyl groups, which enter the mitochondrion and undergo the two-stage transformation diagrammed in Figure 6–6. In the first stage the acetyl groups, linked to *acetyl coenzyme A* (CoA), are taken into the Krebs cycle. Among the products of this cycle are $CO_2$ molecules and hydrogen atoms or protons (H$^+$). In the second stage, which takes place after this cycle, the H$^+$ are taken up by the *respiratory chain* and eventually combined with $O_2$ to yield $H_2O$. This process also generates ATP by the phosphorylation of ADP.

As illustrated in Figure 6–6, the first step of the Krebs cycle involves the condensation of the acetyl group of acetyl CoA with *oxaloacetate* (4 carbons) to form *citrate*, a 6-carbon compound. In the following steps, two molecules of $CO_2$ are released and *succinate* (4 carbons) is formed. Succinate continues through the series of reactions and is eventually transformed into oxaloacetate in preparation for a new cycle.

An important consequence of the Krebs cycle is that at each turn four pairs of hydrogen atoms are removed from the substrate intermediates by enzymatic dehydrogenation. These hydrogen atoms (or equivalent pairs of electrons) enter the respiratory chain, being accepted by either *nicotinamide adenine dinucleotide* (NAD$^+$) or *flavin adenine dinucleotide* (FAD). Three pairs of hydrogens are accepted by NAD$^+$, reducing it to NADH, and one pair by FAD, reducing it to FADH$_2$ (this last pair comes directly from the succinic dehydrogenase reaction) (Fig. 6–6). Since it takes two turns of the cycle to metabolize the two acetate molecules that are produced by glycolysis

tricarboxylic acid (Krebs) cycle—first sequence of reactions occurring in a common pathway for fuel molecule degradation, in which the oxidation of acetyl groups provides energy that is stored in phosphate bonds

acetyl coenzyme A—high energy intermediate in fuel molecule metabolism, bringing suitably modified acetyl groups into the tricarboxylic acid cycle

respiratory chain—second sequence of reactions in the fuel molecule degradation pathway, in which electrons are accepted by NAD$^+$ or FAD and eventually combined with $O_2$ to yield $H_2O$

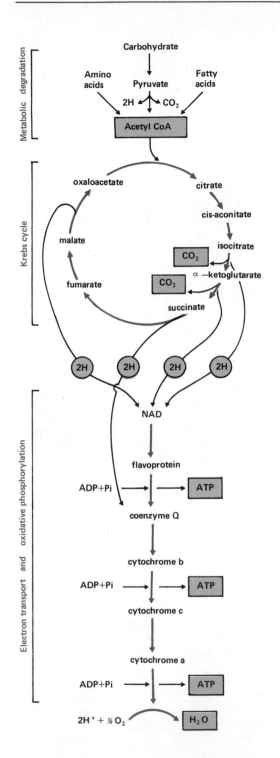

**Figure 6-6** General diagram of aerobic respiration showing the Krebs cycle, the respiratory chain, and its coupling with oxidative phosphorylation. (From A. L. Lehninger, *Biochemistry*. Worth, New York, 1975.)

from one molecule of glucose, a total of six molecules of NADH and two of $FADH_2$ are formed at the starting points of the respiratory chain.

## Electron Transport and Oxidative Phosphorylation

The hydrogen atoms (or equivalent electrons) coming from the Krebs cycle, after being accepted by the coenzymes $NAD^+$ and FAD, are conveyed through the respiratory chain by a group of electron carriers called *cytochromes*. The cytochromes undergo successive oxidations and reductions (reactions in which electrons are transferred from an electron *donor* or *reductant* to an electron *acceptor* or *oxidant*) along a gradient of *redox potential* (i.e., a measure in volts of the reducing capacity).

The cytochromes are iron-containing proteins in which the iron atom is contained within a chemical structure called a *porphyrin ring*. This portion of the molecule is a *heme*, a prosthetic group that is bonded to the protein by sulfur bridges. (A similar heme group is present in molecules of hemoglobin.) During the electron transfer, the iron atom passes from the ferrous to the ferric state, releasing an electron. This reaction is the basis of all cellular oxidation-reduction processes.

$$Fe^{2+} \rightleftharpoons Fe^{3+} + e^- \qquad (4)$$

The inner mitochondrial membrane contains the cytochromes b, c, $c_1$, a, and $a_3$, aligned to form a kind of chain in which the redox potential becomes more and more positive. At each cytochrome the iron may be in the reduced or oxidized state, and the electrons flow along the gradient of redox potential. As shown in Figure 6–7, the

cytochrome—one of several electron carrier enzymes whose prosthetic groups undergo successive oxidations and reductions

reductant—an electron donor

oxidant—an electron acceptor

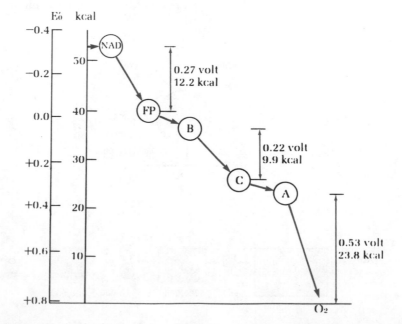

**Figure 6-7** Diagram showing the decline of free energy as electron pairs flow down the respiratory chain from NAD to $O_2$. $E'_0$, oxidoreduction potential; *kcal*, free energy. At the three points indicated there is enough energy drop to generate a molecule of ATP from ADP and phosphate. (Courtesy of A. L. Lehninger. From A. L. Lehninger, *Biochemistry*. 2nd Ed. Worth, New York, 1975, p. 516.)

couple NAD⁺/NADH has the most negative redox potential (designated $E_0$, in this case $-0.32$ volts) of the chain and the highest free energy (designated $\Delta G$, in this case $-52.7$ kcal/mole). This free energy is released upon oxidation. The reversible NAD⁺/NADH system is followed in the chain by a *flavin mononucleotide* (FMN)- and FAD-containing flavoprotein. A lipid-soluble protein called coenzyme Q (CoQ) or *ubiquinone* then acts as a kind of shuttling system between the flavoproteins and the series of cytochromes beginning with cytochrome b. The cytochrome series ends with $O_2$, which has the most positive redox potential ($E_0 = 0.82$) and no free energy to be utilized ($\Delta G = 0$). Thus the most reduced members of the electron transport chain are found at the beginning and the most oxidized members at the end.

All the components of the respiratory chain just described, as well as those involved in the mechanism of phosphorylation (discussed below), are integrated within the molecular structure of the inner mitochondrial membrane. It is important to remember that this membrane has a high protein content (60 to 70 percent) and that most of these proteins are of the integral type, being intercalated into the lipid bilayer.

Using mild detergents, it has been possible to dissociate the respiratory chain into four multi-molecular complexes (I to IV), three of which are represented in Figure 6–8. Complex II, called *succinate-Q-reductase*, precedes the electron transport chain and is coupled to succinate by way of FAD (Fig. 6–6). Complex I, *NADH-Q-reductase*, is the largest complex and includes a flavoprotein containing FMN; this is the first step in the electron transport chain. Electrons are taken into this complex by NAD⁺, which is located at the *matrix side* (M side) of the membrane. Complex III, *$QH_2$-cytochrome c-reductase*, contains cytochromes b, c, and $c_1$. Complex IV, *cytochrome c-oxidase*, contains cytochromes a and $a_3$, and is the one in which the final oxidation of hydrogen, resulting in $H_2O$, takes place.

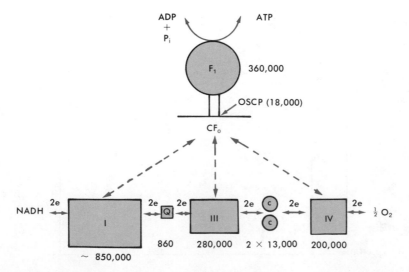

**Figure 6-8** Diagram of the electron-transferring complexes (i.e., respiratory chain) and the ATPase complex ($F_1$) present in the mitochondrial inner membrane; *complex I,* NADH-Q-reductase; *complex III,* $QH_2$-cytochrome c-reductase; *complex IV;* cytochrome c-oxidase; $CF_0$, $F_0$ coupling factor; *OSCP,* oligomycin-sensitive protein; *Q,* ubiquinone; *c,* cytochrome. (See description in text.) (Courtesy of E. C. Slater.)

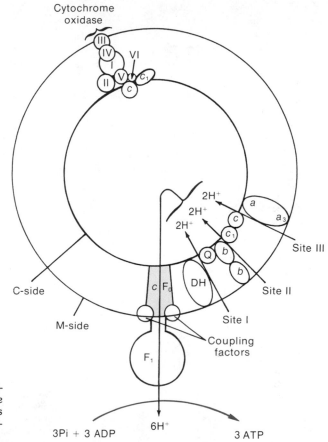

**Figure 6-9** The topography of a vesicle originated from the inner membrane of mitochondria, the transmembrane distribution of the various complexes integrating the respiratory chain, and ATPase are indicated. (Courtesy of E. Racker, 1977.)

The important point to remember is that all these complexes and the phosphorylating system are organized within the inner mitochondrial membrane in a highly asymmetrical arrangement (Fig. 6-9). The electron transport system is only accessible to NADH and succinate from the M side, while cytochrome c is reached from the *cytoplasmic side* (C side) of the membrane. This molecular organization is compatible with the transfer of protons (H+) across the membrane from the M side to the C side.

Figure 6-9 shows that the respiratory chain is coupled at three points with the system in which phosphorylation of ADP to ATP takes place. The six protons that originated in the respiratory chain are translocated across the inner mitochondrial membrane from the M side to the C side, and these six protons will give rise to three molecules of ATP through the intervention of the mitochondrial ATPase. We mentioned above that the phosphorylating system is partly localized in the F1 particles that are present at the M side of the mitochondrial crests (Fig. 6–3). This system is very complex; in addition to the F1 particle (made of five subunits), it has a hydrophobic protein component (F0 in Fig. 6–9) that serves as a proton channel across the membrane. The F0 and F1 components are connected by a protein stalk.

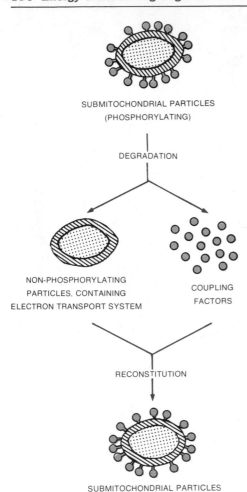

SUBMITOCHONDRIAL PARTICLES
(PHOSPHORYLATING)

DEGRADATION

NON-PHOSPHORYLATING
PARTICLES, CONTAINING
ELECTRON TRANSPORT SYSTEM

COUPLING
FACTORS

RECONSTITUTION

SUBMITOCHONDRIAL PARTICLES
(PHOSPHORYLATING)

**Figure 6-10**  Diagram showing the experiment by which the submitochondrial particles, corresponding to the inner mitochondrial membranes, are submitted to urea to remove the coupling factor $F_1$, thus leaving nonphosphorylating particles. The lower part of the figure shows the subsequent reconstitution of a phosphorylating submitochondrial particle. (From E. Racker, The membrane of the mitochondrion. *Sci. Am.*, *218*:32, 1968. Copyright © 1968 by Scientific American, Inc. All rights reserved.)

**F1 particle—membrane-bound structure with ATP synthetase activity; located on the matrix side of the cristae**

The best demonstration that the F1 particles are involved in phosphorylation is provided by the elegant experiment of Racker, shown in Figure 6–10. If isolated mitochondria are disrupted, fragments of the mitochondrial crests form vesicles with an "inside-out" organization. As shown in the diagram, in these vesicles the F1 particles protrude on the surface at the M side while the C side faces the interior. These vesicles are able to carry out cell respiration and phosphorylation. If the F1 particles are removed by treatment with urea, however, the vesicles no longer phosphorylate. Racker was able to reconstitute the system by mixing the F1 particles with the vesicles. After isolation, the F1 particle functions as an ATPase, causing the hydrolysis of ATP to ADP and phosphate; but when the F1 particle is attached to the membrane, it works as an *ATP synthetase* and makes ATP from ADP and phosphate.

Biochemical and electron microscopic studies demonstrate that the respiratory chain and the phosphorylating system form a kind of huge macromolecular assembly which may reach $5 \times 10^6$ daltons and cover a surface of 20 × 20 nm on the mitochondrial crests. It has been calculated that a mitochondrion from a liver cell may contain about 15,000 such assemblies, and that one from insect flight muscle can contain as many as 100,000. It is evident that all the components involved in oxidation and phosphorylation are organized not only across the membrane but also laterally in a mosaic arrangement, and that knowledge of this molecular structure is indispensable to an understanding of mitochondrial functions.

## The Chemiosmotic Theory

A fundamental problem in the study of cellular energy interrelationships has been that of determining the nature of the link between the respiratory chain and the phosphorylating system. In 1961 Mitchell proposed the *chemiosmotic theory*, which is, for the most part, currently in favor. It postulates that the link between the systems is essentially electrochemical: the translocations of protons across the membrane create a difference in pH and electrical potential. These chemical and electrical gradients would then be the driving force for the synthesis of ATP.

According to this theory, the electron transport system is organized in "redox loops" which translocate protons from the M side to the C side of the membrane (Fig. 6–11, **A**). This vectorial movement of protons gives rise to a *pH gradient* because the concentration

chemiosmotic theory—theory that the pH and electrical gradients across the mitochondrial membrane provide the electrochemical link between oxidation and phosphorylation, and drive the synthesis of ATP

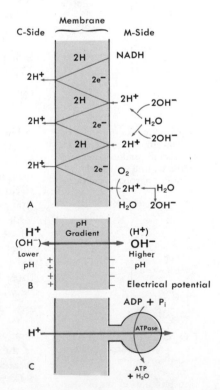

**Figure 6-11**  Diagram of chemiosmotic coupling according to P. Mitchell. **A,** during electron transport H+ ions are driven to the C side of the inner mitochondrial membrane. **B,** this process produces a pH gradient and an electrical potential across the membrane. **C,** this gradient drives the proton pump of the ATPase, and ATP is synthesized from ADP and Pi.

of $H^+$ becomes higher at the C side and that of $OH^-$ increases at the M side. An *electrical gradient* is also established at the same time (Fig. 6–11, **B**). These two gradients would then drive the phosphorylating system through the proton pump of the ATPase. The result is the synthesis of ATP and water (Fig. 6–11, **C**).

$$ADP + Pi \rightleftharpoons ATP + H_2O \qquad (5)$$

A special feature of this theory, as opposed to various others that have been postulated, is that short-range interactions are not needed. The respiratory chain and the ATPase may operate independently of one another, and the coupling does not require direct contact of the molecular components; instead, the coupling is brought about by electrochemical fields that can operate over longer distances.

## Overall Energy Balance: Glucose and the P/O Ratio

*anaerobic glycolysis— degradation of glucose into lactic acid in the absence of oxygen, yielding two ATP molecules*

To determine an overall balance of the energy liberated by the degradation of nutrients, let us consider a molecule of glucose. Glucose is first degraded into lactic acid by a process called *anaerobic glycolysis*, producing two ATP molecules. This is achieved in eleven successive steps, each one catalyzed by a different enzyme. The glycolytic enzymes are soluble in the cytoplasmic matrix, and as shown in Figure 6–12, in this chain of reactions the product of one enzyme serves as a substrate for the next reaction. The figure also shows that during this series of reactions four ATP molecules are produced. Since two ATP were used previously, however, the yield is only two ATP molecules. The general reaction may be written:

$$\underset{\text{glucose}}{C_6H_{12}O_6} + Pi + 2\ ADP \rightarrow \underset{\text{lactic acid}}{2\ C_3H_6O_3} + 2\ ATP + 2\ H_2O \qquad (6)$$

*pyruvate—key product of glycolysis; converted to lactate under anaerobic conditions, and to acetyl CoA under aerobic conditions*

The fate of *pyruvate*, a key product of glycolysis, depends on whether oxygen is available. Under anaerobic conditions, it is used as a hydrogen acceptor for the two NADH generated during glycolysis, and it is converted into lactate. Under aerobic conditions, pyruvate is converted into acetyl coenzyme A and $CO_2$ is released. At this point it is possible for acetyl CoA to enter the Krebs cycle.

Of the 686,000 calories contained in a mole of glucose, less than 10 percent (i.e., 58,000) can be released by anaerobic glycolysis. Much more of this energy is released by oxidative phosphorylation. As described above (see Fig. 6–6), the formation of ATP occurs at three steps of the electron chain. The equation for this process can be written as follows:

$$NADH + H^+ + 3\ ADP + 3\ Pi + 1/2\ O_2 \rightarrow NAD^+ + 4\ H_2O + 3\ ATP \qquad (7)$$

*P/O ratio—measure of ATP yield from oxidative phosphorylation, expressed as moles of inorganic phosphate used per oxygen atom consumed*

One way of indicating the ATP yield from oxidative phosphorylation is the *P/O ratio*, which is expressed as the moles of inorganic phosphate (Pi) used per oxygen atom consumed. [In Equation (7) the P/O ratio is 3 because 3 Pi and $1/2\ O_2$ are used.]

The energy balance of aerobic respiration shows that 36 ATP molecules are produced from each glucose molecule. The overall equation can be written:

$$C_6H_{12}O_6 + 36 \text{ Pi} + 36 \text{ ADP} \rightarrow 6 \text{ CO}_2 + 36 \text{ ATP} + 42 \text{ H}_2\text{O} \quad (8)$$

The cell will store 40 percent of the chemical energy liberated by the combustion of glucose in the form of ATP. The rest of the energy is dissipated as heat or used for other cell functions.

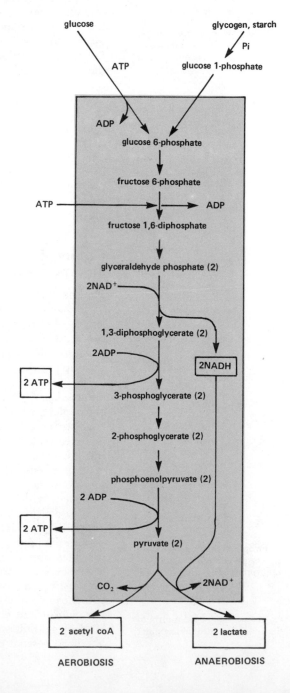

**Figure 6-12**   Diagram of the degradation of glucose or glycogen by anaerobic glycolysis.

## 6-2 THE CHLOROPLAST: STRUCTURE AND FUNCTION

chloroplast—chlorophyll-containing plastid in which photosynthesis occurs

photosynthesis—process of trapping the energy of sunlight, transforming it into chemical energy, and storing it in the chemical bonds of carbohydrates

Chloroplasts, which are found mainly in the cells of the leaves of higher plants and algae, are the most common and the most biologically important of the plastids. By the process of *photosynthesis*, they produce oxygen and most of the chemical energy used by living organisms on our planet. From a human point of view we may say that each molecule of oxygen used in respiration and each carbon atom in our bodies have at one time passed through a chloroplast. Furthermore, it has been calculated that each $CO_2$ molecule from the atmosphere is incorporated into a plant every 200 years, and that all the oxygen in the atmosphere is renewed by plants every 2000 years. The numbers presented here simply help to illustrate the concept that without plants there would be no atmospheric oxygen and that life as we know it would be impossible.

The first living organisms were anaerobic and appeared at least one billion years before the process of photosynthesis evolved. This process appeared first in prokaryotes and then in eukaryotes. With the release of oxygen, a by-product of photosynthesis, the atmosphere favored aerobic organisms, and anaerobic organisms became a small fraction of all living forms.

### Chloroplast Structure and Molecular Organization

The shape and size of chloroplasts vary in different cells and species. In higher plants they are generally disc-shaped, with an average length of 4 to 6 $\mu$m (Fig. 1-1). Sometimes they have a colorless center, due to the presence of starch granules; in general, chloroplasts of plants grown in the shade are larger and contain more chlorophyll than those of plants grown in sunlight.

The number of chloroplasts is relatively constant in different plants. Algae such as *Chlamydomonas* often contain a single huge chloroplast, while in higher plants there are usually 20 to 40 chloroplasts per cell. When the number of chloroplasts is insufficient, it is increased by division (i.e., elongation and constriction at the central portion); when it is excessive, it is reduced by degeneration.

Chloroplasts exhibit both passive and active movements. Changes in shape and volume caused by the presence of light have been observed in chloroplasts isolated from spinach. The volume decreases considerably after the chloroplasts are struck by light and photophosphorylation is initiated; this effect is reversible.

chloroplast envelope—double membrane lacking chlorophyll and cytochromes, and regulating cytosol-chloroplast exchanges

stroma—gel-fluid matrix of the chloroplast

With the light microscope it is possible to distinguish many structural features of the chloroplast, although of course the electron microscope has revealed its components in much greater detail. The *envelope* is a double limiting membrane that lacks chlorophyll and cytochromes, and across which all molecular interchanges between the cytosol and chloroplast take place. The *stroma* or *matrix* is a kind of gel-fluid phase enclosed within the envelope. It contains about 50 percent of the chloroplast proteins, mostly of the soluble

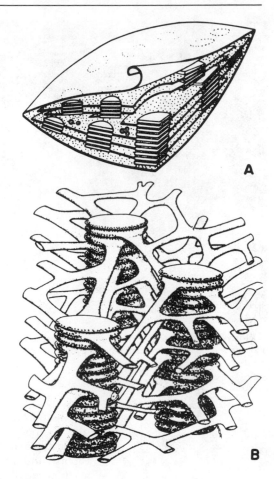

**Figure 6-13** **A,** diagram of a chloroplast showing the inner structure with the grana disposed in stacks perpendicular to the surface. (From G. A. Erickson, E. Kahn, B. Wallis and D. von Wettstein.) **B,** diagram of the ultrastructure of three grana showing the tubules that join some of the membranous compartments of the grana. (From T. E. Weier, C. R. Stocking, W. W. Thompson and H. Drever, *J. Ultrastruct. Res., 8*:122, 1963.)

type; it also has ribosomes, DNA, and the machinery for protein synthesis. We will see that the fixation of $CO_2$ occurs in the stroma, as does the synthesis of sugar, starch, fatty acids, and some proteins.

The *thylakoids* are flattened vesicles arranged as a membranous network within the stroma. The outer surface of the thylakoid is in contact with the stroma, and its inner surface encloses an *intrathylakoid space*. Thylakoids may be stacked like a pile of coins, forming the *grana* (Fig. 6–13), or they may be unstacked *(stroma thylakoids)*, forming a system of tubules that are joined to the grana thylakoids (Fig. 6–14). The number of thylakoids per granum may vary from a few to 50 or more. The thylakoids contain about 50 percent of the chloroplast protein and all the components involved in the essential steps of photosynthesis.

The lipid-protein mosaic model, now accepted for the plasma and mitochondrial membranes, has been accepted for the thylakoids as well. As we have seen, the main characteristics of this model are: (1) fluidity, allowing the free lateral movement of molecular components; (2) asymmetry with regard to the composition of the two lipid half-leaflets, thus preventing lipids and proteins from moving

thylakoid—flattened vesicle found in chloroplasts, in which the light-dependent reactions of photosynthesis take place

granum—stack of thylakoids

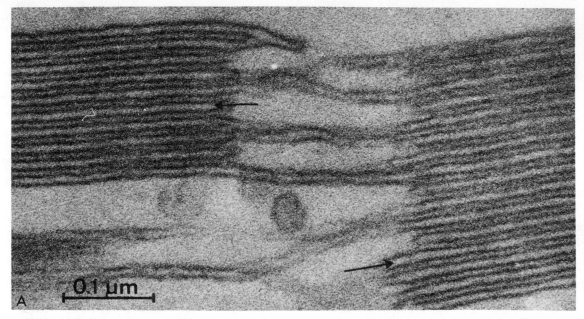

**Figure 6-14** Electron micrograph of a section of two chloroplast grana and their connecting tubules. The arrows indicate the cavity of the membranous compartment of the grana (i.e., thylakoid). ×240,000. (Courtesy of I. Nir and D. C. Pease.)

**chlorophyll—green photosynthetic pigment found in chloroplast thylakoids and prokaryotic photosynthetic membranes**

from one leaflet to the other; and (3) economy, meaning that since the membrane is extremely thin, random movements of its molecular components are only possible in two dimensions.

The thylakoid membrane contains half of the lipids in the chloroplast, and interspersed among them are important molecules such as chlorophyll, carotenoids, and plastoquinone, which are involved in photosynthesis. *Chlorophyll* is an asymmetrical molecule having a hydrophilic head made of four pyrrole rings bound to each other to form a porphyrin ring (Fig. 6–15). This part of the molecule is similar to some animal pigments such as hemoglobin and the cytochromes. In chlorophyll, however, there is a Mg atom forming a complex with the four rings. (In animal pigments the Mg is replaced by Fe.) Chlorophyll has a long hydrophobic chain attached to one of the rings. In higher plants there are two types of chlorophyll, *a* and *b*. In chlorophyll *b* there is a —CHO group in place of the —CH₃ group (circled in Fig. 6–15). Pigments that belong to the group called *carotenoids* are masked by the green color of chlorophyll. In autumn, the amount of chlorophyll decreases and the other pigments become visible. These are the *carotenes* and the *xanthophylls*, which are both related to vitamin A.

**carotenoids—group of yellow and red pigments including the carotenes and xanthophylls; usually masked by chlorophyll**

Chlorophyll molecules are disposed within the thylakoid membrane in close association with integral proteins, forming several complexes that have important functions in photosynthesis. The main complexes are: (1) *Photosystem I* or *PS I complex*, which, in addition to chlorophyll, contains another pigment called P700 (because it absorbs light at 700 nm). (2) *Photosystem II* or *PS II complex*,

**photosynthetic complex—one of the combinations of chlorophyll molecules and integral proteins found in the thylakoid membrane**

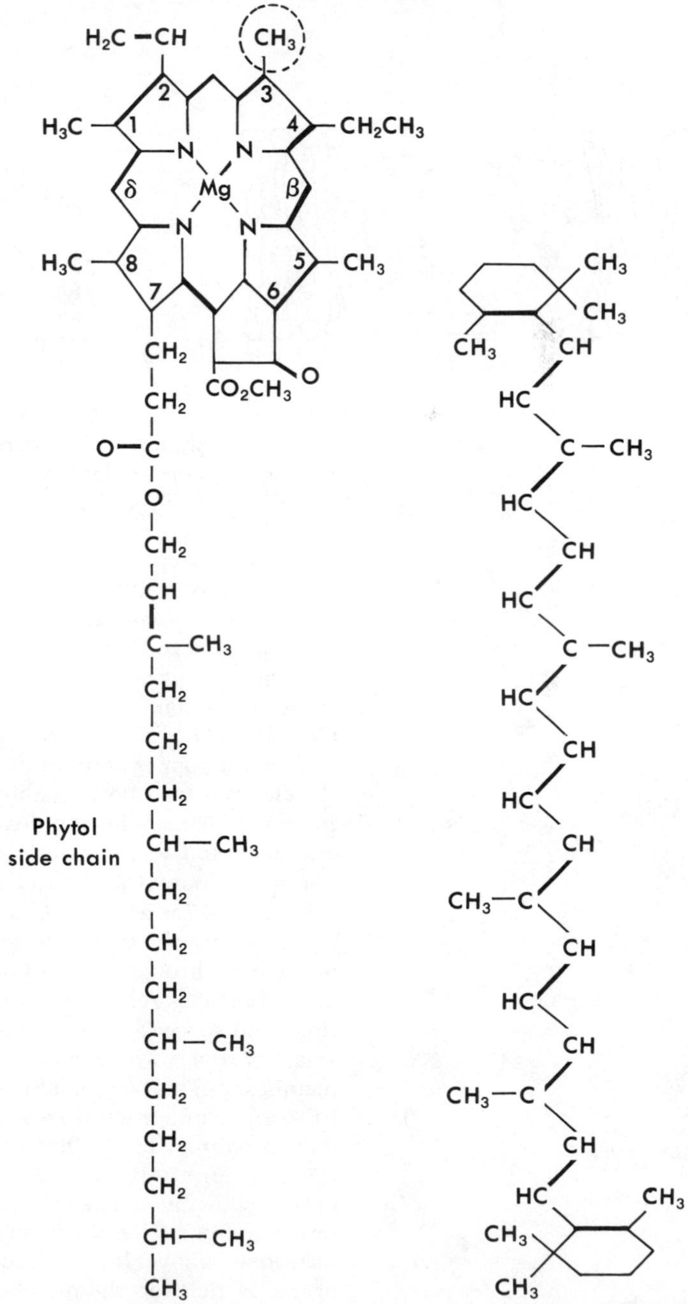

**Figure 6-15** Structural formula of chlorophyll a and β-carotene.

Chlorophyll a

β-Carotene

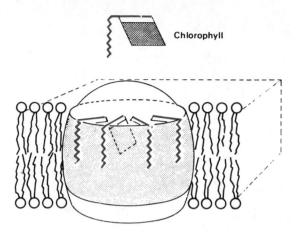

Chlorophyll

**Figure 6-16** Diagram of a thylakoid in cross-section showing an intrinsic protein extending across the membrane. It is proposed that the light-harvesting chlorophyll molecules are part of the boundary lipids. (Courtesy of J. M. Anderson.)

containing chlorophyll and pigment P680. P700 and P680 are the reaction centers of the respective photosystems. (3) *Light-harvesting complex (LHCP)*, which in the mature chloroplast contains about half of the chlorophyll molecules. In contrast with PS I and PS II, the LHCP complex is photochemically inactive, but apparently it transfers captured light energy to the photosystems, particularly to PS II. A possible molecular model of LHCP is shown in Figure 6–16.

In addition to these chlorophyll-protein complexes, which can be separated by treating the membranes with mild detergents, other macromolecular complexes lacking chlorophyll can be isolated. (1) *Cytochrome f-b$_6$ complex* contains cytochromes f and b$_6$ and plasto-cyanine, a copper-containing protein. This complex is involved in the electron transport chain of the chloroplast. (2) *CF0 or HF0* is a hydrophobic protein complex that functions in proton translocation and binds to CF1 to form the ATPase complex. (This system is comparable in many ways with the F0 and F1 systems in mitochondria.)

Figure 6–17 is a diagram of the possible distribution of these complexes within the thylakoid membrane. The figure indicates the line of fracture through the middle of the bilayer, emphasizing that the larger particles remain associated with the leaflet toward the intra-thylakoid space (*loculus* in the diagram), while the more numerous small particles are attached to the outer or stromal leaflet of the membrane. The two leaflets, when fractured, are referred to as the EF (exoplasmic fracture) and PF (protoplasmic fracture) to indicate their proximity to the intrathylakoid space and the stroma, respectively. Figure 6–18 is an electron micrograph of a fractured membrane, showing a smooth continuum corresponding to the lipid phase and particles of various sizes corresponding to the complexes mentioned above. It can be seen that there is a selective segregation of the particles in the membranes, corresponding to the unstacked (u) or to the stacked (s) regions of the thylakoid. (Remember that the stacked regions correspond to the grana.)

We mentioned above that chloroplasts contain a coupling factor, CF1, that is similar to the F1 particle of mitochondria and is involved in the photophosphorylation of ADP to ATP. The electron

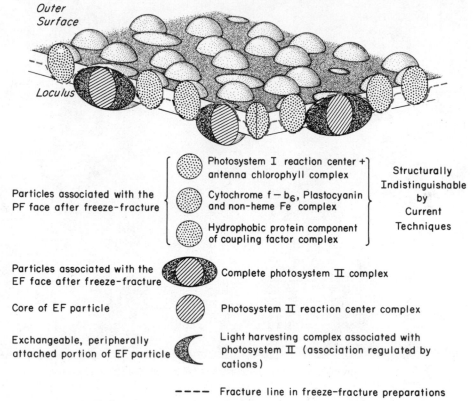

Figure on page showing:

Outer Surface

Loculus

Particles associated with the PF face after freeze-fracture {
- Photosystem I reaction center + antenna chlorophyll complex
- Cytochrome f – $b_6$, Plastocyanin and non-heme Fe complex
- Hydrophobic protein component of coupling factor complex
} Structurally Indistinguishable by Current Techniques

Particles associated with the EF face after freeze-fracture — Complete photosystem II complex

Core of EF particle — Photosystem II reaction center complex

Exchangeable, peripherally attached portion of EF particle — Light harvesting complex associated with photosystem II (association regulated by cations)

– – – – Fracture line in freeze-fracture preparations

**Figure 6-17**   Model of the thylakoid membrane in which the various photosystems and protein complexes are embedded in a lipid bilayer. The diagram shows the cleavage plane of the freeze-fracturing in blue and the relative number, position, and size of the particles in the outer surface and the loculus. (Courtesy of C. J. Arntzen.)

microscope has shown that the CF1 particles are localized in the outer surface of the thylakoid membrane, especially in the region of the stroma thylakoids, and are excluded from the stacked regions or grana thylakoids. The CF1 particles are attached to the membrane by way of the CF0 segment, which represents a proton channel made of hydrophobic proteins.

We will see that not only the photophosphorylating system has a vectorial disposition across the membrane; the components of the photosystems are asymmetric as well. In general, both PS I and PS II have the electron acceptors on the outer or stromal side, while the electron donors are inside, toward the intrathylakoid space. These concepts will be easier to understand after we study the process of photosynthesis.

CF1 particle—structure involved in the photophosphorylation of ADP to ATP; located on the outer surface of the thylakoid membrane

## Photosynthesis

Photosynthesis is one of the most fundamental biological reactions. By means of the chlorophyll contained in the chloroplasts, green plants trap the energy of sunlight emitted as photons and transform

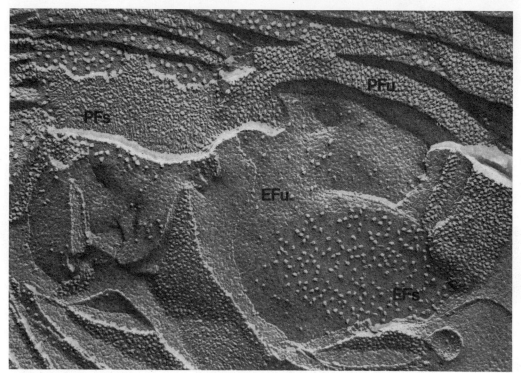

**Figure 6-18** Electron micrograph of a freeze-fractured chloroplast of *Pisum sativum l.* showing the large particles observed in EFs.

it into chemical energy. This energy is stored in the chemical bonds that are produced during the synthesis of various nutrients. We mentioned at the beginning of this chapter that the process of photosynthesis is in some ways opposite to that of oxidative phosphorylation. It will be helpful at this point to refer back to Table 6-1, which outlines the main differences between these two processes.

The overall reaction of photosynthesis can be written:

$$n\ CO_2 + n\ H_2O \xrightarrow[\text{chlorophyll}]{\text{light}} (CH_2O)_n + n\ O_2 \qquad (9)$$

to indicate that photosynthesis is essentially the combining of carbon dioxide and water to form various carbohydrates with loss of oxygen. In Equation (9), water is the donor of $H_2$. The $O_2$ that is liberated also comes from water and not from $CO_2$, as has been confirmed by experiments using "heavy oxygen" ($^{18}O$, rather than the more common $^{16}O$) in the reaction precursors and then examining the products.

$$n\ C^{16}O_2 + n\ H_2{}^{18}O \xrightarrow{\text{light}} (CH_2{}^{16}O)_n + n\ {}^{18}O_2 \qquad (10)$$

$$n\ C^{18}O_2 + n\ H_2{}^{16}O \xrightarrow{\text{light}} (CH_2{}^{18}O)_n + n\ {}^{16}O_2 \qquad (11)$$

Thus, in this reaction, $CO_2$ acts as an electron acceptor (A) and $H_2O$ as an electron donor (D), and an even more general equation for photosynthesis would be

$$H_2D + A \xrightarrow{\text{light}} H_2A + D \qquad (12)$$

The visible light portion of the entire electromagnetic radiation spectrum corresponds only to the very small region between 400 and 700 nm. The energy contained in these wavelengths is transmitted in discrete packets called *photons*. A photon contains one *quantum* of energy. This is expressed mathematically by the equation derived in 1900 by Max Planck:

$$E = \frac{hc}{\lambda} \qquad (13)$$

photon—a quantum of light energy

where $h$ is Planck's constant ($1.585 \times 10^{-34}$ cal-sec), $c$ is the speed of light ($3 \times 10^{10}$ cm/sec), and $\lambda$ is the wavelength of the radiation. This equation shows that photons with shorter wavelengths have higher energy.

Pigments such as chlorophyll and the reaction centers P680 and P700 present in the photosystems are specifically adapted to be excited by light. The photons that are absorbed produce excitation by displacing electrons (in the orbitals of the atoms) from a *ground state* to an *excited state* at a higher energy level. The excitation caused by photons may be dissipated as heat or emitted as radiation (fluorescence), but it may also be converted into chemical energy. We will return to the excitation process and its function later in this section.

Photosynthesis involves a complex series of reactions, some of which take place only in the presence of light while others can also be carried out in the dark. Hence we refer to *light* or *photochemical reactions* and *dark reactions*. In 1937, Hill was the first to provide evidence that the photochemical reactions of plant cells took place in chloroplasts. Later it was discovered that the oxidized form of the coenzyme $NADP^+$ was the normal hydrogen acceptor of the photochemical reaction, producing NADPH. Another important finding was made by Arnon (1954), who showed that, in the presence of light, chloroplasts could make ATP from ADP and Pi (this is the process called photophosphorylation).

photochemical (light) reactions—those photosynthetic processes that take place in chloroplasts and occur only in the presence of light, such as photophosphorylation

At this point it is helpful to recall the process of oxidative phosphorylation in mitochondria, in which the flow of electrons is from NADH to $O_2$ along the gradient of redox potential. In photosynthesis the opposite reaction occurs, and the electrons flow from $H_2O$ to NADPH. The photochemical reaction takes place in the thylakoid membranes. When these are illuminated, there is a transfer of electrons from water ($E'_0 = 0.81$ volts) to the final acceptor ($E'_0 = -0.6$ volts). Transfer against this negative electrochemical gradient uses the energy provided by the photons of light. The light energy is collected by the two photosystems, PS I and PS II, which act as collector antennae.

The diagram in Figure 6–19 shows that the reaction centers of both photosystems (P680 and P700) are excited by photons (2 H$\mu$).

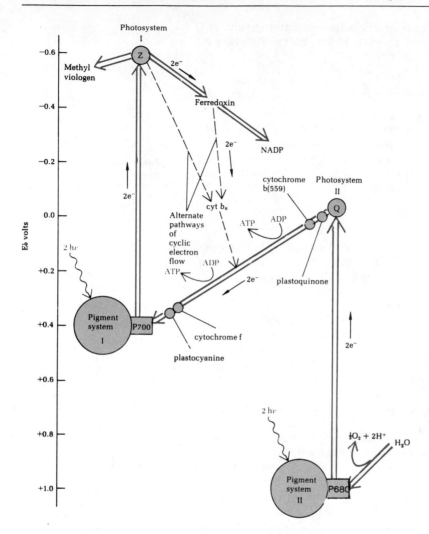

**Figure 6-19** Diagram showing the arrangement of photosystems I and II. These two systems of photosynthesis are connected by an electron transport chain situated between the acceptor Q and P700, a pigment in photosystem I. Observe that starting from $H_2O$ and with the release of $O_2$, electrons are boosted from photosystem II toward photosystem I, finally reaching NADPH + $H^+$. On the left the redox potential of the different components of the system is indicated. (See description in text.) (From A. L. Lehninger, *Biochemistry*. Worth, New York, 1975.)

P680 and P700 are thought to be special molecules of chlorophyll *a* with a higher absorption maximum. The photosystems operate in a sequential manner. In photosystem II the quanta of light remove electrons from the hydrogen of water and boost them to a higher energy level, thus reducing plastoquinone. From there the electrons are brought back "down" to photosystem I (see Fig. 6–19) by way of an electron transfer system. When photosystem I is excited, electrons are boosted to a higher energy level, thereby reducing ferredoxin. This compound is then reoxidized by the transfer of electrons to $NADP^+$, forming $NADPH_2$. Two photons of light appear to be needed to produce one molecule of $NADPH_2$. This reducing component, having high energy electrons, can be used in a variety of biochemical reactions, particularly in the synthesis of carbohydrates. Theoretically, eight quanta of energy are needed to produce one molecule of $O_2$ from $H_2O$.

# Electron Transport and Photophosphorylation

As there is an electron transport chain in mitochondria, there are several electron carrier molecules in the chloroplast thylakoids; the most important of these are the *cytochromes, plastoquinones,* and *plastocyanine.* Our knowledge of the exact sequence of these carriers, however, is not as clear as in mitochondria. One possible sequence is that shown in Figure 6–19. From photosystem II to I the transfer of electrons begins with an unidentified carrier Q, which is followed by the sequence plastoquinone → cytochrome b559 → cytochrome f → plastocyanine. The plastoquinones, being very mobile lipophilic molecules, could establish a dynamic link between the two photosystems by flowing within the membrane. *Ferredoxin* is a protein that transfers electrons from PS I to $NADP^+$. In summary, Figure 6–19 shows that the excitation by photons of photosystem II will cause the release of $O_2$ and electrons. These are boosted to a higher reducing potential (i.e., from 0.81 to 0 volts) to reach the acceptor Q. Then the electrons move in a downhill sequence to reach photosystem I. From there they are again boosted to a higher reducing level (−0.6 volt) from which they finally reach $NADP^+$, reducing it to NADPH.

As in mitochondria, the electron transport system of chloroplasts is coupled to the phosphorylation of ADP to ATP. In this case, too, the coupling is assisted by an ATPase consisting of the CF1 coupling factor, located on the outer surface of the thylakoid membrane, and a CF0 hydrophobic segment that acts as a proton channel. Of the several theories proposed to explain the coupling, the chemiosmotic theory of Mitchell is again favored. As mentioned earlier, this theory suggests that coupling is effected by an electrochemical link in which a gradient of electrical potential and pH is produced across the membrane.

In contrast to the oxidative phosphorylation in mitochondria, $O_2$ is not used in chloroplast photophosphorylation. Green plants can produce 30 times as much ATP by photophosphorylation as by oxidative phosphorylation in their own mitochondria. In addition, these plants contain many more chloroplasts than mitochondria.

In the overall photosynthetic pathway, for every single molecule of $O_2$ liberated, eight light quanta are needed, two molecules of $NADPH_2$ are formed, and three ATP molecules are generated along the electron transport chain. The efficiency of this process is close to 25 percent.

# The Photosynthetic Carbon-Reduction Cycle

The synthesis of carbohydrates and the accompanying reduction of $CO_2$ can take place after the plant is moved from the light to the dark. This is why reactions included in this *photosynthetic carbon-reduction* (PCR) cycle are called *dark reactions.* In the PCR cycle the molecules of ATP and NADPH produced by the photochemical reaction provide the energy needed to fix the $CO_2$ and to synthesize carbohydrates (Fig. 6–20).

electron carriers—in chloroplast thylakoids, these include cytochromes, plastoquinones, and plastocyanine, all of which undergo oxidation–reduction reactions

photosynthetic carbon reduction (PCR) cycle— phase of photosynthesis occurring in the stroma and involving the synthesis of carbohydrates from $CO_2$ and $H_2O$; also Calvin cycle

dark reactions—those photosynthetic processes that can occur in the dark, such as carbohydrate synthesis and $CO_2$ reduction

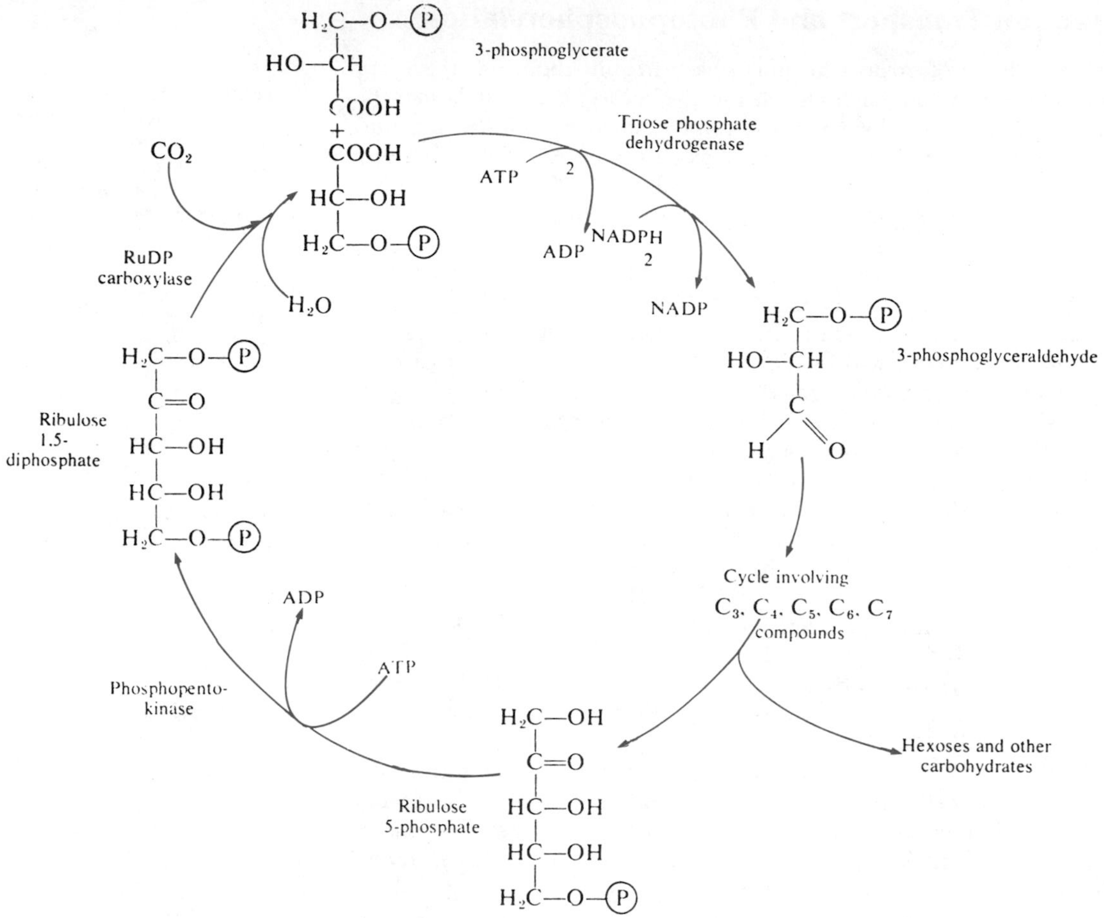

$$H_2C-O-\textcircled{P}$$
$$HO-CH$$
$$COOH$$

3-phosphoglycerate

$$+$$
$$COOH$$
$$HC-OH$$
$$H_2C-O-\textcircled{P}$$

$CO_2$

Triose phosphate
dehydrogenase

ATP   2

RuDP
carboxylase

$H_2O$

ADP   NADPH
2

NADP

$$H_2C-O-\textcircled{P}$$
$$HO-CH$$
$$C$$
$$H \quad O$$

3-phosphoglyceraldehyde

Ribulose
1,5-
diphosphate

$$H_2C-O-\textcircled{P}$$
$$C=O$$
$$HC-OH$$
$$HC-OH$$
$$H_2C-O-\textcircled{P}$$

Cycle involving
$C_3, C_4, C_5, C_6, C_7$
compounds

ADP

ATP

Phosphopento-
kinase

Hexoses and other
carbohydrates

Ribulose
5-phosphate

$$H_2C-OH$$
$$C=O$$
$$HC-OH$$
$$HC-OH$$
$$H_2C-O-\textcircled{P}$$

**Figure 6-20**   Diagram of the Calvin cycle of photosynthesis in which $CO_2$ is reduced and carbohydrates are synthesized.

This phase of photosynthesis takes place in the stroma of the chloroplast by the intervention of numerous enzymes that work in a cycle, the sequence of which was determined mainly by Calvin and his associates (and hence is called the *Calvin cycle*) with the use of $^{14}CO_2$. The reactions involved are so rapid that they occur one second or less after the introduction of $CO_2$. As shown in Figure 6–20, the initial reaction in which $CO_2$ and $H_2O$ enter the cycle is carried out by the enzyme *ribulose-1,5-diphosphate carboxylase*. This is a huge enzyme (500,000 daltons) that represents about half of the stromal proteins. Through its action, *ribulose-1,5-diphosphate* (a pentose) is integrated with $CO_2$ and $H_2O$ to produce two molecules of *3-phosphoglycerate*. These triose sugars are phosphorylated by ATP and form an activated molecule that is able to accept hydrogens and electrons from NADPH. *3-Phosphoglyceraldehyde* is then reduced to form more complex carbohydrates and ribulose-1,5-diphosphate is regenerated. Among the intermediates formed, one of the most important is *fructose-6-phosphate*, which will eventually lead to the formation of glucose.

The energy balance of photosynthesis is

$$6\,CO_2 + 12\,H_2O \xrightarrow{\text{light}} C_6H_{12}O_6 + 6\,O_2 + 6\,H_2O \qquad (14)$$

which represents a storage of 686,000 calories per mole. This energy is provided by a total of 12 NADPH and 18 ATP molecules, representing 750 kcal. The efficiency reached by the PCR cycle is thus as high as 90 percent ($686/750 \times 100 = 90\%$).

In summary, photons absorbed by chlorophyll and other light-sensitive pigments are first converted into chemical energy as NADPH and ATP. During this photochemical reaction, water is oxidated and molecular oxygen is released into the atmosphere as a by-product. The reduction of $CO_2$ takes place even in the dark, provided that NADPH and ATP are present. The initial product of this cycle is a triose which is finally processed to fructose-6-phosphate, glucose, and other carbohydrates. Note that the reactions that result in the synthesis of glucose (Fig. 6–20) and those involved in its breakdown by glycolysis (Fig. 6–12) and aerobic respiration (Fig. 6–6) follow entirely different pathways.

## 6-3 BIOGENESIS OF MITOCHONDRIA AND CHLOROPLASTS

Mitochondria and chloroplasts are remarkable organelles in the way they originate and develop. Both of them are formed by fission of preexisting organelles, and we will see below that two genetic systems are involved in the synthesis of their protein components.

During plant development, *proplastids* appear which are limited by a double membrane. The stages of development that lead to a mature chloroplast are diagrammed in Figure 6–21, **A.** In the presence of light, the inner membrane grows and gives off vesicles into the matrix that are transformed into discs. These intrachloroplastic membranes are the thylakoids which, in certain regions, pile closely to form the grana. In the mature chloroplast the thylakoids are no longer connected to the inner membrane, but the grana remain united by intergranal thylakoids. It is interesting that if a plant is put under low light intensity, the reverse sequence of changes takes place. This process is called *etiolation*, and results in the disorganization of the membranes. The same phenomenon occurs if the plant is grown from the beginning in low light intensity. In this case the vesicles of the proplastid aggregate to form one or more *prolamellar bodies*, which can develop into grana if the plant is subsequently exposed to light.

Both chloroplasts and mitochondria are distributed between the daughter cells during mitosis, and their number increases at interphase.

proplastid—immature chloroplast containing prolamellar bodies that can develop into grana

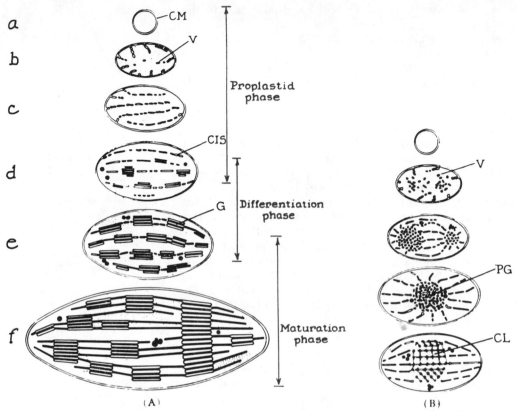

**Figure 6-21**  **A,** phases in the development of a proplastid into a chloroplast in the presence of light. **B,** same, but in the dark, showing the formation of the primary granum *(PG)*, or prolamellar body. *CIS,* flattened cisternae; *CL,* crystal lattice; *CM,* double chloroplast membrane; *G,* granum; *V,* vesicles. (Modified from D. von Wettstein, *J. Ultrastruct. Res.,* 3:235, 1959.)

## Limited Autonomy in Mitochondria and Chloroplasts

New light was thrown on the problem of mitochondrial and chloroplast biogenesis when it was demonstrated that these organelles contain DNA, RNAs, and other components involved in protein synthesis. These two organelles are thus semiautonomous, and their functioning depends on the cooperation between their own genetic systems plus that belonging to the rest of the cell.

Mitochondrial DNA (mtDNA) is contained in the matrix and is probably attached to the inner membrane. mtDNA is circular, 5.5 $\mu$m long, and generally occurs in one copy per mitochondrion (Fig. 6-22). Apparently all the mtDNAs in an individual (about $10^{17}$ molecules in a human) are identical and contain single copies of genes. mtDNA codes for the ribosomal RNAs of mitochondria and for about 19 transfer RNAs for the 20 or so proteins that are incorporated into the inner membrane. (A striking characteristic of these proteins is that they are all hydrophobic.) Although ribosomal *RNAs* originate from mtDNA, all ribosomal *proteins* come from the cytosol.

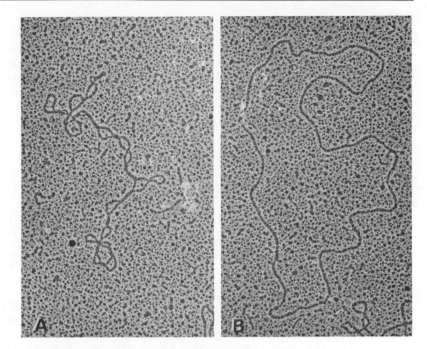

**Figure 6-22** DNA extracted from rat liver mitochondria and observed by the spreading technique. **A,** configuration in twisted circle ("supercoiled"); **B,** configuration in open circle. (Courtesy of B. Stevens.)

Chloroplasts contain DNA in the stroma; this DNA is much longer than mtDNA. Ribosomes and polyribosomes are also present, although the genetic information of the chloroplasts is evidently limited, and their proteins are synthesized by several mechanisms involving (1) chloroplast DNA and ribosomes, (2) nuclear DNA and cytoplasmic ribosomes, and (3) nuclear DNA and chloroplast ribosomes.

## The Symbiont Hypothesis

At the end of the last century, cytologists such as Altmann and Schimper speculated on purely morphological grounds that mitochondria and chloroplasts might be intracellular parasites that had established a *symbiotic* relationship with the eukaryotic cell. Bacteria were thought to have given rise to the mitochondria and blue-green algae to the chloroplasts. The *symbiont hypothesis* is based on many similarities between prokaryotes and mitochondria and chloroplasts, such as the presence of DNA and ribosomes. Another similarity is evident in the location of the respiratory chain and ATPase in the bacterial plasma membrane; and in certain bacteria, the membrane forms projections called *mesosomes*, which are comparable to the mitochondrial crests.

Support for the possible prokaryotic origin of mitochondria and chloroplasts is also found in the fact that *intracellular symbiosis* can be found in nature. Thus *Paramecia* may contain certain bacteria, and blue-green algae may occur in simple animals. The endosymbiosis of a photosynthetic prokaryote confers upon the host the ability to capture light energy with which to synthesize various prod-

symbiont hypothesis—proposal that mitochondria and chloroplasts might have originated as prokaryotic intracellular parasites that established a symbiotic relationship with the eukaryotic cell

ucts; at the same time, it provides the prokaryote with a constant environment in which to grow and reproduce. In evolutionary terms it is possible to conceive that a symbiotic relationship could have given rise to the present situation, in which the organelles have only limited autonomy and depend on the nucleus and the cytosol of the cell for most of their specific components.

# SUMMARY

## 6-1 THE MITOCHONDRION: STRUCTURE AND FUNCTION

Energy transformation in cells is carried out by two main transducing systems represented by mitochondria and chloroplasts, which in some respects function in opposite directions. By the process of photosynthesis, chloroplasts capture light energy and transduce it into chemical energy; by the process of oxidative phosphorylation, mitochondria release the chemical energy stored in nutrients by plant cells.

Mitochondria are cylindrical structures about 0.5 $\mu$m in diameter and several $\mu$m in length. In the living cell they display passive and active motion and show changes in volume that are related to their activity. These organelles consist of an outer and an inner membrane and two separate compartments in which different biochemical activities take place. The cristae formed by the inner membrane are covered with mushroom-like projections and contain all the components of the respiratory chain as well as those related to phosphorylation. The mitochondrial matrix contains all the soluble enzymes involved in the Krebs cycle as well as DNA and the machinery for protein synthesis.

The chemical energy of nutrients is stored in the covalent bonds between the atoms of each molecule. This energy is not released suddenly, but is made available in a stepwise and controlled manner. It can then be used to synthesize new molecules, perform mechanical work, carry out active transport, maintain membrane potentials, or produce radiant energy. The common link in all these processes is the compound ATP, which is distinguished by having two terminal bonds of very high potential energy.

The Krebs or tricarboxylic acid cycle, which takes place in the mitochondrial matrix, is the first step in a common pathway for the degradation of fuel molecules. These molecules are acted upon metabolically (in the cytoplasm) to produce acetyl groups, which are taken into the cycle by acetyl coenzyme A. Each turn of the cycle generates two $CO_2$ molecules and four pairs of hydrogen atoms. The hydrogen atoms then enter the respiratory chain: three pairs are accepted by $NAD^+$, reducing it to NADH, and one pair is accepted by FAD, reducing it to $FADH_2$. It takes two turns of the cycle to metabolize the two acetate molecules that are produced by glycolysis from one molecule of glucose.

The hydrogen atoms coming from the Krebs cycle are conveyed through the respiratory chain by a group of electron carriers called cytochromes, and progress along a gradient of increasingly positive redox potential ending with $O_2$. All the components of the respiratory chain, as well as those involved in phosphorylation, are integrated within the mitochondrial inner membrane. The former can be dissociated into four multi-molecular complexes (I–IV), which are arranged within the membrane in a highly asymmetrical fashion compatible with the transfer of protons from the M side to the C side of the membrane. These protons will give rise to ATP molecules through the intervention of mitochondrial ATPase and the phosphorylating system, which are located primarily in the F1 particles attached to the M side of the mitochondrial crests.

The chemiosmotic theory has been postulated to explain the nature of the link between the respiratory chain and the phosphorylating system. This theory suggests that the link is essentially electrochemical: the translocations of protons across the membrane create a difference in pH and electrical potential that would drive the process of ATP synthesis.

Less than 10 percent of the energy in a mole of glucose can be released by anaerobic glycolysis. Much more is released by oxidative phosphorylation. One way of indicating ATP yield from the latter process is by the P/O ratio, which is expressed as the moles of inorganic phosphate used per oxygen atom consumed.

## 6-2 THE CHLOROPLAST: STRUCTURE AND FUNCTION

Chloroplasts in higher plants are generally disc-shaped and have an average length of 4 to 6 $\mu$m; they tend to be larger and contain more chlorophyll in those plants grown in the shade. The main components of the chloroplast are the envelope; the stroma (or matrix); and the thylakoids, which may be found either stacked (in units called grana) or unstacked. The thylakoid membranes, which are thought to be constructed along the lines of the fluid mosaic model, contain many proteins and all the components involved in the essential steps of photosynthesis. These include chlorophyll, carotenoids, and plastoquinone. The chlorophyll molecules occur in the membrane in conjunction with integral proteins, forming several important complexes including PS I, PS II, and LHCP. Other complexes lacking chlorophyll have also been isolated. The CF1 particles, which are similar to the F1 particles in mitochondria, are localized in the outer surface of the stroma thylakoids. Again, as in mitochondria, these components have a vectorial disposition and are asymmetric with respect to the membrane.

Photosynthesis is essentially the light-catalyzed combination of carbon dioxide and water to produce various carbohydrates and molecular oxygen. The energy contained in visible light is transmitted in packets called photons. Pigments such as chlorophyll are specifically adapted to be excited by light photons, and respond by displacing electrons to higher energy levels. Recall that in oxidative phosphorylation the flow of electrons is from NADH to $O_2$; in photosynthesis, the electrons flow in the opposite direction, from $H_2O$ to NADPH. Electron transfer against this negative gradient uses the energy provided by light photons, which is collected by PS I and PS II. Two photons of light appear to be needed to produce one molecule of $NADPH_2$.

The electron carrier molecules in the thylakoids include cytochromes, plastoquinones, and plastocyanine. As in mitochondria, the chloroplast electron transport system is coupled to the phosphorylation of ADP to ATP. The coupling is assisted by the CF1 coupling factor located on the outer surface of the thylakoid membrane. The chemiosmotic theory is also thought to explain this instance of coupling. Green plants can produce 30 times as much ATP by photophosphorylation as by oxidative phosphorylation in their own mitochondria, and the efficiency of the overall photosynthetic pathway is close to 25 percent.

Carbohydrate synthesis and $CO_2$ reduction can take place in the dark, fueled by the ATP and NADPH produced by the photochemical reaction and catalyzed by the enzymes of the Calvin cycle. The initial product of this cycle is processed to glucose and other carbohydrates.

## 6–3 BIOGENESIS OF MITOCHONDRIA AND CHLOROPLASTS

Both mitochondria and chloroplasts are produced by the fission of preexisting organelles, and both contain their own DNA, RNAs, and other components involved in protein synthesis. These two organelles are thus semiautonomous, and their functioning depends on the cooperation between their own genetic systems and that of the cell nucleus. In mitochondria, for example, the ribosomal RNAs originate from the mtDNA whereas all ribosomal proteins come from the cytosol. Similarly, chloroplast proteins are synthesized from both chloroplast and nuclear DNA, using both chloroplast and cytoplasmic ribosomes.

The symbiont hypothesis suggests that mitochondria and chloroplasts may originally have been prokaryotic intracellular parasites that established a symbiotic relationship with eukaryotic cells. This hypothesis is supported by a good deal of structural and functional evidence, and by the existence of intracellular symbiosis in nature.

# STUDY QUESTIONS

- A molecule of glucose has been ingested by a eukaryotic cell. Explain in detail how it will be metabolized under aerobic conditions, beginning with its breakdown in the cytoplasm, and suggest ways in which its potential energy might be utilized by the cell.
- What is meant by a positive redox potential?
- Describe the arrangement of the respiratory chain complexes I–IV and the phosphorylating system within the mitochondrial inner membrane.
- What is the chemiosmotic theory?
- Describe the structure and function of the chloroplast thylakoids.
- Using Figure 6–19 as a guide, explain how electrons are transported from $H_2O$ to NADPH. How is this transport fueled?
- Explain how carbohydrates are synthesized during the photosynthetic carbon reduction cycle. Why is light unnecessary when this process takes place?
- How does light intensity affect the development of chloroplasts?
- In what respects are mitochondria and chloroplasts semiautonomous?
- What is the symbiont hypothesis?

# SUGGESTED READINGS

**Hinkle, D. C., and McCarty, R. E.** (1978) How cells make ATP. *Sci. Am., 238:*104.

**Lehninger, A. L.** (1977) *Biochemistry.* Worth, New York.

**Miller, K. R.** (1979) The photosynthetic membrane. *Sci. Am., 241:*102.

**Mitchell, P.** (1979) Keilin's respiratory chain concept and its chemiosmotic consequences. *Science,* 206:1148.

**Racker, E.** (1970) *Membranes of Mitochondria and Chloroplasts.* Van Nostrand Reinhold, New York.

**Srere, P. A.** (1980) The infrastructure of the mitochondrial matrix. *Trends Biochem. Sci.,* 5:120.

**Tzagoloff, A., and Macino, G.** (1979) Mitochondrial genes and translation products. *Annu. Rev. Biochem., 48:*419.

# The Nucleus, Chromatin, and the Chromosomes

About a century ago, in 1876, Balbiani described rod-like structures that were formed in the nucleus before cell division. In 1879 Flemming used the word "chromatin" (Gr., *chroma*, color) to describe the substance that stained intensely with basic dyes in interphase nuclei. He suggested that the affinity of chromatin for basic dyes was due to its content of "nuclein," a phosphorus-containing compound isolated from pus cells by Meischer in 1871 and now called DNA. In 1888 Waldeyer used the word "chromosome" to emphasize the continuity between the chromatin of interphase nuclei and the rod-like objects observed during mitosis. In this chapter we will analyze the molecular organization of chromatin and of the chromosomes, as revealed by more recent studies.

Like other cell organelles, the nucleus is self-contained, includes specific components related to its activity (in this case, chromatin, nucleoli, RNAs, and various proteins), and exchanges materials with the rest of the cell. We will begin this chapter by considering the element that separates the nucleus from the cytoplasm: the nuclear envelope.

## 7-1 THE NUCLEAR ENVELOPE

We mentioned in Chapter 1 that one of the main differences between prokaryotes and eukaryotes is that the latter have a distinct nucleus in their cells whereas the former do not. The light microscope provided little information about the structure of the *nuclear envelope*

**nuclear envelope**—flattened sac or double membrane surrounding the nucleoplasm and genetic material

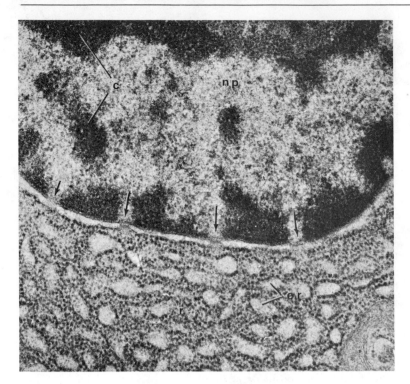

**Figure 7-1** Electron micrograph of a portion of the nucleus and cytoplasm from a pancreatic cell of the mouse. The pores in the nuclear envelope are indicated by arrows within the interchromatin channels; *c*, chromatin; *er*, endoplasmic reticulum; *np*, nucleoplasm, ×48,000. Chromatin blocks are attached to the inner nuclear membrane but not to the nuclear pores. Ribosomes are attached to the outer nuclear membrane. (Courtesy of J. Andrè.)

that separates the eukaryotic chromosome(s) from the cytoplasm; however, with the use of the electron microscope it became apparent that this envelope is yet another differentiation of the cytoplasmic membrane system.

## Nuclear Pores and Pore Complexes

The nuclear envelope consists of two concentric membranes separated by a perinuclear cisterna 10 to 15 nm in width. Biochemical analyses have revealed that these membranes have a basic unit structure similar to that of the plasma membrane, and consist of flattened cisternae of the endoplasmic reticulum having ribosomes only on the outer surface.

As shown in Figure 7-1, at certain points the nuclear envelope is interrupted by structures called *pores*. Both membranes of the envelope are in continuity around the margins of these pores. At their nuclear side the pores are generally aligned with channels of nucleoplasm situated between more condensed lumps of chromatin attached to the inner membrane (Fig. 7-1).

The nuclear pores are very large, usually 50 to 80 nm in diameter. It has been calculated that in the nuclei of *Mammalia* these pores account for 5 to 15 percent of the surface area of the nuclear membrane; and in amphibian oocytes, certain plant cells, and protozoa the surface area they occupy may be as high as 20 to 36 percent. The pores are not wide-open channels, however, and the electron microscope has revealed that they are occluded by an electron-dense material and enclosed by circular structures which are called *an-*

**nuclear pore—interruption in the nuclear envelope that allows the exchange of material between the nucleus and cytoplasm**

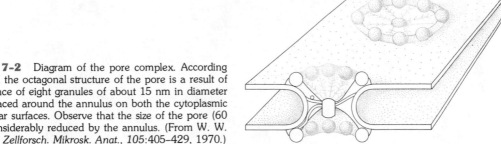

**Figure 7-2**  Diagram of the pore complex. According to Franke, the octagonal structure of the pore is a result of the presence of eight granules of about 15 nm in diameter evenly spaced around the annulus on both the cytoplasmic and nuclear surfaces. Observe that the size of the pore (60 nm) is considerably reduced by the annulus. (From W. W. Franke, *Z. Zellforsch. Mikrosk. Anat.*, 105:405–429, 1970.)

*nuli*. The annuli are octagonal in shape and about 60 nm in diameter. Each annulus consists of eight *granules* present at both the nuclear and cytoplasmic surfaces (Fig. 7-2) and a less defined, amorphous *annular material* in the opening itself. Biochemical studies indicate that this material is some type of protein. The pores and annuli together are referred to as the *pore complex*. This is apparently a rather rigid structure present in a fixed number according to cell type. In certain physiological stages, however, this number may vary. In general, cells of high transcriptional activity have more pore complexes (frog oocytes have 60 pores/$\mu$m$^2$) than those cells that synthesize little RNA (frog erythrocytes have 3 pores/$\mu$m$^2$).

**annulus**—circular structure enclosing the nuclear pore, consisting of eight surrounding granules and a proteinaceous material in the pore opening

**pore complex**—structure formed by the nuclear pores and annuli

## Nuclear Proteins Accumulate in the Nucleus

The cell nucleus contains a specific subset of proteins, including (for example) DNA polymerase, RNA polymerase, and histones. All these proteins are synthesized in the cytoplasm and must subsequently be transported into the nucleus. The correct localization is achieved because mature nuclear proteins contain a signal that enables them to accumulate selectively in the nucleus. (It follows that this signal is absent in proteins which do not accumulate in the nucleus.)

This selectivity has been demonstrated with labeled nuclear proteins microinjected into the cytoplasm of an oocyte, which not only entered the nucleus but accumulated in it. Figure 7-3 shows the results of an experiment in which the cytoplasm of frog oocyte was injected with a preparation of radioactive proteins. Most of the proteins in this preparation were nucleus-specific, but others were common to both the nucleus and cytoplasm. Actin, for example, is a main component of both cellular compartments. Twenty-four hours after injecting the labeled proteins into the cytoplasm, the oocyte nuclei were manually separated from the cytoplasm (this is possible because both the oocytes and their nuclei are extremely large), and the proteins were analyzed on two-dimensional gels. It was found that nucleus-specific proteins were concentrated in the nucleus, while microinjected actin was found in both the nucleus and cytoplasm.

This experiment provides additional confirmation that cells have the capacity to place specific proteins into the appropriate cell compartments, as occurs with the mitochondria or chloroplast proteins

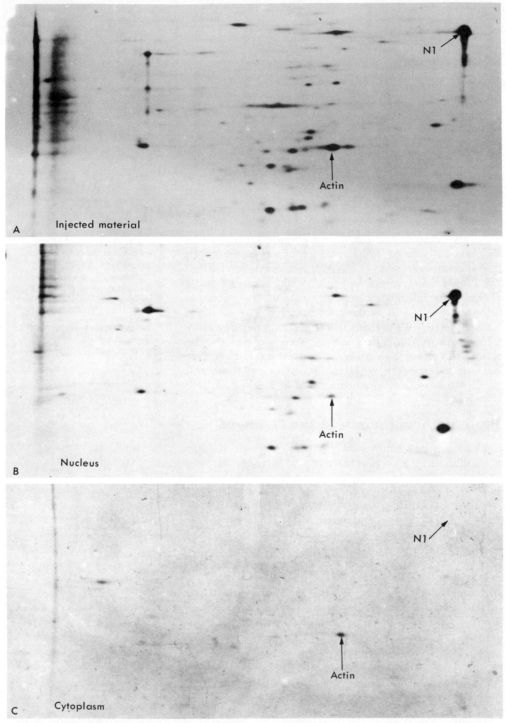

**Figure 7-3** Selective migration of nuclear proteins. Two-dimensional gel analysis of $^{35}$S-methionine-labeled nucleoplasmic proteins microinjected into *Xenopus* oocytes. **A,** a preparation of $^{35}$S-*Xenopus* nucleoplasmic proteins that was injected into oocytes. Note that it contains actin and many nuclear proteins. **B,** the nuclear fraction reisolated from unlabeled oocytes injected 24 hours earlier into the cytoplasm with the $^{35}$S proteins shown in **A.** Note that the nuclear proteins re-enter the nucleus, while actin decreases in relative amount. **C,** cytoplasmic fraction from the same microinjected oocytes. Note that actin is the most prominent protein, while the nuclear proteins are not detectable in the cytoplasm. (From E. M. De Robertis.)

that are synthesized in the cytoplasm, or with proteins that are exported from the cell. It is important to emphasize that the mechanism by which proteins are secreted into the endoplasmic reticulum (Section 5–3) is very different from that used for the accumulation of proteins inside the nucleus. Recall that the signal peptide enables the secreted proteins to traverse the ER membrane only as they are being synthesized, and that this peptide is quickly removed thereafter. The nuclear proteins, in contrast, accumulate in the nucleus independent of protein synthesis, and this is not a transient property but is intrinsic to the structure of the mature protein.

# 7–2  CHROMATIN

DNA is the main genetic material of cells, carrying information in a coded form from cell to cell and from organism to organism. Within the eukaryotic cell the DNA is not free, but is complexed with proteins in a structure called *chromatin*.

**chromatin**—complex of DNA and several basic proteins, constituting the eukaryotic genetic material

## Nuclei Contain a Constant Amount of DNA

The demonstration that nuclei contain a constant amount of DNA, performed by the Vendrelys and by Mirsky and Ris in 1948, was a landmark in cell biology because it confirmed that DNA was the genetic material and that genetic information is not lost during the differentiations of the various somatic tissues.

Each species has a characteristic content of DNA, sometimes called the *C value*. All the *diploid* cells (normal non-gametes) in an organism have the same DNA content, designated *2C*. Gametes are haploid and therefore have half the DNA content or *1C*. Some tissues, such as liver, contain occasional *polyploid* cells whose nuclei have a higher DNA content (e.g., 4C or 8C). Eukaryotes vary greatly in DNA content but always contain much more DNA than prokaryotes. Lower eukaryotes generally have less DNA; for example, the nematode *C. elegans* has only 20 times more DNA than *E. coli*, and the fruit fly *D. melanogaster* has only 40 times more. Vertebrates have a higher DNA content, generally about 700 times greater than that of *E. coli*.

**C value**—characteristic DNA content of the cells of a given species

The DNA in metaphase chromosomes is usually compacted between 5000- and 10,000-fold, and DNA content tends to be proportional to the size of the chromosome. The largest human chromosome (1), which is 10 $\mu$m long, has been calculated to accommodate about 7.2 cm of DNA in a tightly packed form, thus giving a compaction value of about 7000-fold. This degree of compaction is a remarkable feature of the genetic material, and we will now analyze how this folding is achieved.

## Chromatin: DNA Plus Histones

In eukaryotic cells the DNA is complexed with small proteins called *histones*, which are bound to it in a 1:1 ratio by weight. Figure 7–4 shows that there are five different histone proteins. All of them are

**histones**—group of five basic proteins that associate with DNA in the cell nucleus to form chromatin

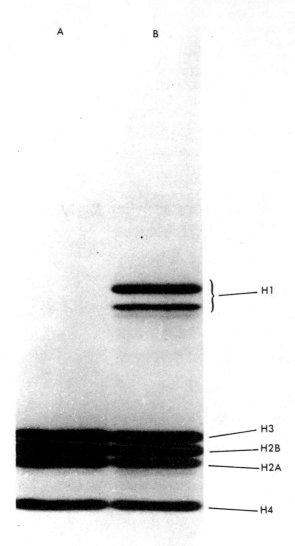

**Figure 7-4**  Histone proteins after electrophoresis in polyacrylamide gels containing sodium dodecyl sulfate. **A,** histones from nucleosome "cores" which do not contain H1. **B,** histones from total chromatin. Note that the four fundamental histones are present in about equal amounts. (Courtesy of Jean O. Thomas.)

basic because they have a high content (10 to 20 percent) of the basic amino acids arginine and lysine. Being basic, histones bind tightly to DNA, which is an acid.

The four main histones, H2A, H2B, H3, and H4, are very similar in different species. (The sequence of histone H3 from the rat, for example, differs in only two amino acids from that of peas.) These four histones are present in equimolar amounts, two of each being present every 200 base pairs of DNA. Histone H1 is not conserved between species and has tissue-specific forms. It is present only once per 200 base pairs of DNA and is rather loosely associated with chromatin (it can be eluted from DNA by adding low concentrations of salt); it is not a component of the DNA-histone structural unit called the *nucleosome core* (see below), but instead is bound to the linker segments of DNA that join neighboring nucleosomes.

When eukaryotic nuclei were spread on electron microscopic specimen holders, it was found that chromatin has a repeating structure of "beads" about 10 nm in diameter connected by a "string" of DNA. Most if not all of the DNA is present in this form, as shown in Figure 7–5. The beads-on-a-string appearance is not the true structure of chromatin but is an artifact arising from the way in which the sample is prepared, which results in a loss of histone H1. If milder treatments which do not remove H1 are used, the beads are not stretched but can be observed as a 10 nm fiber, with the beads touching each other. The 10 nm fiber therefore represents the first level of organization of chromatin within cells. A possible model of this organization is shown in Figure 7–6.

**10 nm fiber**—structure that represents the first level of chromatin organization within cells

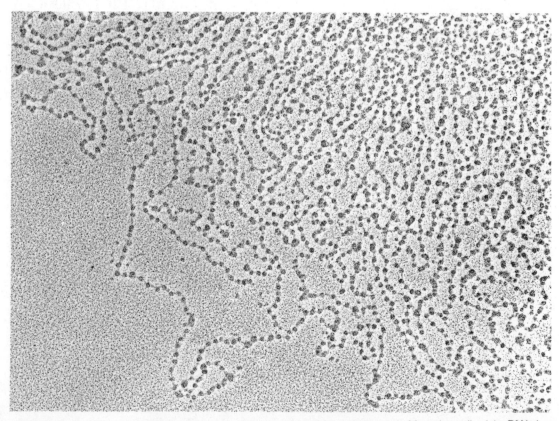

**Figure 7-5**  A *D. melanogaster* embryonic nucleus spread for electron microscopy. Most, if not all, of the DNA, has a beaded structure. ×50,000. (Courtesy of O. L. Miller, Jr.)

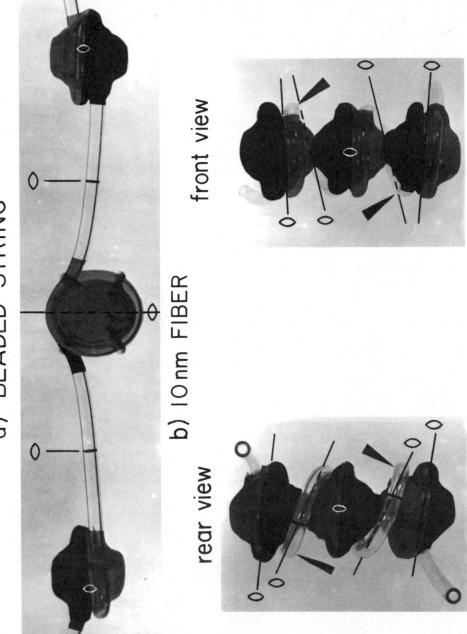

## a) BEADED STRING

## b) 10 nm FIBER

rear view      front view

**Figure 7-6** A model showing the two possible appearances of chromatin in the electron microscope. The nucleosome is represented by tennis balls, and the DNA by plastic tubing. It is important to note that the DNA is wound on the *outside* of the nucleosome. **A,** the beads-on-a-string configuration results from the unwinding of part of the DNA during sample preparation. **B,** the 10 nm fiber, in which nucleosomes are in close contact, is the natural configuration of chromatin. (Courtesy of A. Worcel, from *Cold Spring Harbor Symp. Quant. Biol., 42:313,* 1977.)

# The Nucleosome and the 20 to 30 nm Fiber

The existence of a repeating unit of chromatin, called the *nucleosome*, was also predicted from biochemical studies independent of the EM results described above. Studies of chromatin digestion by nucleases showed that the DNA was cut into multiples of a unit size, which was later found to be about 200 base pairs of DNA in length; that is, the DNA was cut at intervals of 200, 400, 600, 800 (and so forth) base pairs. Other studies showed that histones H3 and H4 tend to associate in solution, forming tetramers consisting of two each. Since the four histones are present in equimolar amounts (two of each per 200 base pairs of DNA), in 1974 R. Kornberg proposed a model for the nucleosome in which the four histones H2A, H2B, H3, and H4 are arranged in octamers containing two of each protein, located every 200 base pairs of DNA. The histone octamers are in close contact (as in the 10 nm fiber), and the DNA is coiled on the *outside* of the nucleosome.

When nucleosomes are in close apposition in the 10 nm filaments, the packing of DNA is about five- to seven-fold, i.e., five to seven times more compact than free DNA. This is still 1000-fold lower than the packing ratio of DNA in metaphase chromosomes, which is achieved by further folding of the 10 nm fiber.

nucleosome—repeating unit of chromatin consisting of 200 base pairs of DNA coiled around a histone octamer (two each of histones H2A, H2B, H3, and H4)

**Figure 7-7** A possible model of the coiling of the 10 nm fiber into a 20 to 30 nm fiber. The model stresses the fact that the DNA (plastic tubing) is wound on the outside of the nucleosomes (tennis balls). Note that the histone H1 (indicated by arrowheads) is located in the inner part of the solenoid. (Courtesy of A. Worcel, from *Cold Spring Harbor Symp. Quant. Biol.*, 42:313, 1977.)

thick fiber—second-level
chromatin structure of 20–30
nm diameter, probably
representing inactive
chromatin

Electron microscopic studies of chromatin fibers in interphase nuclei and mitotic chromosomes have revealed a *thick fiber* having a diameter that varies from 20 to 30 nm. This fiber probably represents the structure of inactive chromatin. Figure 7–7 shows a possible model of how the thick fiber could arise by coiling the 10 nm fiber into a solenoid. The DNA of a 20 to 30 nm solenoid is compacted about 40-fold, and to achieve the packing ratio of a metaphase chromosome the thick fiber must be folded yet another 100-fold. At present we know very little about how this feat is achieved.

# 7-3 THE CHROMOSOMES

chromosome—structure
consisting of chromatin
strands that become
extremely compact during
cell division

At the time of cell division, chromatin becomes condensed into the chromosomes. Since chromosomes can be observed clearly with the optical microscope, they were the subject of much study soon after their discovery in 1876, and by 1910 it became clear that genetic phenomena could be explained in terms of chromosome behavior. The study of chromosomes allows one to learn about the behavior of genes and DNA molecules in the microscope, and for this reason it remains important even though more powerful (and complicated) techniques for genetic studies are now available. In this section we will analyze the general morphology and composition of chromosomes; their behavior during mitosis and meiosis will be discussed in Chapter 9.

Chromosomes can be studied in tissue sections, but they are best observed in *squash* preparations. Fragments of tissues are stained with basic dyes (such as *orcein* or *Giemsa*) and then squashed between a slide and coverslip by gentle pressure. Sometimes hypotonic solutions are used prior to squashing to make the nucleus swell and produce a better separation of the individual chromosomes. Chromosomal morphology can be studied to best advantage during anaphase and metaphase, which are the periods of maximal contraction.

Chromosomes are classified into four types according to their shape, which is determined by the position of the *centromere* (the point of attachment to the mitotic spindle). As shown in Figure 7–8, *telocentric* chromosomes have the centromere located on one end; *acrocentric* chromosomes have a very small short arm; *submetacentric* chromosomes have arms of unequal length; and *metacentric* chromosomes have equal or almost equal arms.

karyotype—characteristic
chromosome set of a given
species

The characteristics of the chromosome set, called the *karyotype*, vary considerably between different species. For example, the mouse has acrocentric chromosomes; many amphibia have only metacentric chromosomes; and humans have a characteristic assortment of the latter three types. Some species are also more favorable than others for genetic analysis. Salamanders and grasshoppers, for example, have a high DNA content and very large chromosomes which produce beautiful meiotic preparations.

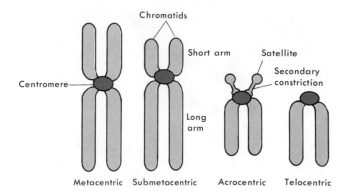

**Figure 7-8** Chromosomes are classified according to the position of the centromere.

## Chromosome Components

Several elements can be distinguished in condensed chromosomes:

**CHROMATID.** At metaphase each chromosome consists of two symmetrical structures, the *chromatids* (Fig. 7–8), each of which contains a single DNA molecule. The chromatids are attached to each other only by the centromere and become separated at the start of anaphase, when the sister chromatids migrate to opposite poles. Anaphase chromosomes thus have only one chromatid, whereas metaphase chromosomes have two.

chromatid—in the metaphase chromosome, one of two symmetrical structures each containing a single DNA molecule

**CENTROMERE.** This is the region of the chromosome that becomes attached to the mitotic spindle (Fig. 7–8). The centromere lies within a thinner segment of the chromosome called the *primary constriction*. The regions flanking the centromere often contain highly repetitive DNA (heterochromatin; see below) and may stain more intensely with basic dyes. The microtubules of the mitotic spindle attach to a protein disc, the *kinetochore*, which is bound to the centromeric heterochromatin.

centromere—point of attachment of sister chromatids, and site of chromosome attachment to the mitotic spindle

kinetochore—proteinaceous region of the centromere to which spindle microtubules attach

**TELOMERE.** This term applies to the tips of the chromosomes which contain the ends of the long DNA molecule that makes up each chromatid. Telomeres have special properties; for example, when chromosomes are broken by x-rays, the free ends (those without telomeres) will become "sticky" and fuse with other broken chromosomes, although they will not fuse with a normal telomere.

**NUCLEOLAR ORGANIZER.** These areas are present in certain *secondary constrictions* which contain the genes coding for 18S and 28S ribosomal RNA, and they induce the formation of nucleoli (which, in turn, assemble ribosomes). The secondary constriction arises because the rRNA genes are transcribed very actively, interfering with chromosomal condensation. In humans, the nucleolar organizers are located in the secondary constrictions of chromosomes 13, 14, 15, 21, and 22, all of which are acrocentric and have a small rounded body (called a chromosome *satellite*) that is sepa-

nucleolar organizer—region that contains the genes for ribosomal RNAs and induces formation of the nucleolus

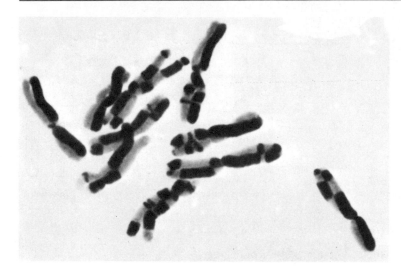

**Figure 7-9** Chromosomes of onion root tips that had been treated two generations earlier with bromodeoxyuridine and that had experimentally induced sister chromatid exchanges. Note that each chromosome has two chromatids. The DNA molecule that incorporated bromodeoxyuridine does not stain with Giemsa. Observe that each change in one chromatid is accompanied by a reciprocal one in the other. These images are consistent with the idea that each chromatid contains a single DNA molecule. (Courtesy of J. B. Schvartzman.)

rated from the rest of the chromosome by the nucleolar secondary constriction (Fig. 7–8).

## The Unineme Theory

**unineme theory—theory that each chromatid represents a single linear DNA molecule with its associated proteins**

Each chromatid represents a single linear DNA molecule with its associated proteins. This concept, sometimes called the *unineme theory,* is supported by several lines of evidence. As will be discussed further in Chapter 8, the replication of chromatids is consistent with the semiconservative replication of a single DNA molecule (see Fig. 8–6). Furthermore, exchanges between sister chromatids can be induced experimentally, and in such cases every change in a chromatid is accompanied by a reciprocal change in the other (Fig. 7–9), a fact which is also consistent with there being only one DNA molecule per chromatid. Additional evidence is provided by the finding, mentioned above, that chromosomes which have not yet replicated their DNA have only one chromatid, while those that have done so have two (see Fig. 8–3).

The unineme organization of the chromatid is supported directly by the isolation of DNA molecules that are long enough to contain all the DNA from a single chromatid. The length of these giant molecules can be estimated by viscoelastic measurements.

The two sister chromatids are mirror images one of the other. The morphologic characteristics of one chromatid, such as chromomeres, secondary constrictions, and satellites, always have their counterparts in the other (Fig. 7–8). The reason for this symmetry is that sister chromatids contain identical DNA molecules, and the morphologic features of chromosomes are ultimately determined by the DNA sequence.

The solution to the problem of higher levels of DNA folding may come from further studies of chromatid structure. When histones are removed from metaphase chromosomes, the chromosomes adopt the configuration shown in Figure 7–10. Histone-depleted

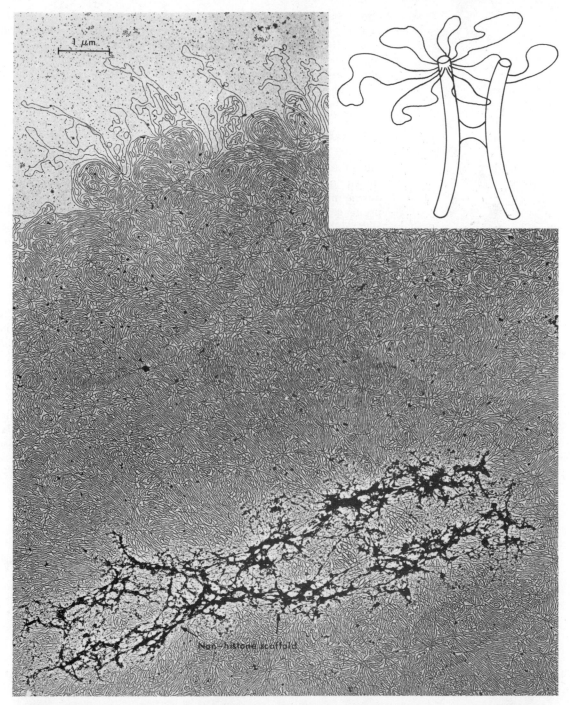

**Figure 7-10** Electron micrograph of a histone-depleted human chromosome. The nonhistone proteins form two scaffolds, one per chromatid, which are joined at the centromere. The scaffold retains the shape of an intact chromosome while the naked DNA fibers form a halo around it. The inset shows a model of chromosome organization in which loops of DNA emerge from the nonhistone protein scaffold. (Courtesy of U. K. Laemmli. From J. R. Paulson and U. K. Laemmli, *Cell, 12*:817, 1977.)

chromosomes have a central core or *scaffold*, surrounded by a halo of loops of DNA. The scaffold is made of nonhistone proteins and retains the general shape of the metaphase chromosome. Each chromosome has two scaffolds, one for each chromatid, connected together at the centromere region. The loops of DNA are about 75,000 base pairs long (see Fig. 7–10, inset), and in nondepleted chromosomes they would be folded 40 times in the 30 nm thick fiber.

## 7-4 HETEROCHROMATIN

Chromatin condenses and decondenses throughout the cell cycle. As shown in Figure 7–11, interphase chromatin becomes condensed during mitosis—the chromatids becoming increasingly shorter from prophase to metaphase—and decondensed at telophase. In 1928 Heitz defined *heterochromatin* as those regions of the chromosome that remain condensed during interphase. The rest of the chromosome, which remains in a noncondensed state, is called *euchromatin* (Gr., *eu*, true). The heterochromatic segments tend to show preferential localization in the centromeres of most plants and animals, at the telomeres (especially in plants), or intercalated in the chromosomal arms (frequently adjacent to the nucleolar organizers). In other cases, whole chromosomes become heterochromatic. It is thought that in heterochromatin the DNA remains tightly packed in the 20 to 30 nm fiber, which as we have mentioned probably represents the configuration of transcriptionally inactive chromatin.

　Two types of heterochromatin are generally recognized: *constitutive* heterochromatin, which is permanently condensed in all types of cells, and *facultative* heterochromatin, which is condensed only in certain cell types or at special stages of development. Frequently in the case of facultative heterochromatin, one chromosome of the pair becomes either partially or totally heterochromatic. The best known case is that of the X chromosomes in the mammalian female, one of which is active and remains euchromatic, whereas the other

**euchromatin**—uncondensed, genetically active regions of the chromatin

**constitutive heterochromatin**—condensed and permanently inactive chromatin

**facultative heterochromatin**—chromatin that is condensed and inactive only in certain cell types or at particular stages of development

**sex chromatin**—an X chromosome randomly selected and inactivated at an early stage of development in the cells of mammalian females; also called a Barr body

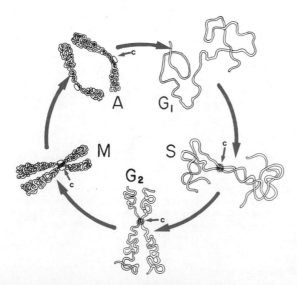

**Figure 7-11**　The condensation–decondensation cycle of chromosomes **G₁,** chromosomes are completely dispersed; **S,** duplication occurs; and **G₂,** condensation starts. At metaphase, **M,** and anaphase, **A,** the condensation is maximal and the two centromeres are clearly visible.

is inactive and forms the *sex chromatin*, or *Barr body*, at interphase.

Constitutive heterochromatin is the most common type of heterochromatin. Most chromosomes contain large blocks of heterochromatin at the centromeres, frequently comprising 5 to 10 percent of the total chromosomal DNA. In some extreme cases, such as in *Drosophila virilis*, it can constitute up to 40 percent of the chromosome. This type of heterochromatin contains highly repeated DNA sequences called *satellite DNAs*, which might have a structural role in chromosomes (see Chapter 13).

satellite DNA—region of highly repeated, constitutive heterochromatin which may have a structural role in chromosomes

A property common to all types of heterochromatin is *late replication*. When cells are given a brief pulse of $^3$H-thymidine late during the phase of DNA synthesis, the label is incorporated only into the heterochromatic segments, indicating that they replicate after the bulk of the DNA.

It is generally agreed that condensed chromatin is inactive in RNA synthesis, and there is good genetic evidence indicating that genes contained in heterochromatic segments are not expressed. Here we will examine three of these lines of evidence.

Some cats have a striking spotted black and yellow coat coloring. Because of the peculiar colored patches, these cats are known as *tortoiseshells* (Fig. 7–12). Tortoiseshell cats are always female. The reason for this became clear when Lyon suggested the hypothesis that the patchy coloring is produced by a gene contained on the X chromosome, which becomes heterochromatic and inactive in some groups of cells but not in others. One of the X chromosomes in each cell of a mammalian female becomes inactivated early in embryogenesis, but this occurs on a random basis, so that adult animals are *mosaics* in which 50 percent of the cells have an active paternal X chromosome and the other 50 percent have an active maternal X.

mosaic—tissue or organism whose cells exhibit more than one genotype, for example as a result of random X chromosome inactivation

Another striking example of facultative heterochromatin is found in the mealy bug *(Planococcus citri)*, an insect in which the males have a haploid set of chromosomes that are entirely heterochromatic. The heterochromatic set is the paternal one. The genes in these paternal chromosomes are switched off, as demonstrated by the fact that it is not possible to induce lethal mutations in them, even by high doses of radiation, whereas dominant lethals can readily be obtained by irradiation of the euchromatic chromosome set. In female mealy bugs both sets of chromosomes are euchromatic and both are genetically active, so lethal mutations can be induced in both the paternal and the maternal set.

It is also possible to inactivate specific genes by translocating them into heterochromatic regions of the chromosomes. This was first shown in *Drosophila melanogaster*, in which a gene coding for eye color can be inactivated by translocating it close to the centromeric heterochromatin.

The biological significance of the switching off of genes in condensed chromatin may be more universal than indicated in the limited examples discussed here. Multicellular organisms have many specialized tissues, and in each one of them a particular set of genes is active and others are inactive. Globin genes, for example, are expressed in red blood cells but are not expressed in other tissues. We do not know how entire sets of genes are switched off, but it is widely believed that variable degrees of chromatin condensation could be involved in the maintenance of the differentiated state.

cells with
an active
black gene

cells with
an active
orange gene

**Figure 7-12**  A tortoiseshell cat. This cat has a white background and spots that have black or orange patches. The genes for the spot color are located in the X chromosomes, and this cat is heterozygous (one X chromosome has the orange allele; the other has the black allele). The appearance is patchy because one of the chromosomes becomes heterochromatic at random during early development; only one of the X chromosomes is active in any given cell. An orange patch represents the cells descended from a single cell in which the X chromosome containing the gene for black color became heterochromatic. Tortoiseshell cats are always female (XX). Male cats (XY) can have either orange or black coats, but never display the tortoiseshell patches. (From Mange, A. P., and Mange, E. J.: *Genetics: Human Aspects.* Saunders College Publishing, Philadelphia, 1980, p. 150.)

# SUMMARY

## 7-1 THE NUCLEAR ENVELOPE

The nuclear envelope consists of two concentric membranes belonging to flattened cisternae of the ER and having ribosomes only on the outer (cytoplasmic) surface. The envelope is covered with relatively large and numerous pores which are occluded by an electron-dense material and enclosed by circular structures called annuli. The pores and annuli together make up the pore complexes.

The cell nucleus contains a specific set of proteins; these are synthesized in the cytoplasm and selectively transported to the proper location. This is possible because mature nuclear proteins contain a signal that enables them to accumulate in the nucleus, as has been demonstrated with labeled nuclear proteins microinjected into the cytoplasm of an oocyte.

## 7-2 CHROMATIN

Each species has a characteristic DNA content, called the C value. Eukaryotes usually have more DNA than prokaryotes, and higher eukaryotes generally have more than lower eukaryotes. The DNA in metaphase chromosomes is usually compacted between 5000- and 10,000-fold, and this degree of compaction is a remarkable (and so far not well understood) feature of the genetic material.

Eukaryotic DNA is complexed with five small, basic histone proteins: H1, H2A, H2B, H3, and H4. The latter four are very similar in different species and are present in equimolar amounts at regular intervals along the nucleic acid molecule; H1 is not conserved between species and is bound rather loosely to the DNA. The association of an octamer of H2A, H2B, H3, and H4 (two of each) with the DNA, plus H1 in its typical configuration, constitutes a 10 nm fiber, representing the first level of chromatin organization within cells.

The nucleosome is a repeating unit of chromatin about 200 bp of DNA in length and containing a histine octamer. In the 10 nm fiber, the octamers are in close contact; the DNA is coiled on the outside of the nucleosome. A still higher level of organization is the 20 to 30 nm fiber, which could arise by coiling of the 10 nm fiber into a solenoid and probably represents the structure of inactive chromatin.

## 7-3 THE CHROMOSOMES

Chromosomes can be observed most easily in squash preparations made at the anaphase or metaphase stage of the mitotic cycle. They are classified into four types according to their shape, which is determined by the position of the centromere. The characteristics of the chromosome set or karyotype vary considerably between species.

Among the elements of the chromosome that can be distinguished in condensed preparations are the chromatid(s), the centromere, the telomeres, and the nucleolar organizers (present in the secondary constrictions).

Several lines of evidence support the unineme theory, which postulates that each chromatid represents a single linear DNA molecule with its associated proteins. More information about higher levels of DNA folding may come from studies of chromatids and particularly histone-depleted chromosomes.

## 7-4 HETEROCHROMATIN

Heterochromatin has been defined as those regions of the chromosome that remain condensed during interphase; euchromatin remains in a noncondensed state. Constitutive heterochromatin is permanently condensed in all kinds of cells and is the more common type; facultative heterochromatin is condensed only in some kinds of cells or at certain developmental stages, as is true of one X chromosome in each cell of the mammalian female. A property common to all types of heterochromatin is late replication, and there is good evidence indicating that genes contained in heterochromatic segments of DNA are not expressed. The ability of the cells to switch genes off in condensed chromatin probably has an important role in the maintenance of the differentiated state.

# STUDY QUESTIONS

- Describe the structure of the nuclear envelope. How might materials be transferred through the pores?
- How do nuclear proteins accumulate in the nucleus? Do you think the ribosomes on the outer surface of the nuclear envelope may have any part in this process? Why or why not?
- What is chromatin?
- Describe the nucleosome, the 10 nm fiber, and the 20 to 30 nm fiber. How much is the DNA folded within them?
- What are telocentric, acrocentric, submetacentric, and metacentric chromosomes? Define centromere, telomere, and secondary constrictions.
- What is a chromatid, and what is the unineme theory?
- Given the two X chromosomes in any female mammalian somatic cell, one is euchromatin and the other is facultative heterochromatin. What exactly does this mean with regard to the morphology of the chromosomes, the time of replication, and the expression of genes located on each?
- Why are there no male tortoiseshell cats?

# SUGGESTED READINGS

**Bostock, C. J., and Summer, A. T.** (1978) *The Eukaryotic Chromosome.* North-Holland, Amsterdam.

**Brown, S. W.** (1966) Heterochromatin. *Science, 151:*417.

**DeRobertis, E. M., Longthorne, R. F., and Gurdon, J. B.** (1978) Intracellular migration of nuclear proteins in *Xenopus* oocytes. *Nature, 272:*254.

**Felsenfeld, G.** (1978) Chromatin. *Nature, 271:*115.

**Kornberg, R. D.** (1974) Chromatin structure: A repeating unit of histones and DNA. *Science, 184:*868.

**Laskey, R. A., and Earnshaw, W. C.** (1980) Nucleosome assembly. *Nature, 286:*763.

**Lewin, B.** (1980) *Eukaryotic Chromosomes,* 2nd ed., volume 2: *Gene Expression.* John Wiley & Sons, New York.

**Lyon, M.** (1974) Review lecture. Mechanisms and evolutionary origins of variable X chromosome activity in mammals. *Proc. R. Soc. London [Biol.], 187:*243.

**Marsden, M. P. F., and Laemmli, U. K.** (1979) Metaphase chromosome structure: Evidence for a radial loop model. *Cell, 17:*849.

**Rattner, J. B., Goldsmith, M., and Hamkalo, B. A.** (1980) Chromatin organization during meiotic prophase of *Bombyx mori. Chromosoma, 79:*215.

# The Cell Cycle and DNA Replication

The ability to reproduce is a fundamental property of cells. The magnitude of cell multiplication can be appreciated by realizing that an adult person is made up of $10^{14}$ cells, all derived from a single cell, the fertilized ovum. Even in fully grown adults, the amount of cell multiplication is impressive. A man contains $2.5 \times 10^{13}$ red blood cells (5 liters of blood, with $5 \times 10^6$ red blood cells per mm³), the average life span of which is $10^7$ seconds (or 120 days). To maintain a constant blood supply, about 2.5 million new cells are required per second. Furthermore, cell reproduction is precisely regulated so that the production of new cells exactly compensates for the loss of cells in adult tissues. In this chapter we will analyze the cell cycle and DNA replication.

## 8-1 THE CELL CYCLE

A growing cell undergoes a cell cycle that consists essentially of two periods: *interphase* and *division*. For many years cytologists were concerned mainly with the period of division, in which dramatic chromosomal changes were visible under the light microscope, whereas interphase was considered a "resting" phase. However, it has since been learned that cells spend most of their life span in interphase, during which intense biosynthetic activity takes place as the cell doubles in size and duplicates precisely its chromosome complement. Cell division is only the final and microscopically visible phase of an underlying change that has already occurred at the molecular level.

interphase—interval between eukaryotic cell divisions, during which growth and synthetic activities take place

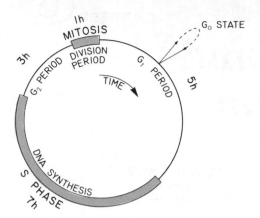

**Figure 8-1**   The cell cycle. The duration of each phase corresponds to a mammalian cell growing with a generation time of 16 hours. (After D. M. Prescott.)

## Stages of Interphase

synthetic (S) period—specific part of interphase during which DNA synthesis occurs

$G_1$ period—interval between the end of mitosis and the start of S phase

$G_2$ period—interval between the end of S phase and the start of mitosis

DNA synthesis occurs only during a restricted portion of interphase called the *synthetic* period or *S* period, which is preceded and followed by two *gap* periods of interphase ($G_1$ and $G_2$) in which there is no DNA synthesis. This finding led investigators to divide the cell cycle into four successive intervals: $G_1$, S, $G_2$, and mitosis. $G_1$ is the interval between the end of mitosis and the start of DNA synthesis, S is the period of DNA synthesis, and $G_2$ is the interval between the end of DNA synthesis and the start of mitosis (Fig. 8–1). During $G_2$ a cell contains two times (4C) the amount of DNA present in the original diploid cell (2C). After mitosis the daughter cells again enter the $G_1$ period and have a DNA content equivalent to 2C (Fig. 8–2).

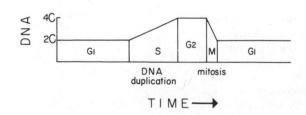

**Figure 8-2**   Life cycle of a cell showing the changes in DNA content during the various periods as a function of time. *2C* corresponds to the diploid content of DNA; *4C* corresponds to the tetraploid content.

The duration of the cell cycle varies greatly from one type of cell to another. For a mammalian cell growing in culture with a generation time of 16 hours, the lengths of the different periods would be $G_1$ = 5 hours, S = 7 hours, $G_2$ = 3 hours, and mitosis = 1 hour. Generally speaking, the S, $G_2$, and mitotic periods are relatively constant in the cells of the same organism. The $G_1$ period is the most variable in length. Depending on the physiological condition of the cells, it may last for days, months, or years. Those tissues that normally do not divide (such as nerve cells or skeletal muscle) or that divide rarely (such as circulating lymphocytes) contain the amount of DNA present in the $G_1$ period. Cultured cells that stop multiplying because of contact inhibition also stop at $G_1$.

The regulation of length of the cell cycle occurs primarily by arresting it at a specific point of $G_1$, and the cell in the arrested condition is said to be in the $G_0$ state. In the $G_0$ state the cell is considered to be withdrawn from the cell cycle. When conditions change and growth is resumed, the cell re-enters the $G_1$ period.

During interphase the chromosomes are decondensed and cannot be distinguished under the microscope. However, using an experimental trick, it is possible to induce the condensation of chromosomes in all three stages of interphase. Mitotic cells fused to interphase cells (using inactivated Sendai virus—see Fig. 14-13) are able to induce premature chromosome condensation in the interphase nuclei. As shown in Figure 8-3, the prematurely condensed chromosomes from $G_1$ nuclei show only one chromatid, while the ones from $G_2$ nuclei have two chromatids. This illustrates that the $G_1$ phase corresponds to the period of interphase prior to DNA replication and that $G_2$ is the period of interphase that follows DNA replication.

**$G_0$ state**—condition of a cell whose division has been arrested at $G_1$, and which is considered to be withdrawn from the cell cycle

## Biochemical Activity during the Cell Cycle

RNA synthesis stops during mitosis, as shown dramatically in Figure 8-4. The rate of RNA synthesis declines rapidly in late prophase and stops in metaphase and anaphase (these stages of mitosis are described in detail in Chapter 9). As in other instances (Section 7-4), highly condensed chromatin cannot be transcribed, perhaps because the DNA cannot be reached by the enzymes that transcribe it.

We have already mentioned that DNA synthesis occurs only in S phase. During this period many units of replication are activated sequentially; the more condensed, heterochromatic regions of the chromosomes (such as the centromeric heterochromatin and the inactivated X chromosome) replicate late during S phase in all cells.

The most important regulatory event occurs in the $G_1$ phase, during which the cell must either start a new cell cycle or become arrested in the $G_0$ state. Once this $G_1$ checkpoint has been passed, the cell will go on to complete a new cycle. Unfortunately, we still know very little about the regulation of this fundamental step in cell proliferation.

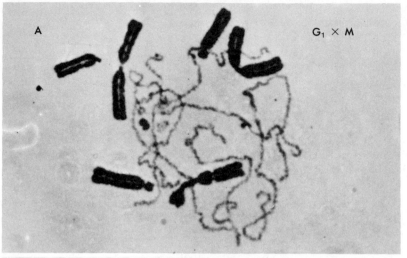

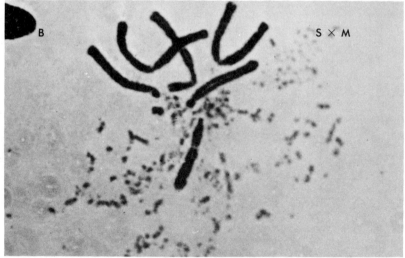

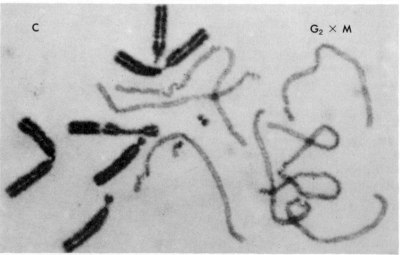

**Figure 8-3** Premature chromosome condensation of $G_1$, S, and $G_2$ chromosomes, induced by fusion to mitotic cells. All cells are from the Indian muntjac, a deer that has a small number of chromosomes. The thick metaphase chromosomes are from the mitotic cells. **A,** $G_1 \times M$, note that the $G_1$ chromosomes have a single chromatid. **B,** $S \times M$, the S phase chromosomes have a fragmented appearance. **C,** $G_2 \times M$, the $G_2$ chromosomes have two chromatids. (Courtesy of D. Röhme.)

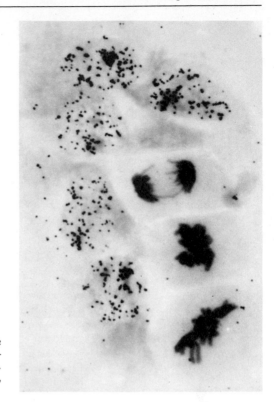

**Figure 8-4**  RNA synthesis stops during mitosis. Chinese hamster ovary (CHO) cells were labeled for 15 minutes with ³H-uridine and then subjected to autoradiography. (Courtesy of D. M. Prescott, from *Reproduction of Eukaryotic Cells*. Academic, New York, 1976.)

## 8-2  DNA REPLICATION

### Replication in Eukaryotes

**SEMICONSERVATIVE.**  The mechanism of DNA replication may be considered a direct consequence of its structure, described in the molecular model proposed by Watson and Crick in 1953 (see Section 2-1). The two strands of the double helix can be separated (recall that they are joined by relatively weak hydrogen bonds), and each polynucleotide chain serves as a template for the synthesis of a new DNA molecule (Fig. 8-5). DNA replication follows the base-pairing rules that A pairs with T and G with C, and consequently each daughter molecule is an exact replica of the parental molecule.

When we say that DNA replication is *semiconservative* we mean that each daughter molecule consists of one parental and one newly synthesized strand (Fig. 8-6). This has been verified by several demonstrations. In their classic experiment with the bacterium *E. coli*, Meselson and Stahl made use of the heavy isotope ¹⁵N. DNA containing ¹⁵N (heavy-heavy, or HH-DNA) is denser than DNA containing the more common isotope ¹⁴N. *E. coli* was grown for several

semiconservative replication—usual process of DNA replication, in which each strand of the parent molecule serves as a template for the synthesis of a new strand (i.e., each daughter molecule contains one parental and one new strand)

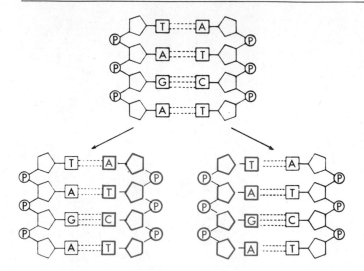

**Figure 8-5** Diagram showing the mechanism of DNA duplication. **Above,** the two standard parent molecules, which separate by opening of hydrogen bonds. **Below,** the two new strands that have been synthesized and that have a complementary base composition with respect to the parent DNA strands are indicated by bold outlines. (From A. Kornberg, *Ciba Lecture in Microbial Biochemistry.* Wiley, New York, 1962.)

generations in a medium containing $^{15}N$ and was then transferred to another medium containing $^{14}N$. The DNA was isolated and its density was determined by ultracentrifugation in a cesium chloride gradient. It was found that after the first division cycle there is only one peak, corresponding to the heavy-light (HL) hybrid molecule in which one strand contains $^{14}N$ and the other $^{15}N$. At the second generation two DNA peaks appear, one in which the two DNA strands

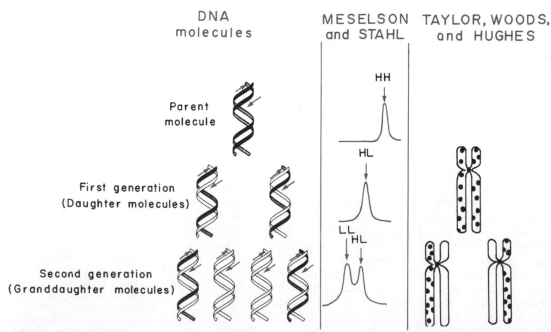

**Figure 8-6** Semiconservative replication of DNA in *E. coli* and in eukaryotic chromosomes. Meselson and Stahl labeled the two chains of the parent DNA molecule with the heavy isotope $^{15}N$ and analyzed the results by CsCl density gradients, while Taylor et al. used $^3H$-thymidine and autoradiography. (See description in text.)

contain $^{14}$N (LL) and the other still corresponding to hybrid (HL) molecules (Fig. 8–6).

Semiconservative DNA replication can also be demonstrated in higher organisms, since (as explained in Chapter 7) each chromatid of a metaphase chromosome represents a single DNA molecule. In another classic experiment, Taylor, Woods, and Hughes labeled *Vicia fava* (broad bean) root tip cells with $^3$H-thymidine and then allowed them to grow in unlabeled medium. When the metaphase chromosomes were analyzed by autoradiography one or two generations later (Fig. 8–6), it was found that both chromatids were labeled after one generation, but only one was radioactive after two cell cycles.

**DISCONTINUOUS.** DNA is synthesized by enzymes called *DNA polymerases*. *E. coli* has three different DNA polymerases, all of which are involved in some aspect of DNA replication. Eukaryotic cells also have three DNA polymerases, but they are less well characterized. (We will see that these enzymes are required not only for DNA replication but also for DNA repair.) One of the features of these polymerases is that they proceed along the DNA molecule only in the 5′ to 3′ direction. However, as discussed in Chapter 2, the two DNA strands in each molecule are oriented in antiparallel fashion. When the two strands unwind (Fig. 8–7), one of them will be facing the DNA polymerase in the correct direction (5′ → 3′), but the other strand will have an unfavorable orientation (3′ → 5′). The solution to this problem is that the DNA is synthesized in *discontinuous segments* on at least one of the strands. It seems likely that DNA synthesis is discontinuous on both strands; in other words, after the unwinding of the DNA, the new strands are made in short segments, always in the 5′ → 3′ direction. These segments are then joined together by the action of another enzyme, *polynucleotide ligase* (Fig. 8–7).

The model of discontinuous synthesis of DNA is supported by experimental evidence. If bacteria are exposed to $^3$H-thymidine for a few seconds while the replication of their genetic material is in progress, only short pieces of labeled DNA (1000 to 2000 nucleotides in

DNA polymerase—one of several enzymes involved in various aspects of DNA synthesis

discontinuous synthesis—the 5′ → 3′ production of short fragments of DNA on the parental template during replication

polynucleotide ligase—enzyme that joins the DNA segments produced discontinuously

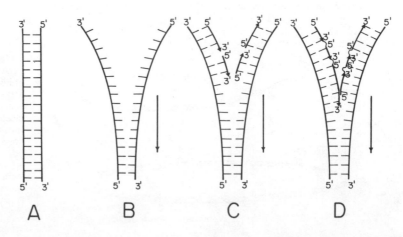

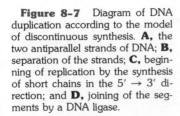

**Figure 8-7** Diagram of DNA duplication according to the model of discontinuous synthesis. **A,** the two antiparallel strands of DNA; **B,** separation of the strands; **C,** beginning of replication by the synthesis of short chains in the 5′ → 3′ direction; and **D,** joining of the segments by a DNA ligase.

length) are found. With longer labeling periods the DNA is already in the form of high molecular weight strands, but some mutations that impair the DNA ligase enzyme result in the accumulation of short fragments even after longer labeling periods. The short DNA pieces are called *Okasaki fragments* after their discoverer. It has been shown that Okasaki fragments sometimes start with a short segment of RNA which acts as a *primer* for DNA synthesis and is later removed.

**Okasaki fragment—a discontinuously produced DNA segment**

**MULTIPLE ORIGINS.** We have seen that eukaryotic chromosomes contain a very large amount of DNA, all of which is contained in only two linear molecules, one for each chromatid. If these huge molecules were copied from a single origin of replication, the S phase would be exceedingly long. Eukaryotic cells solve this problem by having multiple initiation sites in each chromosome (Fig. 8–8).

In 1968, Huberman and Riggs demonstrated that eukaryotic DNA has tandem units of replication called *replicons*. The average replicons are 30 μm long, and there are about 30,000 of them per haploid mammalian genome. Each chromosome may have several thousand origins of replication, from which synthesis proceeds in both directions until the entire molecule has been duplicated (Fig. 8–8).

**replicon—~30 μm "replication unit" of eukaryotic DNA, having its own replication initiation site**

The thousands of eukaryotic replication origins are successively activated over the course of several hours in S phase. If the DNA is to be duplicated accurately, the cell must ensure that no initiation site is used more than once in each cell cycle. Prevention of reinitiation must therefore be a crucial feature of eukaryotic replication. We do not know how this is achieved, but some mechanism must

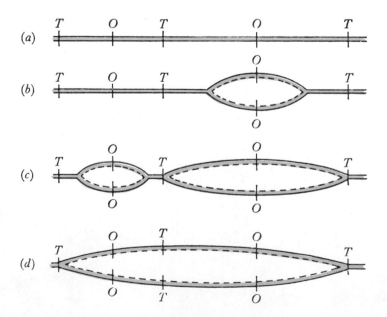

**Figure 8-8** Bidirectional model of DNA replication in eukaryotes. Each pair of horizontal lines represents a segment of a double helical DNA molecule with two strands. The newly formed chains are indicated by broken lines. *O* and *T* indicate sites of origin and termination of replication. **(a)** two adjacent replication units prior to replication; **(b)**, replication started in the unit to the right; **(c)**, replication started in the unit to the left; and **(d)**, replication completed in both units. (From J. A. Huberman and A. D. Riggs, *J. Mol. Biol., 32:*327, 1968.)

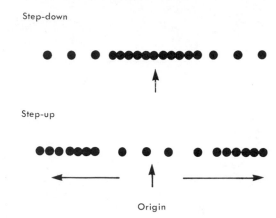

Step-down

Step-up

Origin

**Figure 8-9** Bidirectionality of eukaryotic DNA synthesis, as revealed by DNA fiber autoradiography. Cells were incubated first with highly labeled ³H-thymidine and then with moderately labeled thymidine (step-down), and vice versa (step-up), as described in the text.

exist to distinguish newly replicated chromatin from the unreplicated material; for example, specific modifications in the histones of the newly assembled nucleosomes could be utilized.

**BIDIRECTIONAL.** The *bidirectional* nature of eukaryotic chromosome replication has been demonstrated experimentally using cells labeled with ³H-thymidine of two different specific activities (a measure of relative radioactivity). The labeled cells were lysed with detergent, and the radioactive DNA was attached to nitrocellulose filters and made visible by autoradiography. In this technique, called DNA fiber autoradiography, the replicating DNA molecules can be seen as linear tracks of radioactive grains. As shown in Figure 8-9, cells grown in highly radioactive thymidine and then transferred to less radioactive thymidine ("step down") show two "tails" of diminishing grain density after DNA fiber autoradiography. Conversely, cells receiving the opposite treatment ("step up") show increased grain density in the two tails.

Figure 8-10 shows a replication unit of *D. melanogaster* spread for electron microscopy under conditions that retain the chromosomal

bidirectional synthesis—progress of two growing points in opposite directions along the chromosome after replication has been initiated

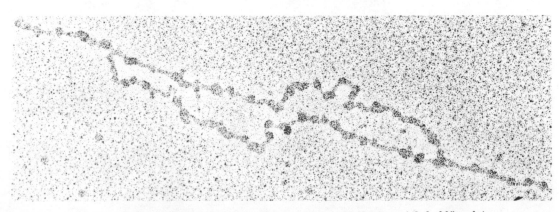

**Figure 8-10** Replicating chromatin (Courtesy of S. L. McKnight and O. L. Miller, Jr.)

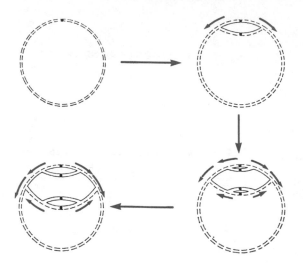

**Figure 8-11** Diagram of the formation of a symmetric chromosome from a circular structure containing a single initiation site. The small arrows indicate the directions of movement of replication forks. (From R. G. Wake, *J. Mol. Biol.*, *68*:501, 1972.)

proteins. It can be observed that the newly replicated DNA rapidly becomes associated with histones, giving a nucleosome configuration. In eukaryotic cells we must therefore think in terms of chromatin replication rather than simply that of DNA.

Unlike eukaryotes, prokaryotes have single origins of DNA replication. *E. coli* has a single circular chromosome about 1.1 mm long (4.2 million base pairs). This DNA molecule has a single origin of replication from which two replication forks proceed bidirectionally (Fig. 8-11). The complete replication of this chromosome takes 30 minutes, but if the bacterium is growing very fast (in a rich culture medium, *E. coli* can divide every 20 minutes or less), new replication forks can start at the same point of origin even before the previous replication cycle is completed, as shown in Figure 8-11.

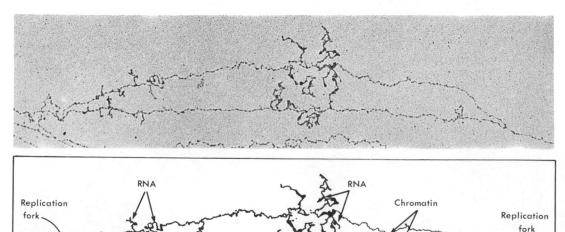

**Figure 8-12** RNA synthesis during DNA replication. (Courtesy of O. L. Miller, Jr. and S. L. McKnight.)

## RNA Synthesis during DNA Replication

Figure 8–12 shows a replication unit of *D. melanogaster* that is being transcribed. It is apparent that RNA synthesis can be initiated almost immediately after replication and that transcription therefore continues during the S phase. The growing RNA chains can be observed on both sister chromatids in a rather symmetrical arrangement. (In a more general sense, this figure also reveals what an enormously powerful and useful tool electron microscopy can provide for cell biologists.)

## DNA Repair

Since DNA is the genetic material it must be passed to future generations in an unaltered form. Although changes in the DNA sequence sometimes occur (i.e., *mutations*, which are of considerable importance in evolution), cells have a DNA repair machinery to prevent excessive changes. Ultraviolet light is a frequent agent of DNA alteration. Its main effect is to produce *thymine dimers* between adjoining T residues in the same strand of DNA. These dimers can be removed and the molecule repaired by a complex series of reactions involving several repair enzymes, as shown in Figure 8–13.

mutation—physical or chemical alteration in the genetic material

thymine dimer—chemical association of adjacent thymines such that they can no longer base-pair with the opposing adenines, causing a distortion in the DNA helix

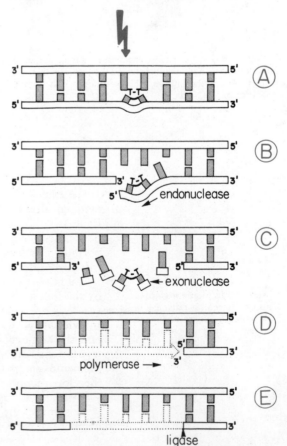

**Figure 8–13**    Diagram showing the effect of ultraviolet light on the DNA molecule and the mechanism of DNA repair. **A,** under the action of ultraviolet light a dimer of T—T is produced; **B,** the affected DNA strand is recognized and incised by a molecule of endonuclease; **C,** the strand segment is excised by a molecule of exonuclease; **D,** the gap is filled by DNA polymerase; and **E,** the synthesized segment is joined by a DNA ligase.

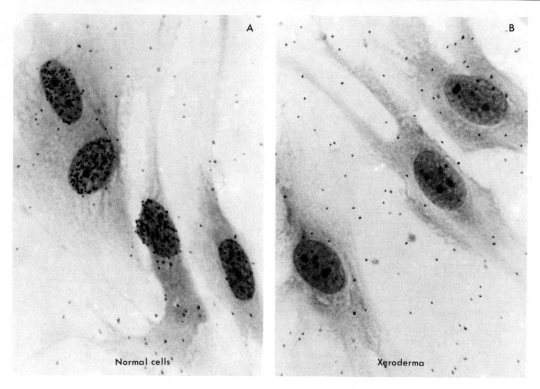

**Figure 8-14** Repair of thymine dimers in normal and xeroderma pigmentosum human cultured cells. Cells were treated with ultraviolet light and then with ³H-thymidine for one hour. The normal cells incorporate low amounts of thymidine in their nuclei, while xeroderma cells do not. None of the cells shown here was in S phase. (Courtesy of Professor Y. Okada.)

xeroderma pigmentosum—
human hereditary disease in
which cells are unable to
repair the damage caused by
ultraviolet light

The excision and repair of thymine dimers can be demonstrated cytologically by irradiating cultured fibroblasts with ultraviolet light and then incubating them with ³H-thymidine for one to two hours. As shown in Figure 8-14, the ultraviolet treatment induces normal cells to incorporate low levels of thymidine. This incorporation represents the repair of the excised thymine dimers.

The importance of DNA repair to cells is perhaps best illustrated by the human hereditary disease *xeroderma pigmentosum*, in which the cells are unable to repair damage caused by ultraviolet light. Individuals with this disease have a severe intolerance to sunlight, and the sun-exposed areas (face and hands) develop fibrosis, pigmentations, and multiple skin cancers. Fibroblasts from xeroderma pigmentosum patients are unable to repair thymine dimers and therefore do not incorporate thymidine after the treatment described above (Fig. 8-14). This disease clearly links human cancer with somatic mutations and shows the benefit of having mechanisms to repair DNA defects.

## Reverse Transcription

The so-called central dogma of molecular biology, discussed in Chapter 2, is:

$$\text{DNA replication} \xrightarrow{\text{transcription}} \text{RNA} \xrightarrow{\text{translation}} \text{Protein}$$

However, there is an exception to this strict unidirectional flow of information, which is that RNA can sometimes be copied or transcribed into DNA. This is done by an enzyme called *reverse transcriptase*, an *RNA-dependent DNA polymerase* that is able to synthesize DNA on an RNA template.

$$\text{DNA} \rightleftharpoons \text{RNA} \rightarrow \text{Protein}$$

reverse transcriptase— enzyme that can transcribe RNA into DNA, found for example in some RNA tumor viruses

Oncogenic (i.e., tumor-producing) viruses containing an RNA genome have been found to act as templates for the synthesis of DNA. The tumor virion contains the reverse transcriptase by which an RNA/DNA hybrid molecule can be produced. In this way RNA viral genes can be copied into DNA and integrated into the genome of the host cell.

# SUMMARY

### 8-1  THE CELL CYCLE

The cell cycle consists of two periods, interphase and division. Interphase can be further subdivided into three stages, $G_1$, S, and $G_2$, with S being the period of DNA synthesis and $G_1$ and $G_2$ being the intervals between DNA replication and mitosis. The $G_1$ period is the most variable in length, ranging from hours to years depending on the cell type. A cell whose growth is arrested in $G_1$ is said to be in the $G_0$ state.

### 8-2  DNA REPLICATION

DNA replication in eukaryotes is semiconservative, discontinuous, and bidirectional, all as a consequence of the structure of the DNA molecule and the enzymes that duplicate it. Synthesis of the replication units or replicons is initiated successively at thousands of origins of replication located at intervals throughout the genome. Some mechanism must exist in the cells to prevent reinitiation at any single origin, although little is known about what this might involve. Newly replicated DNA associates with histones so rapidly that in eukaryotes it is more correct to think in terms of chromatin replication, not just that of DNA. Prokaryotes have only a single origin of DNA replication from which two replication forks proceed bidirectionally. RNA synthesis can be initiated almost immediately after DNA replication.

Ultraviolet light can cause thymine dimers to occur in the DNA molecule. Normal cells contain several enzymes that can repair the DNA by excising the dimers and reconstructing the molecule properly. Cells from individuals with the hereditary disease xeroderma pigmentosum are unable to repair the damage caused by ultraviolet light.

The usual flow of genetic information from DNA to RNA to protein is not always strictly unidirectional. RNA can sometimes be transcribed into DNA by reverse transcriptase, an RNA-dependent DNA polymerase. This enzyme is found in tumor viruses that contain RNA as their genetic material.

# STUDY QUESTIONS

- Summarize the cellular activities that are likely to be occurring during $G_1$, S, and $G_2$.
- What does the experiment shown in Figure 8–3 demonstrate?

- What are some examples of cells that have an exceptionally long $G_1$ period?
- Describe the experiment carried out by Meselson and Stahl. What did it demonstrate?
- What are Okasaki fragments?
- Explain the similarities and differences between prokaryotic and eukaryotic DNA replication.
- Explain how a cell would repair a thymine dimer.
- What conclusions can be drawn from Figures 8–4, 8–11, and 8–12?

# SUGGESTED READINGS

**Blumenthal, A., Kriegstein, J., and Hogness, D.** (1974) The units of DNA replication in *Drosophila melanogaster. Cold Spring Harbor Symp. Quant. Biol., 38:*205.

**DePamphilis, M. L., and Wassarman, P. M.** (1980) Replication of eukaryotic chromosomes: A close-up of the replication fork. *Annu. Rev. Biochem., 49:*627.

**Hand, R.** (1978) Organization of the genome for replication. *Cell, 15:*317.

**Harland, R. M., and Laskey, R. A.** (1980) Regulated replication of DNA microinjected into eggs of *Xenopus laevis. Cell, 21:*761.

**Kornberg, A.** (1980) *DNA Replication.* Freeman, San Francisco.

**Meselson, M., and Stahl, F. W.** (1958) The replication of DNA in *Escherichia coli. Proc. Natl. Acad. Sci. USA, 44:*671.

**Ogawa, T., and Okasaki, T.** (1980) Discontinuous DNA replication. *Annu. Rev. Biochem., 49:*421.

**Prescott, D. M.** (1976) *Reproduction of Eukaryotic Cells.* Academic, New York.

**Ringertz, N. R., and Savage, R. E.** (1976) *Cell Hybrids.* Academic, New York.

# Mitosis, Meiosis, and Heredity

We have seen in earlier chapters that cell division is a complex phenomenon involving the separation of cellular material between daughter cells, and that this process is the final and microscopically visible phase of an underlying change that began at the molecular level with DNA replication. In this chapter we will discuss the two main types of cell division, mitosis and meiosis.

*Mitosis* is characteristic of all somatic cells in both plants and animals, and leads to the equal separation of the chromosomes between the daughter cells. This type of division also occurs in protozoa, algae, and fungi that reproduce asexually; in such cases the progeny have a *uniparental inheritance*.

*Meiosis* is a special type of cell division that takes place in the germ cells of organisms that reproduce sexually (i.e., have *biparental inheritance*). To produce sexual cells or *gametes* (eggs and sperms) there is a complex mechanism consisting of two successive divisions, by which the number of chromosomes is reduced to half. Thus the four cells resulting from a meiotic division have a *haploid number* of chromosomes.

The study of meiosis is directly related to that of heredity, since the laws that rule the transmission of the hereditary characters (the *genes*) are based on the behavior of chromosomes during meiosis.

## 9–1 MITOSIS

The general mechanism of mitosis (Gr., *mitos*, thread) is similar in all eukaryotic cells, although there are differences between animal

**mitosis**—process by which the genetic material is precisely duplicated and two new chromosome sets identical to the original are generated

237

and plant cells. In Chapter 4 we described the mitotic apparatus as including the spindle and asters, and mentioned that in plant cells mitosis is *anastral*, lacking asters and centrioles. In this chapter our description of mitosis will be based mainly on animal cells (see Fig. 9–1).

The series of events that make up the *mitotic cycle* begins at the end of the G2 period of interphase and terminates at the beginning of G1 of a new interphase. The main phases of mitosis are *prophase*, *metaphase*, *anaphase*, and *telophase*. The process of separation of the two cytoplasmic territories, called *cytokinesis*, can be simultaneous with anaphase and telophase, or it can occur at a later stage.

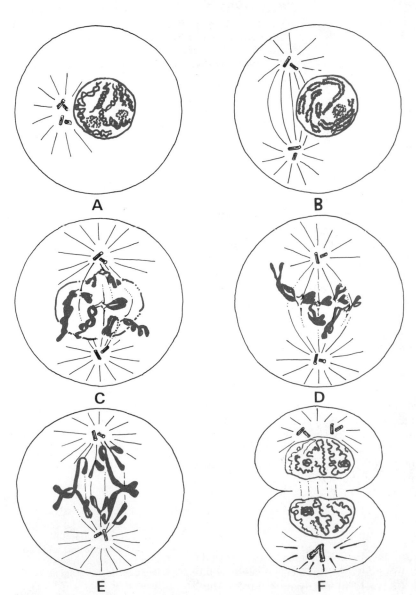

**Figure 9–1** General diagram of mitosis, **A,** *prophase,* the nucleoli and chromosomes, shown as thin threads; in the cytoplasm the aster with the pairs of centrioles are shown. **B,** *prophase,* a more advanced stage of this phase in which the chromosomes have shortened. The primary constriction with the centromere is shown; in the cytoplasm the spindle is formed between the asters. **C,** *late prophase or prometaphase,* the nuclear envelope disintegrates and the chromosomes become attached to the spindle fibers. **D,** *metaphase,* the chromosomes are arranged along the equatorial plane. **E,** *anaphase,* the daughter chromosomes, preceded by the centromeres, are moving toward the poles. **F,** *telophase,* the daughter nuclei are in the process of reconstitution; cell cleavage has started.

# General Description of Mitosis

At *prophase* each chromosome is composed of two chromatids, which have resulted from the duplication of the DNA during the S period (Fig. 9–1, A). As prophase progresses, the chromatids become shorter and thicker by a coiling process, and the primary constrictions become visible (Fig. 9–1, B). At the end of prophase the chromosomes appear as two cylindrical parallel elements (the daughter chromatids). At this time the nucleolus and the nuclear envelope disintegrate, disappearing from view, and the nucleoplasm mixes with the cytoplasm. While these changes are occurring in the nucleus, asters are formed around the centriole pairs at each cell pole, and the spindle begins to form between the asters. The elongation of the spindle helps to situate the asters at the two antipodal positions (Fig. 9–1, C).

prophase—stage of mitosis during which the chromosomes condense and become visible within the nucleus, followed by the dissolution of the nuclear envelope

The transition between prophase and metaphase (Gr., *meta*, between) is sometimes called *prometaphase*. This is the short period in which the nuclear envelope disintegrates and the chromosomes are in apparent disorder. At the beginning of *metaphase* the spindle microtubules invade the central area of the cell, and the chromosomes become attached to them by their kinetochores. The chromosomes then become radially oriented in the equatorial plane, forming the *equatorial plate* (Fig. 9–1, D). This is the high point of metaphase.

metaphase—stage of mitosis during which the chromosomes become attached to spindle fibers and arrayed in the equatorial plane of the cell

During *anaphase* (Gr., *ana*, back) the equilibrium of forces that characterizes metaphase is broken by the separation of the centromeres, which move apart; the sister chromatids separate and begin their migration toward the poles (Fig. 9–1, E). The centromere leads the rest of the chromatid or daughter chromosome, being pulled by the spindle fibers attached to the kinetochore. This pulling causes the chromosomes to assume their characteristic V (metacentric) or L (submetacentric) anaphase shape. During anaphase the microtubules attached to the chromosomes apparently shorten from one third to one fifth of their original length; at the same time the microtubules situated between the poles (the *continuous fibers*) elongate and form the *interzonal fibers*.

anaphase—stage of mitosis during which the centromeres and daughter chromosomes separate and begin moving toward opposite poles of the cell

*Telophase* (Gr., *telo*, end) begins at the end of the polar migration of the daughter chromosomes. The chromosomes uncoil in a process comparable to a reversal of prophase, and gather into masses of chromatin that are then surrounded by cisternae of the endoplasmic reticulum, which fuse to form two new nuclear envelopes. During the final stages, nucleoli are formed by the intervention of the *nucleolar organizers* present in certain chromosomes.

telophase—stage of mitosis during which the chromosomes uncoil and become surrounded by two new nuclear envelopes

*Cytokinesis* is the process of cleavage and separation of the cytoplasm. In animal cells a constriction forms in the equatorial region which deepens until the cell divides (Fig. 9–1, F). During this period there are signs of active movement at the cell surface *(blebs)* and ameboism. This can be observed most clearly in films of dividing cells and in scanning electron micrographs (Fig. 9–2). Cytokinesis is the time during which the cytoplasmic components and organelles are distributed between the daughter cells. Each cell receives a pair of centrioles that will replicate (to form two pairs) during the S period of interphase.

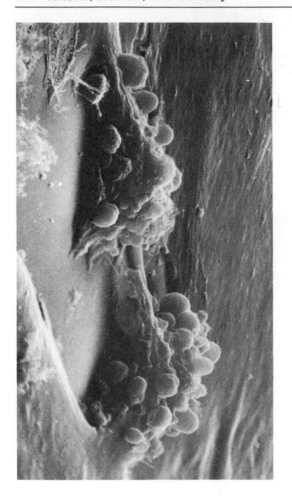

**Figure 9-2**  Scanning electron micrograph of cultured normal hamster kidney cells. Two cells showing many blebs on the surface are in late cytokinesis, while the interphase cells remain flattened. ×1820. (Courtesy of R. D. Goldman.)

## Kinetochores and Chromosome Movement

centromere—point of attachment of sister chromatids, and site of chromosome attachment to the mitotic spindle

kinetochore—proteinaceous region of the centromere to which spindle microtubules attach

In Chapter 4 we described the mitotic apparatus formed by the asters and the spindle, and mentioned that these structures consist of microtubules. At the primary constriction of each chromosome there is a special differentiation called the *centromere*, which is associated with a proteinaceous structure called the *kinetochore*. This is the site of implantation of microtubules that are designated kinetochoric or chromosomal, and we have seen that such microtubules are considered to be functionally related to chromosome movement. Usually there are two kinetochores, one in each chromatid, but in some cases a large number may be found. In such cases this assembly is called a *diffuse* centromere; it is found in the cells of certain plants, insects, and some algae.

In the electron microscope the kinetochore appears as a plate or cuplike disc, 0.20 to 0.25 μm in diameter, plastered upon the primary constriction (Fig. 9–3). It consists of three protein layers: (1) an outer layer of dense material; (2) a middle one of lower density;

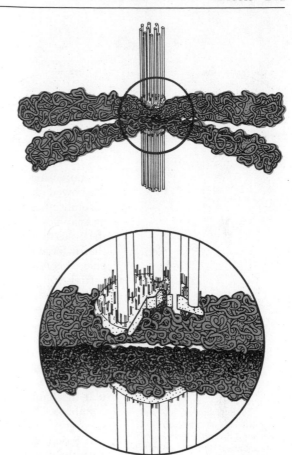

**Figure 9-3** Diagram of a metaphase chromosome showing the folded-fiber structure and the centromere with implanted microtubules. **Below,** an inset at higher magnification, showing the convex electron-dense layer and the fibrillar material forming the "corona" of the kinetochore. Several microtubules of the spindle are shown penetrating the various layers reaching the chromosome fibers.

and (3) an inner dense layer in contact with the chromatin fibers. From the convex surface of the outer layer, in addition to the implanted microtubules, emanate fine filaments that form a kind of "corona." We mentioned in Chapter 4 that kinetochores contain tubulin and that they serve as centers for microtubular polymerization.

Current hypotheses about the role of the mitotic apparatus in the movement of chromosomes and other cell components are based on the idea that the microtubules can generate some sort of mechanical force. The chromosomes can either be pushed or pulled, involving either the elongation or shortening of microtubules. During prophase, centriole migration toward the poles is probably caused by the "pushing" of microtubules. (In fibroblasts this separation occurs at a rate of 0.8 to 2.4 $\mu$m per minute.)

The motion of chromosomes toward the metaphase plate may be the result of lateral interaction between polar and kinetochoric microtubules. Figure 9-4 is a diagram showing the equilibrium of forces that are probably at work in metaphase. There are pole-directed forces that tend to pull the sister kinetochores apart, but these are cancelled by forces that join the sister chromosomes until the beginning of anaphase. This equilibrium of forces, together with

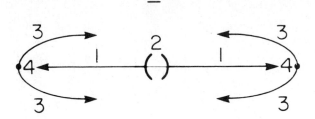

**Figure 9-4** Diagram indicating the equilibrium of forces at metaphase: *1,* pole-directed forces that pull sister kinetochores to opposite poles. *2,* forces that link the two kinetochores until anaphase. *3,* forces that pull the poles inward. *4,* resistance to the inward force. (From J. R. McIntosh, Z. W. Cande, and J. Synder, In *Molecules and Cell Movement,* p. 31. Inoué, S., and Stephens, R. E., eds. Raven, New York, 1975.)

counteracting ones that tend to pull the poles inward, could explain the bowed shape of the spindle and the fact that at metaphase the chromosomes temporarily remain still. This equilibrium is broken at anaphase, during which the chromosomes usually move toward the poles at a rate of about 1 $\mu$m per minute. It has been calculated that a force of $10^{-8}$ dyne is needed to move a chromosome to the pole; this may require the use of about 30 ATP molecules.

High-voltage electron microscopy has made it possible to examine thick sections of the mitotic spindle. The images suggest that two processes are taking place: (1) motion of chromosomes toward the pole in each half-spindle and (2) separation of the two half-spindles by elongation of the free microtubules in the interzonal region. In this area there is a dense zone classically known as the *stem body,* representing a high concentration of interdigitating microtubules.

Two main hypotheses have been postulated to explain the molecular mechanisms responsible for the movement of chromosomes at anaphase. The *dynamic equilibrium hypothesis* was first proposed by Ostergren in 1946, but was established more definitively by Inoué and Sato (1967) from their studies of living mitotic cells examined with the polarization microscope. This hypothesis suggests that the polymerization-depolymerization of microtubules is directly responsible for the displacement of the chromosomes. The *sliding hypothesis* of McIntosh postulates that the motive force originates from the interaction of tubulin molecules with other molecules, such as dynein in the case of cilia and flagella (see Chapter 4). The demonstration that actin and myosin are present in the spindle raises the possibility that these contractile proteins could also play a role in mitosis. In this case, mitosis would be yet another example of the mechanism of cell motility discussed in Chapter 4. For the moment it must be said that the actual nature of the molecular mechanisms underlying the anaphasic movement of chromosomes is still unknown.

**dynamic equilibrium hypothesis**—suggestion that the polymerization/depolymerization of spindle microtubules is responsible for chromosome displacement during anaphase

**sliding hypothesis**—suggestion that anaphase chromosome displacement is caused by the force originating from the interaction of spindle tubulin with other molecules

## Cytokinesis

**cytokinesis**—process of cleavage and separation of the cytoplasm; the final stage of mitosis

The separation of the daughter nuclei at anaphase-telophase usually but not always ends in cytokinesis or cell cleavage. (For example, the eggs of most insects undergo division and form a multi-nucleated *plasmodium* without separation of the cytoplasmic territories.) In animal cells, cytokinesis is first evident with the appearance of a dense material around the microtubules at the equator of the spindle, which later forms part of the *midbody* (Fig. 9–5). Current

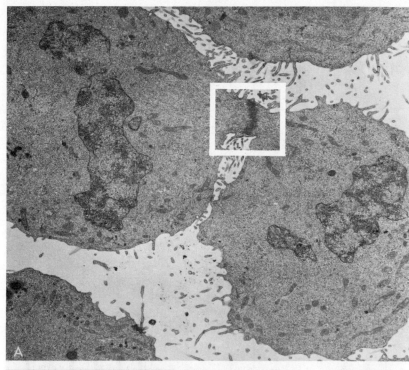

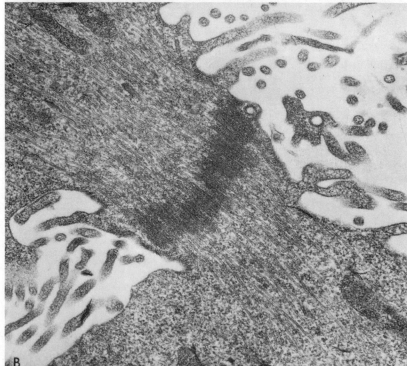

**Figure 9-5** Electron micrograph of a HeLa cell at the completion of cytokinesis. **A,** the two daughter cells are still joined by a small bridge which contains the interzonal microtubules and the electron-dense midbody. **B,** boxed area in **A** shown at higher magnification. **A,** ×10,000; **B,** ×30,000. (Courtesy of B. R. Brinkley.)

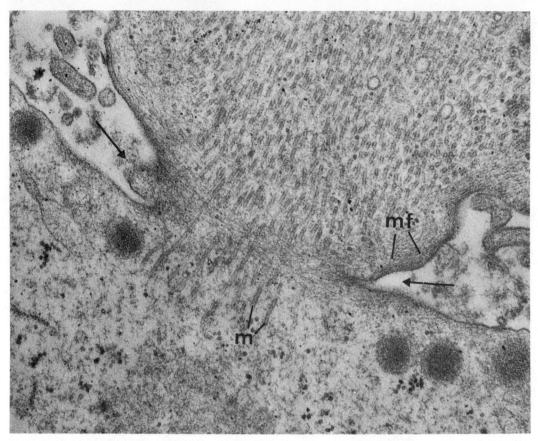

**Figure 9-6**    Electron micrograph of the advancing furrow *(arrows)* in a cleaving rat egg. Microtubules *(m)* of the interzonal fibers of the spindle are observed in the bridge between the daughter cells. Below the plasma membrane a network of fine microfilaments *(mf)* is observed. (See description in text.) ×34,000. (Courtesy of D. Szollosi.)

hypotheses suggest that cleavage by furrowing results from the action of a contractile ring present in the cell cortex (Fig. 9-6). This ring contains actin microfilaments and myosin with an ATPase activity. The existence of this mechanism is supported by evidence from cells treated with cytochalasin B: cell cleavage is inhibited and the contractile ring of microfilaments is dispersed.

## 9-2  MEIOSIS

*Meiosis* (Gr., *meioum,* to diminish) occurs in the cells that produce gametes in sexually reproducing organisms. In these organisms, the fusion of an egg and a sperm cell produces a *zygote,* which develops into a new organism. Such cases of biparental inheritance are based on *fertilization* or *nuclear fusion* of the two gametes.

Figure 9-7 diagrams the essentials of meiosis in humans. The human diploid karyotype consists of 46 chromosomes (44 + XY in the

meiosis—process occurring during gamete formation and involving a reduction division, whereby each daughter cell receives one of each pair of homologous chromosomes, thus reducing the number of chromosomes in each cell to one half

zygote—product of the union of two gametes, which develops into a new organism

male and 44 + XX in the female; see Chapter 10). If mitosis were the only type of cell division, each gamete would carry 46 chromosomes, and after fertilization there would be 92 chromosomes. This progression would be repeated in the next generation, leading to an increase in chromosome number according to the square of the number of generation(!). Meiosis is the mechanism that has evolved to prevent this from occurring. By a series of two divisions the number of chromosomes in the gametes is reduced to half (producing haploid cells), thus allowing recovery of the diploid number at fertilization.

Gametes are produced in the gonads (testicles and ovaries) by processes called *spermatogenesis* and *oogenesis* (Fig. 9-7). This type of meiosis, found in animals and a few lower plants, is called *terminal* or *gametic* because it occurs just before the formation of the

terminal (gametic) meiosis—meiosis occurring just before gamete formation

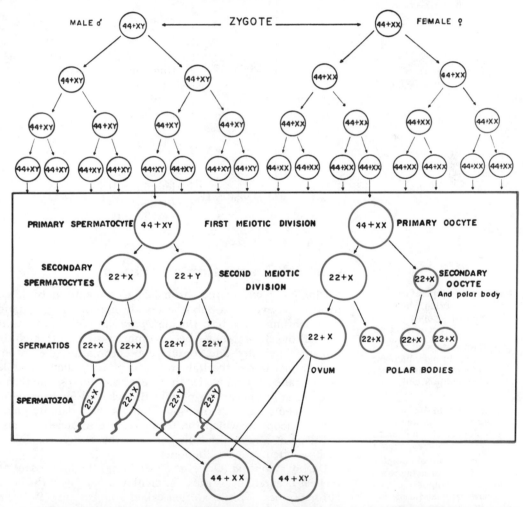

**Figure 9-7** Diagram of spermatogenesis and oogenesis in the human. **Above,** mitosis of spermatogonia and oogonia. **Middle** (within the box), the meiotic divisions. **Below,** fertilization and zygote. Notice the 44 autosomes and the XY or XX sex chromosomes.

gametes. In most plants, meiosis is *intermediary* or *sporic*, since it takes place some time between fertilization and the formation of gametes; in such cases meiosis is followed by mitosis, producing large numbers of gametes. Cells undergoing meiosis are sometimes called *meiocytes*.

Figure 9–8 shows that after several mitoses of the spermatogonia or oogonia, meiotic division starts. At a certain moment of the $G_2$ period there is a decisive change that directs the cell toward meiosis. Meiosis involves two cell divisions with only one duplication of the chromosomes. In division I there is a long prophase, in which the designated stages of mitosis do not suffice to describe all the complexities. The successive stages of meiosis can be outlined as shown:

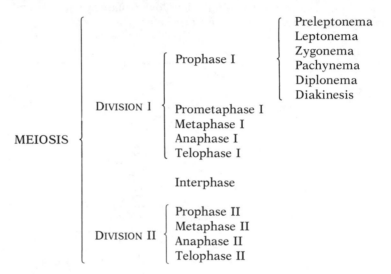

MEIOSIS

DIVISION I

Prophase I
- Preleptonema
- Leptonema
- Zygonema
- Pachynema
- Diplonema
- Diakinesis

Prometaphase I
Metaphase I
Anaphase I
Telophase I

Interphase

DIVISION II

Prophase II
Metaphase II
Anaphase II
Telophase II

## General Description of Meiosis

During *preleptonema* the chromosomes are extremely thin and difficult to observe, and only the sex chromosomes may stand out as heterochromatic bodies. During *leptonema* (Gr., *leptos*, thin; *nema*, thread) the chromosomes become more visible. Although they have already duplicated and contain two chromatids, the chromosomes look single. Under the light microscope they show beadlike thickenings called *chromomeres*. In the electron microscope the unit fiber of chromatin appears to be folded back and forth at the chromomeres. Leptonemic chromosomes may show a definite polarization, forming loops in which the telomeres are attached to the nuclear envelope at a region pointing toward the centrioles. This arrangement is often called a "bouquet."

During *zygonema* (Gr., *zygon*, adjoining) the first essential step of meiosis occurs. This is the alignment and pairing of the homologous chromosomes, a process often called *synapsis* (Fig. 9–8). Pairing is highly specific, involving the formation of a special proteinaceous structure called the *synaptonemal complex* (SC). Figure 9–9 shows

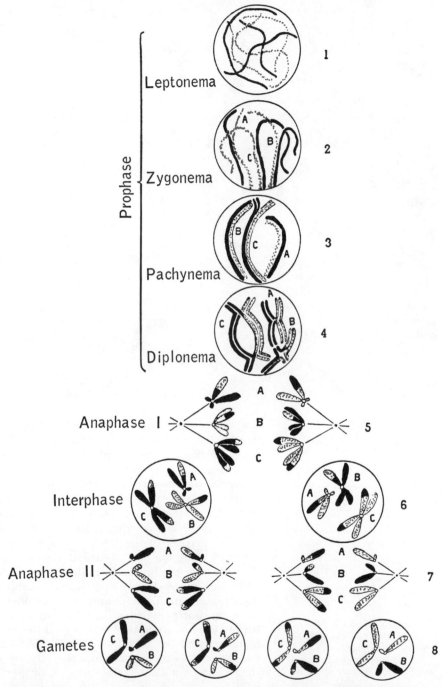

**Figure 9-8**   General diagram of meiosis, illustrating the union, separation, and distribution of the chromosomes.

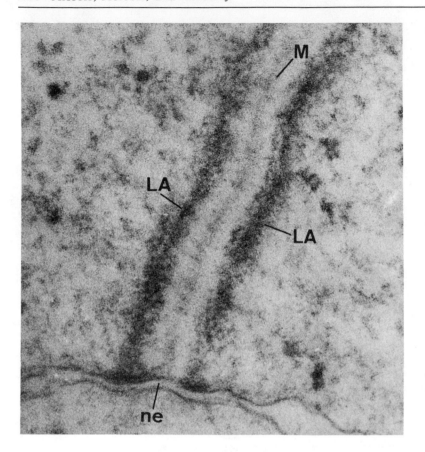

**Figure 9-9** Electron micrograph of the synaptonemal complex in a dog spermatocyte. The two lateral arms *(LA)*, corresponding to the homologous chromosomes, are shown parallel to each other and extending into the nuclear envelope *(ne)*. The medial element *(M)* is simpler than in certain invertebrates. ×125,000. (Courtesy of J. R. Sotelo.)

pachynema—stage of meiosis during which two chromatids belonging to different homologs exchange segments of genetic material

tetrad—a pair of homologs at pachynema, consisting of four chromatids; also bivalent

that this complex, first described by Moses in 1956, is composed of two lateral arms and a medial element. The SC is interposed between pairing homologs and can be considered the structural basis of pairing. It is important to emphasize that the lateral components of the SC start to form at leptonema, while the medial component appears with the pairing at zygonema. Another important point is that at zygonema each chromosome (and each chromatid) has undergone a 300:1 packing of its DNA. For this reason, at this stage only 0.3 percent of the DNA of the homologous chromosomes is matched along the SC. To explain this situation it is assumed that the unit fiber of chromatin makes loops which join the SC at one point (see Fig. 9-10). Pairing is remarkably exact and specific. It takes place point for point and chromomere for chromomere in each homolog. During pairing, the two homologs remain separated by a 0.15 to 0.2 $\mu$m space that is occupied by the SC. A study of all the homologs and the corresponding SCs has been carried out by the use of serial sections and, more recently, by a squashing technique that makes it possible to observe all of them at once, as in Figure 9-11.

At *pachynema* (Gr., *pachus*, thick) the pairing process has been completed and the chromosomes are shorter and thicker. Each one is now a *bivalent* or *tetrad*, composed of two homologs (i.e., four

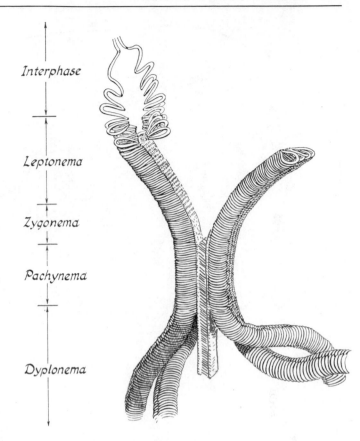

Interphase

Leptonema

Zygonema

Pachynema

Dyplonema

**Figure 9-10**   Three-dimensional diagram of the synaptonemal complex (SC) at different stages of meiosis. In pachynema the homologous chromosomes, now fully paired, are joined by the synaptonemal complex. This complex separates at diplonema; the chromatids show a relational coil. (From T. F. Roth, *Protoplasma*, 61:346, 1966.)

chromatids). The two chromatids of each homolog are called *sister chromatids*. During pachynema the characteristic phenomenon is the exchange of segments; that is, the recombination of chromosomal segments between two chromatids belonging to different homologs. As shown in Figure 9–12, transverse breaks occur in the two *homologous chromatids*, followed by the fusion of the segments. Electron microscopy has revealed dense nodules (called *recombination nodules*) that are related to the SC and may represent the sites of genetic exchange. The function of the SC during pachynema may involve stabilizing the pairing to facilitate recombination. This process occurs at the molecular level, and it is thought that to achieve recombination the broken nucleotide sequences of the two homologs search for each other within the central component of the SC.

At *diplonema* the paired chromosomes begin to separate but remain united at the points of interchange or *chiasmata* (Gr., *chiasma*, crosspiece) (Fig. 9–12.) The number of chiasmata per chromosome varies and there may be one, two, or several, depending on the length of the chromosome. At this time the four chromatids of the tetrad become visible (Fig. 9–13) and the SC disappears. Diplonema is a long-lasting period. (In the fifth month of prenatal life, for example, human oocytes have reached the stage of diplonema and

recombination—exchange of genetic material between homologs

diplonema—stage of meiosis during which paired homologs begin to separate

chiasma—point of chromosomal interchange that becomes visible when the homologs begin to separate at diplonema

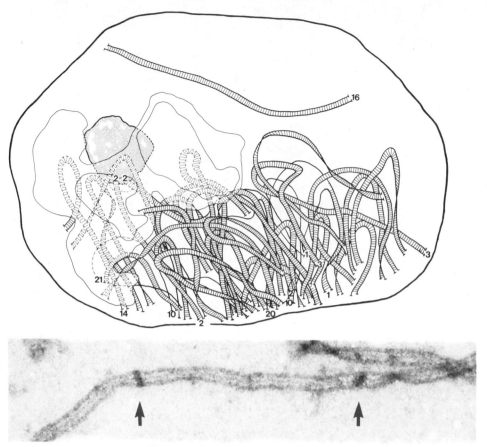

**Figure 9-11**  Reconstruction of the late zygonema stage of meiosis in the *Bombyx mori* female. It can be observed that all homologous chromosomes are paired, with the exception of pairs 1, 2, 5, 10, and 17. In those that are paired, a full synaptonemal complex is present; in those with unpaired regions, the lateral arms of the SC are present. Note that most of the telomeres are attached to a region of the nuclear envelope, forming a *bouquet* figure. ×11,500. (Courtesy of S. W. Rasmussen.) **Inset.** Electron micrograph of a synaptosomal complex from human spermatocyte, showing with arrows two recombination nodules that appear as a dense bar across the complex. ×17,500. (Courtesy of A. J. Solari.)

remain in it until many years later, when ovulation occurs.) In most species the chromosomes uncoil to some extent during diplonema; in fish, amphibian, reptilian, and avian oocytes, however, the uncoiling becomes so marked that the greatly enlarged nucleus assumes an interphase appearance. In these cases the bivalent chromosomes may attain a special configuration known as the *lampbrush chromosome*, in which the chromatids uncoil into loops that converge upon a more highly coiled axis. We will see later that the presence of these lampbrush chromosomes is related to intensive RNA synthesis and to the enormous growth of the oocyte (Chapter 14).

At *diakinesis* (Gr., *dia*, across) the contraction of the chromosomes is accentuated and the number of chiasmata becomes reduced by a process formerly called terminalization (see Fig. 9–8). By the end of diakinesis the homologs are held together only at the chiasmata.

The phases of division I that follow diakinesis are somewhat sim-

**diakinesis—stage of meiosis during which chromosome contraction increases; at the end of this stage the homologs are attached only at the chiasmata**

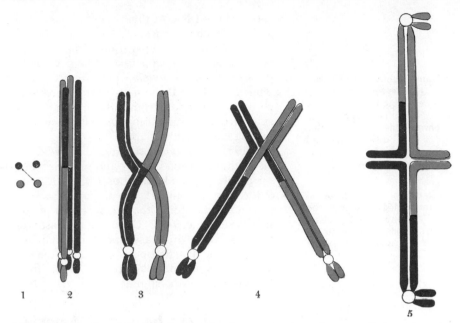

Figure 9-12 **1** and **2,** diagrams showing the process of crossing over; **3,** formation of a chiasma; **4,** terminalization; **5,** rotation of the chromatids of one bivalent.

ilar to those of mitosis. In prometaphase I, chromosome condensation reaches its maximum. The nuclear envelope disappears and the spindle microtubules attach to the kinetochore of the homologous centromeres. As a result the two sister chromatids behave as a functional unit and move together toward one pole. During metaphase I the homologs are still attached at their chiasmata while the centromeres are pulled toward opposite poles (anaphase I). Following telophase I, there is a short interphase in which there is no repli-

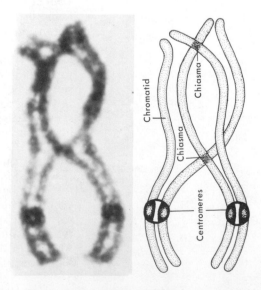

Figure 9-13 **Above,** photomicrograph of a single bivalent at diplonema in a salamander spermatocyte. **Below,** diagram interpreting the photomicrograph. The four chromatids are clearly apparent. Two chiasmata and the position of the centromeres are also visible. Observe that the two sister centromeres are side by side. (Courtesy of J. Kezer.)

cation of the DNA. At this time the number of chromosomes is haploid, but each chromosome has two chromatids.

At division II the separation of sister chromatids and of the corresponding centromeres takes place. A very important point to emphasize is that in division I the *homologous* centromeres separate, while in division II the *sister* centromeres separate. In each case the chromosomes and chromatids have mixed segments that have resulted from recombination; we will discuss the consequences of such mixing later in this chapter.

In 1951 Ostergren postulated that the differences in chromosome distribution during mitosis and meiosis were the result of different orientations of the kinetochores. In fact, in mitosis the kinetochores of the sister chromatids are arranged back-to-back (Fig. 9–3), while in meiotic division I the kinetochores of the sister chromatids lie side by side (Fig. 9–13). The use of microsurgery has made it possible to demonstrate in grasshopper spermatocytes that a key role in chromosome distribution is played by the orientation of the kinetochores. This orientation determines the initial interaction with the microtubules of the spindle.

homologous chromatids— nonidentical chromatids belonging to two different homologs; homologous chromatids separate during meiotic division I

sister chromatids—identical chromatids belonging to the same homolog; sister chromatids separate during meiotic division II

## Biochemical Changes in Meiosis

In certain organisms it is possible to isolate meiocytes at different stages and in sufficient quantities to carry out biochemical analyses. A favorable material is provided by *Lillium* anthers, but mammalian spermatocytes are also suitable. The special biochemical events that occur in meiotic cells are known to include the following:

**PREMEIOTIC DNA REPLICATION.**  Before the start of meiosis there is a premeiotic S phase which is 100–200 times longer than the S phase of a normal cell cycle. This long duration is due to a reduced number of initiation points (see Chapter 8).

**UNREPLICATED DNA.**  In microsporocytes of *Lillium* it has been found that 0.3 to 0.4 percent of the DNA remains unreplicated during the premeiotic S phase.

**Z-DNA.**  During zygonema the unreplicated DNA replicates in conjunction with chromosomal pairing. This late-replicating DNA seems to be essential if synapsis is to occur properly. Z-DNA is rich in GC and represents highly repeated nucleotide sequences. Protein synthesis also seems to be essential to maintain pairing. If it is inhibited at zygonema, the chromosome pairs fall apart.

**P-DNA.**  During pachynema there is a small amount of DNA synthesis that is probably related to the process of recombination. Figure 9–14 shows a molecular model of recombination based on DNA synthesis and repair at pachynema. It is thought that at the end of zygonema, an endonuclease is activated that produces nicks in the two DNA molecules that are aligned for recombination (Fig. 9–14, A). After recombination (Fig. 9–14, B), the overlaps of nucleotide sequences can be excised by exonucleases (Fig. 9–14, C). Then the

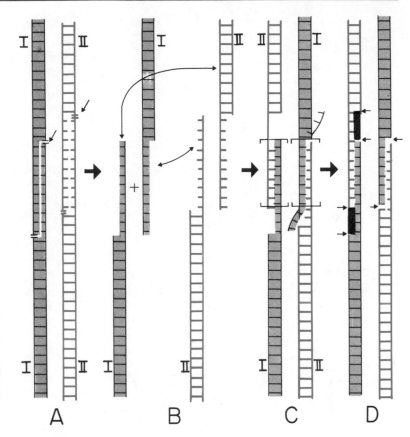

**Figure 9-14** Molecular model of recombination during meiotic prophase. **A,** the DNA of the pairing chromatids (I and II) undergoing the effect of endonucleases, producing "nicks" on each of the strands; **B,** unraveling of the strands; **C,** rejoining of the opposite chromatids; and **D,** elimination of excess pieces of DNA and filling of gaps by a process similar to that of DNA repair. (Modified from a diagram of C. A. Thomas, Jr.)

gaps are filled in by DNA polymerase and ligase in a typical mechanism of DNA repair (Fig. 9-14, D) (Chapter 8). This recombination-repair mechanism accounts for the small amount of DNA synthesis at pachynema.

## Meiosis in Plants

We mentioned above that intermediary or sporic meiosis is characteristic of certain plants. As shown in Figure 9-15, meiosis takes place at some intermediate time between fertilization and the formation of the gametes. In higher plants the reproductive organs—anthers in the male and ovary or pistil in the female—produce *microspores* and *megaspores*, respectively. Microspores are produced by microsporocytes (pollen mother cells). The cells that undergo meiosis to produce megaspores are called megasporocytes. Each microsporocyte gives rise, by meiosis, to four functional microspores. Each megasporocyte produces four megaspores by meiosis, of which three degenerate. The remaining megaspore develops into the female gametophyte, which gives rise to the egg cell.

The microspores and megaspores are not the final gametes of the plants. Before fertilization, they undergo two mitotic divisions in the anther or three in the ovary to produce the male and female

microspore—gamete precursor produced by the anther in (male) higher plants

megaspore—gamete precursor produced by the ovary or pistil in (female) higher plants

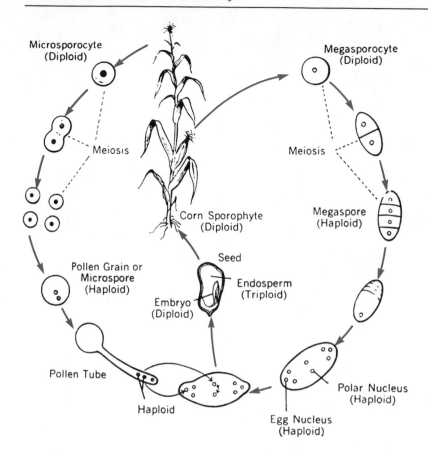

Microsporocyte
(Diploid)

Meiosis

Megasporocyte
(Diploid)

Meiosis

Corn Sporophyte
(Diploid)

Megaspore
(Haploid)

Pollen Grain or
Microspore
(Haploid)

Seed

Embryo
(Diploid)

Endosperm
(Triploid)

Pollen Tube

Haploid

Polar Nucleus
(Haploid)

Egg Nucleus
(Haploid)

**Figure 9-15**   Life cycle of a
plant. (Modified from E. W. Sin-
nott, L. C. Dunn, and T. Dob-
zhansky, *Principles of Genetics.*
5th Ed. McGraw-Hill, New York,
1958.)

gametophytes, respectively. Fertilization in plants is also a complex phenomenon. Each pollen grain or microspore (haploid) carries two sperm nuclei. One of them fertilizes the egg nucleus to produce a *diploid zygote,* which will eventually form the embryo of a new plant. The other sperm fuses with two polar nuclei to form a *triploid endosperm nucleus,* which by mitotic divisions will give rise to the *endosperm,* which contains the nutritive material for the embryo. Thus the plant seed is a mosaic of tissues consisting of the diploid zygote, the triploid endosperm, and the diploid integuments which are of maternal origin (Fig. 9–15).

## Meiosis in Humans

gonocyte—primary germ cell
found in the human embryo

The primary female germ cells (called *gonocytes*) appear in the human embryo in the wall of the yolk sac at about 20 days of gestation and migrate to the gonadal ridges during the fifth week. By mitosis, they form oogonia and become surrounded by follicular cells, forming the primary follicles. At the end of the third month of prenatal development the oogonia enter meiosis, becoming oocytes I, and are then arrested at the stage of diplonema until sexual maturity at about 12 years. Only about 400 of the estimated 1,000,000 oocytes in a human female reach maturity during the course of her reproductive life; thus in some of the later-appearing cells, meiosis may last as long as fifty years. This may explain the increase in the incidence of chromosomal aberrations with the increasing age of the mother (see Chapter 10).

When the ovum is released into the oviduct, the first meiotic division occurs, producing one polar body. Only when the ovum is fertilized by the spermatozoon does the second meiotic division take place, producing the second polar body. Only one viable egg results from the process of oogenesis.

In the human male the primitive gonad also starts with the migration of the primary gonocytes into the gonadal ridges, and later these are incorporated into the seminiferous tubules. These contain the spermatogonia that enter meiosis at puberty. In contrast to oogenesis in the female, spermatogenesis continues in the male until an advanced age. One cycle of this process is completed in about 24 days, and the meiotic prophase I occurs in 13 to 14 days. Four viable spermatozoa are formed from each meiotic cycle (Fig. 9–7). The transformation of the spermatids into mature sperm cells occurs by a complex process called spermiogenesis.

oogenesis—process by which viable egg cells arise from gonocytes in the female

spermatogenesis—process by which mature sperm cells arise from gonocytes in the male

## Comparison of Mitosis and Meiosis

At this point it is important to make a comparison between the two processes of cell division. Many events are similar in mitosis and meiosis, including the changes in the nucleus and cytoplasm during the stages of prophase, metaphase, anaphase, and telophase; the formation of the spindle apparatus; the condensation cycle of the chromosomes; and the structure and function of the centromeres. There are also several essential differences.

(1) Mitosis occurs in all *somatic* cells of an organism, while meiosis is limited to the *germinal* cells.

(2) In mitosis each DNA replication cycle is followed by one division. The resulting daughter cells have a *diploid* number of chromosomes and the same amount of DNA as the parent cell. In meiosis each DNA replication cycle is followed by two divisions; the four resulting cells are *haploid* and contain half the parental amount of DNA.

(3) In mitosis DNA synthesis occurs in the S period, which is followed by a $G_2$ phase before the onset of division. In meiosis there is a period of *premeiotic DNA synthesis* which is much longer than in mitosis and is followed immediately by meiosis. In other words, the $G_2$ phase is short or nonexistent.

(4) In mitosis every chromosome behaves independently. In meiosis the homologous chromosomes become mechanically related (i.e., undergo *meiotic pairing*) during the first meiotic division.

(5) While mitosis is rather brief, lasting perhaps one or two hours, meiosis is a long process. In the human male, for example, it may last for 24 days, and in the human female it may go on for several years.

(6) In the course of mitosis the genetic material remains constant (i.e., with only rare mutations or chromosomal aberrations), but *genetic variability* is one of the characteristic consequences of meiosis.

There are three general features of meiosis that may have genetic

**segregation—random partitioning of homologous chromosomes (and their genes) into the daughter cells during meiosis**

consequences. These are (1) the *pairing* or synapsis of the homologs, (2) the process of crossing over and *recombination*, and (3) the segregation of the homologous chromosomes. As a result of meiosis there has been not only a reduction in the number of chromosomes, but after the two divisions each chromatid of the tetrad has been

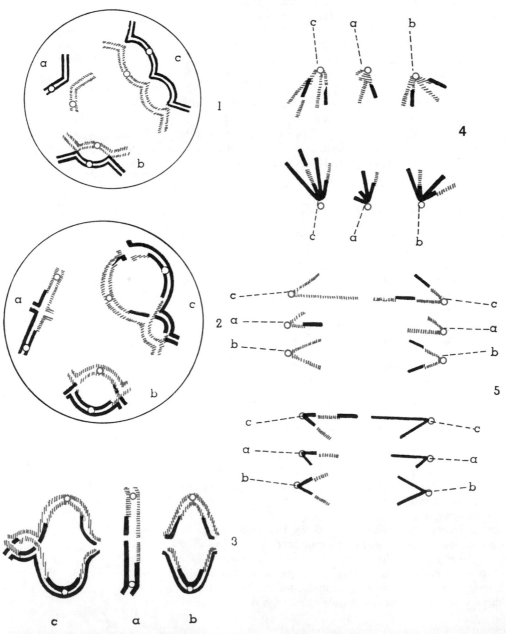

**Figure 9-16** Diagram showing the genetic consequences of the meiosis of three pairs of chromosomes with (a) one chiasma, (b) two chiasmata, and (c) three chiasmata. **1,** diplonema; **2,** advanced diplonema showing the process of terminalization; **3,** metaphase I; **4,** anaphase I; **5,** anaphase II, showing the distribution of the chromosomes in the four nuclei formed. **Solid line,** the paternal chromosomes; **dashed line,** the maternal chromosomes. The centromere is represented by a circle.

segregated into one of the four resulting nuclei. Recall that in division I the homologous centromeres are separated, while in division II the sister centromeres separate. Since the crossing over has mixed the homologous chromatids at pachynema, both meiotic divisions are needed to segregate the genes contained in each chromatid. This can be illustrated schematically as in Figure 9–16 by the distribution of chromosomes a, b, and c. (In this case the chromosomes have undergone one, two, or three crossing-over events, and the corresponding chiasmata are shown.) It is obvious that the four gametes have a different genetic constitution. Due to recombination, the chromosomes may contain alternating segments of maternal and paternal origin. In this figure, however, the random segregation of the chromatids is not apparent. For simplicity, the homologous and sister centromeres are shown to migrate to the same pole. This is not the case in living cells, where the chromosomes segregate at random in both divisions. This means that even in the absence of crossing over, in the case of three pairs of chromosomes, $2^3 = 8$ different chromosome combinations could result. In humans, since each cell contains 23 pairs of chromosomes, the possible chromosome combinations could reach the immense number of $2^{23} = 8,388,608$.

# 9–3  HEREDITY AND CYTOGENETICS

Cytogenetics is a branch of cell biology that emerged from the convergence of cytology and genetics. It concerns the cytological bases of heredity and deals with important problems applicable to, for example, agriculture and medicine. In this chapter we will consider the relationship between chromosomes and heredity in preparation for later discussions of human cytogenetics and the molecular aspects of genetics.

cytogenetics—study of inheritance using the methods of cytology and genetics

## Mendelian Laws of Inheritance

The Mendelian laws that rule the transmission of hereditary characters are based on knowledge about the behavior of the chromosomes during meiosis and the genetic consequences of this mechanism. However, when Gregor Mendel discovered the fundamental laws of inheritance in 1866, nothing was known about chromosomes or meiosis. His findings were based on precise quantitative experiments and logical abstract thinking. He studied crosses between peas *(Pisum sativum)* having pairs of different or contrasting characteristics. For example, he used plants that have white and red flowers, smooth and rough seeds, yellow and green seeds, long and short stems, and so forth. After observing the parental generation ($P_1$), he observed the resulting *hybrids* of the first filial generation, $F_1$. Then he crossed the hybrids among themselves and studied the result in the second filial generation, $F_2$.

hybrid—offspring of parents having different genetic characteristics

For example, in a cross between plants with yellow and green seeds, in the first generation he found that all the hybrids had yellow seeds, and thus resembled only one of the parents. In the second

gene—hereditary unit consisting of a particular sequence of bases in DNA and specifying the production of a distinct protein (e.g., an enzyme)

allele—one of two (or more) genes for a given trait that occurs in a specific position on each homologous chromosome

locus—position of an allele on the chromosome

cross, $F_2$, the characteristics of both parents reappeared in the proportion of 75 percent yellow to 25 percent green, or 3:1.

Mendel's results can now be explained in terms of the behavior of chromosomes and genes. The genes are present in pairs called *alleles*, one allele being located on each homologous chromosome. They are situated at a specific *locus* occurring in the same position on each chromosome. When the plants having yellow and green seeds were crossed, the alleles were distributed in the progeny as diagrammed below, where $A$ stands for the yellow character and $a$ for the green.

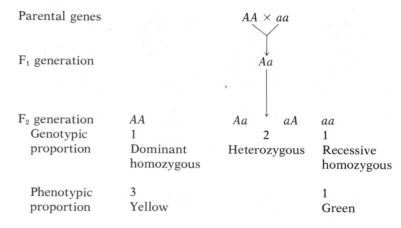

Parental genes $\qquad\qquad AA \times aa$

$F_1$ generation $\qquad\qquad Aa$

| $F_2$ generation Genotypic proportion | $AA$ 1 Dominant homozygous | $Aa$ $\quad$ $aA$ 2 Heterozygous | $aa$ 1 Recessive homozygous |
|---|---|---|---|
| Phenotypic proportion | 3 Yellow | | 1 Green |

genotype—genetic constitution of an organism

phenotype—any visible characteristics of an organism that are ultimately an expression of gene activity

homozygous—having two identical alleles for a given trait

heterozygous—having two different alleles for a given trait

dominance—suppression of one allele in a heterozygous organism by the other (dominant) allele for the same trait

recessiveness—susceptibility to suppression by a dominant allele

It is evident in the diagram that the genetic constitution of the $F_2$ progeny is 25 percent $AA$, 50 percent $Aa$ (or $aA$), and 25 percent $aa$, giving a *genotypic proportion* of 1:2:1. The appearance of the $F_2$ progeny, however, is 75 percent yellow and 25 percent green.

In 1911, Johanssen proposed the term *genotype* for the genetic characteristics of an organism and the term *phenotype* for its visible characteristics. Phenotype is now used to refer to all the characteristics of an individual that are ultimately an expression of gene activity. (In humans, for example, phenotypic characteristics include eye color, baldness, presence of various hemoglobins, and blood group identity.) Thus, in the example above, in the $F_2$ offspring there are two phenotypes (yellow and green) but three genotypes ($AA$, $Aa$, and $aa$) in the proportions described. Of these genotypes, $AA$ is *dominant homozygous*, $Aa$ is *dominant heterozygous*, and $aa$ is *recessive*.

**LAW OF SEGREGATION OF THE GENES.** The first law or principle that Mendel postulated from his experiments was that the genes are distributed independently, without mixing. In the above experiment, in the $F_1$ generation an $Aa$ hybrid is produced in which gene $A$ (yellow) is dominant; gene $a$ (green) is recessive and remains hidden. Meiosis in the $F_1$ hybrid produces gametes, half of which carry the $A$ gene and half of which carry the $a$ gene. Four possible $F_2$ combinations result from the random mixing of these two types of gametes: $AA$, $Aa$, $aA$, and $aa$. As we have seen, these combinations correspond to plants of which 25 percent will have pure yellow seeds, 50 percent will have hybrid yellow seeds, and 25 percent will have pure green seeds.

## LAW OF INDEPENDENT ASSORTMENT OF THE GENES.

Whereas the law of segregation applies to the behavior of a single pair of genes, the law of independent assortment describes the simultaneous behavior of two or more pairs of genes located in different pairs of chromosomes. Genes that are located in separate chromosomes are independently distributed during meiosis. The resulting offspring is a hybrid (also called a *dihybrid*) at two loci.

Figure 9–17 diagrams the cross between a black, short-haired guinea pig *(BBSS)* and a brown, long-haired guinea pig *(bbss)*. The *BBSS* individual produces only *BS* gametes; the *bbss* guinea pig produces only *bs* gametes. The F₁ offspring are heterozygous for hair

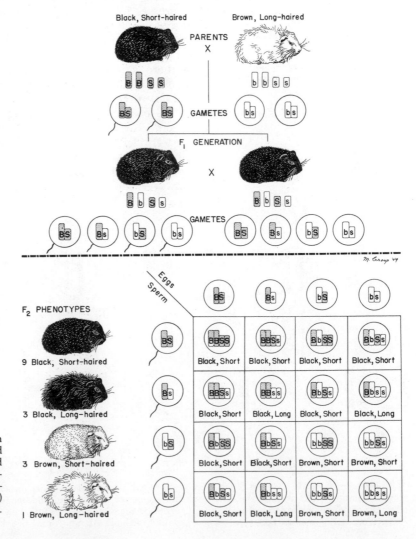

**Figure 9–17** Diagram of a cross between black short-haired (dominant) and brown long-haired (recessive) guinea pigs. The independent assortment of genes is evident. (See description in text.) (From C. A. Villee, *Biology*. 6th Ed. Saunders, Philadelphia, 1972.)

color and hair length *(BbSs)*. Phenotypically they are all black and short-haired. However, when two of the F$_1$ dihybrids are mated, each produces four types of gametes *(BS, Bs, bS, bs)*, which by fertilization result in 16 zygotic combinations as shown in the figure. In the F$_2$ generation there are nine black, short-haired individuals; three black, long-haired; three brown, short-haired; and only one brown, long-haired individual. This phenotypic proportion (9:3:3:1) is characteristic of the second generation of a cross between two allelic pairs of genes.

## Linkage and Recombination

*Drosophila melanogaster—species of fruitfly well suited for genetic studies*

*linked genes—genes present on the same chromosome*

In the cross shown in Figure 9–17, the genes were located in different chromosomes, and during meiosis they segregated randomly in the gametes. However, when Morgan and his collaborators (1910–1915) studied crosses in the fruit fly *Drosophila melanogaster* it became evident that the law of independent assortment could not be applied in every case. Sometimes there are certain limitations on free segregation, resulting in a reduced number of combinations. This is because there is *linkage of genes* within a given chromosome. Since *Drosophila* has four pairs of chromosomes, the possibility of linkage is increased. After studying many crosses in *Drosophila*, Morgan reached the conclusion that all its genes are clustered into four linked groups corresponding to the four pairs of chromosomes.

Figure 9–18, left, shows the case of the segregation of two genes that are in the same chromosome. For these genes, only two types of gametes *(AB* and *ab)* are produced during meiosis. Figure 9–18, right, however, shows the effect of recombination on the linkage of the two genes *(A* and *B)*. In this case, four types of gametes are

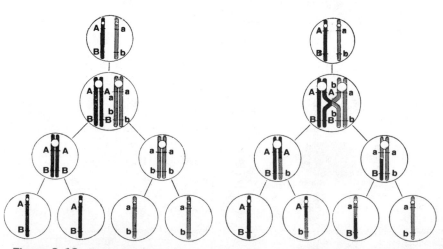

**Figure 9-18   Left,** diagram of the segregation of two pairs of allelic genes localized on the same pair of chromosomes without crossing over. The result is two types of gametes, AB and ab. A case of linkage. **Right,** diagram of the segregation of two pairs of allelic genes on the same chromosome, between which crossing over takes place during meiosis. Four types of gametes result: AB, aB, Ab, ab. A case of linkage with crossing over.

produced and, because of the interchange of homologous chromatid segments, the linkage is ruptured.

There is a correlation between recombination and the number of chiasmata present at diplonema. The frequency of recombination of two linked genes is a function of the distance that separates them along the chromosome. When two genes are close to one another, the probability of crossing over is less than when they are far apart. If the distance between genes is estimated by analyzing the frequency with which linkage occurs, it is possible to construct a map indicating the relative position of each gene along the chromosome.

*Neurospora* is an ideal organism for the study of recombination and gene expression. The advantage it provides is two-fold: (1) it is possible to identify and to follow the fate of each of the four chromatids present in the bivalent meiotic chromosome, and thus to determine whether the crossing over involves two, three, or all four chromatids; and (2) it is possible to make a close correlation between genetic constitution and the biochemical expression (i.e., activity) of genes. As shown in Figure 9–19, the four cells resulting from the two meiotic divisions then undergo a mitotic division, which gives rise to eight haploid ascospores. Each of these ascospores can be isolated by dissection and cultured separately, producing haploid individuals having the genetic constitution of each of the four original chromatids of the bivalent chromosome.

*Neurospora*—genus of fungus well suited for genetic studies

## Chromosomal Aberrations

The normal functioning of the genetic system depends on the constancy of the hereditary material contained in the chromosomes. Under certain conditions, however, the karyotype may change in such a way as to produce abnormal genetic consequences. Such karyotypic changes may involve alterations in the number or structure of the chromosomes, and may occur either spontaneously or as a consequence of ionizing radiation or certain chemicals *(mutagens)* that damage the DNA molecule.

**CHANGES IN CHROMOSOME NUMBER.**    A change in the number of chromosomes in a given karyotype may lead to *euploidy* or *aneuploidy*. Euploidy is an abnormality in which the whole set of chromosomes is decreased or increased, while aneuploidy signifies the presence of abnormal numbers of one or more chromosomes within the set.

Some exceptional plants or animals may be *monoploid* or *haploid*, having a single set of chromosomes. In these organisms, which lack homologs, meiosis may be irregular. When an individual has more than two sets of chromosomes it is said to be *polyploid* (Fig. 9–20). Polyploidy occurs frequently in flowering plants but less often in animals; it can be induced by the drug colchicine, which inhibits the formation of the spindle, allowing the accumulation of additional sets of chromosomes.

Aneuploidy may result from a failure in the separation of one chromosome pair *(nondisjunction)* during meiosis. In *monosomic in-*

euploidy—karyotypic abnormality in which the entire set of chromosomes is increased or decreased

aneuploidy—karyotypic abnormality in which a specific chromosome(s) is present in too many or too few copies

nondisjunction—failure of one or more homologous chromosome pairs to separate properly during cell division

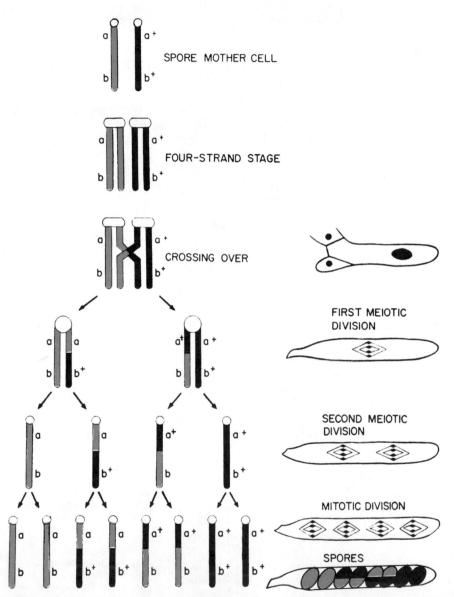

**Figure 9-19** Diagram of the formation of ascospores in *Neurospora crassa*. A single crossing over between genes a and b, the behavior of one pair of chromosomes during the first and second meiotic divisions, and the division by mitosis of each of the four products are shown. The presence of a single chromatid in each spore is indicated diagrammatically.

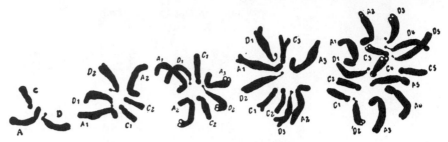

**Figure 9-20**  Polyploid series in the plant *Crepis*. (After Nawashin.)

dividuals, one of the chromosomes was lost because it did not reach the pole. In *trisomics* there is a gain of a single chromosome, and thus there are three of the same type. (As will be discussed in Chapter 10, aneuploidy in humans may lead to severe mental and physical defects.) When nondisjunction occurs at meiosis, all the cells of an individual show the change in chromosome number. However, when nondisjunction takes place in mitosis, only some somatic cells may be aneuploid. This allows the development of *mosaic* individuals in which parts of the body are aneuploid. In cultured cells such changes are common, and may lead to a cancerous transformation. Furthermore, cytogenetic analysis of tumors shows that they frequently contain altered karyotypes.

**CHANGES IN CHROMOSOME STRUCTURE.**  A *chromosomal aberration* is an alteration in the structural organization of the chromosome that can be observed under the microscope. This is in contrast to a *gene mutation* (also called a *point mutation*), in which the changes occur at the molecular level. When a gene mutation occurs the alteration is in the sequence of bases in the DNA; such mutations can only be detected by the subsequent expression of the gene. Chromosomal aberrations often affect many genes, and can be intra- or interchromosomal.

To understand the mechanisms by which chromosomal aberrations are produced, it is useful to recall what was said in Chapter 7 about the molecular structure of chromosomes. First, the chromatids are *mononemic* (i.e., consist of a single DNA molecule). Remember, too, that during the $G_1$ period of the cell cycle there is a single chromatid. Only after the S period (in $G_2$) does the chromosome contain two chromatids. The main chromosomal aberrations are:

(1) *Deficiency* or *deletion* involves the loss of chromosomal material and may be either *terminal* (at the end of a chromosome) or *intercalary* (within the chromosome). The former aberration originates from a single break at the $G_1$ stage and the latter from two breaks (Fig. 9–21, A).

(2) *Duplication* occurs when a segment of a chromosome becomes represented two or more times (Fig. 9–21, B). If the fragment includes the centromere, it may be incorporated into the karyotype as a small extrachromosome.

chromosomal aberration—physical alteration in chromosome structure that can be observed under the microscope

point mutation—change in a single base of the DNA molecule; can only (and not always) be detected upon subsequent gene expression

Intrachromosomal aberrations

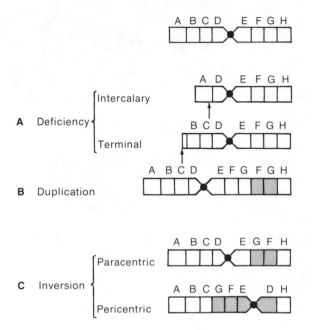

Interchromosomal aberrations

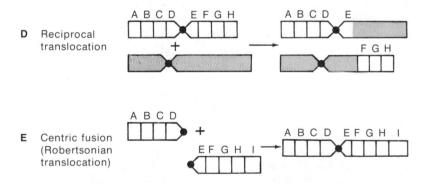

**Figure 9-21** Diagram showing some of the most frequent chromosomal aberrations. See description in text.

(3) *Inversion.* This involves a 180 degree inversion of a chromosomal segment. It may be *pericentric* if it includes the centromere or *paracentric* if it does not (Fig. 9–21, C). At pachynema a loop is formed that allows for the pairing of the inverted segment.

(4) *Translocation.* This interchromosomal aberration is called *reciprocal* when segments are exchanged between nonhomologous chromosomes (Fig. 9–21, D). Translocations may be *homozygotic* when involving segments of two chromosomes of a pair and *heterozygotic* when only one chromosome of a pair is translocated.

When both chromosomes are broken near the centromere, a metacentric chromosome may originate by *centric fusion* or *Robertsonian translocation* (Fig. 9–21, E). This phenomenon has occurred in the phylogeny of many species and has led to a reduction in the number of chromosomes. This mechanism has played a part in the evolution of various species.

(5) *Isochromosomes.* A break at the centromere may give rise to two new chromosomes, each containing part of the original one (Fig. 9–22).

(6) *Sister chromatid exchange* involves a break in the sister chromatids of a chromosome with an interchange of segments. These changes are difficult to observe with common methods because there is no visible morphological alteration. However, they can be detected by labeling the chromosome with $^3$H-thymidine or bromodeoxyuridine (BrdU). BrdU is a thymidine analog which can incorporate into the chromosome in place of thymidine, upon which it produces a change in the staining of the chromatids (Fig. 9–23). It has been observed that *mutagenic* drugs (i.e., those that produce mutations) tend to increase the number of breaks and chromatid exchanges, increasing the susceptibility of the chromosome to other types of damage in the process.

Robertsonian translocation—special case of translocation in which the long arms of two acrocentric chromosomes form a metacentric chromosome; also centric fusion

isochromosome—chromosome formed by the joining of two identical chromosome arms

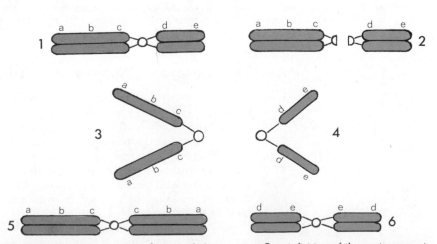

**Figure 9-22** Formation of isochromosomes. **1,** original chromosome; **2,** misdivision of the centromere at the beginning of mitotic anaphase; **3** and **4,** the chromatids unfold into two isochromosomes; **5** and **6,** in the next division, two complete isochromosomes are present. Note that in each isochromosome, the arms exhibit the same genetic constitution.

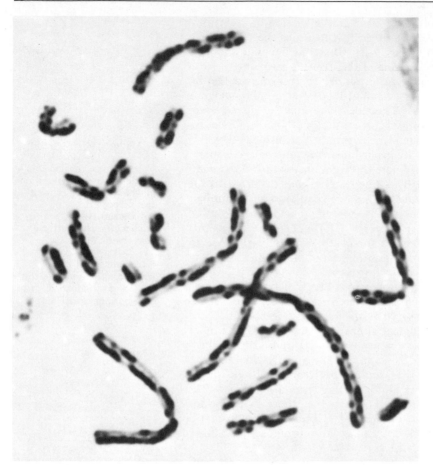

**Figure 9-23**   Cultured cell exposed to the mutagenic chemical agent 8-methoxypsoralen and to bromode-oxyuridine. The regions stained only lightly with Giemsa are those in which the DNA was synthesized in the presence of BrdU. Observe the numerous sister chromatid exchanges in each chromosome. (Courtesy of S. Latt.)

## Chromosomal Changes and Evolution

The development of comparative cytology and cytogenetics has brought about great progress in our understanding of species taxonomy and evolution. The study of the karyotypes of different species has revealed interesting facts about both the plant and the animal kingdoms. It has been demonstrated, for example, that individuals in wild populations are to some extent heterogeneous cytologically and genetically. In some cases, even if the genes are identical they may be ordered in a different way, owing to alterations of the chromosomal segments.

Many plant species originate from an abrupt and rapid change in nature, and aneuploidy and polyploidy are the prime sources of variation. This is not the case in the animal kingdom, where there are two primary mechanisms by which the chromosome number can be decreased and increased during evolution. These two opposite changes in number (and configuration) are (1) *centric fusion*, a process that leads to a decrease in chromosome number, in which two acrocentric chromosomes join together to produce a metacentric chromosome; and (2) *dissociation* or *fission*, a process that leads to an increase in chromosome number, in which a metacentric chromosome (usually large) and a small, supernumerary metacentric fragment become translocated in such a way that two acrocentric or submetacentric chromosomes are produced.

In recent years the karyotypes of the chimpanzee, gorilla, and orangutan have been compared with that of man. It was found that these primates have 48 chromosomes, and attempts were made to correlate each chromosome pair with those of the 46 human chromosomes. Various banding techniques (Chapter 11) have revealed that there is a direct correlation between most human chromosomes and those of the chimpanzee. The difference accounting for 48 versus 46 chromosomes is the fusion of two acrocentric chromosomes in the hominoid apes to make chromosome N°2 of man.

**centric fusion—joining of two acrocentric chromosomes to produce a metacentric chromosome, with a net decrease in chromosome number**

**dissociation—translocation leading to the production of two chromosomes where one existed previously; also fission**

# SUMMARY

### 9-1 MITOSIS

Mitosis is similar in all eukaryotic cells, although it is usually anastral in plant cells and astral in animal cells. During prophase the two chromatids formed during the S phase coil up, becoming shorter and thicker. The nucleolus and nuclear envelope disintegrate, and the nucleoplasm mixes with the cytoplasm. At this time the asters and spindle are beginning to form. During metaphase the spindle microtubules invade the central area of the cell; the chromosomes become attached to them by their kinetochores, and the spindle fibers form the equatorial plate. During anaphase the centromeres and sister chromatids separate and begin to migrate toward the poles. This pulling is thought to involve the shortening of the microtubules attached to the kinetochores. At the beginning of telophase the daughter chromosomes have finished migrating to the poles and have begun to uncoil, at which time each set becomes surrounded by a new nuclear envelope. New nucleoli are formed and cytokinesis takes place, in the process of which the cytoplasmic components and organelles are distributed between the daughter cells.

The centromere, in association with a proteinaceous kinetochore, is located at the primary constriction of each chromosome. Usually there is one kinetochore in each chromatid; it serves as the site of microtubule implantation. Current hypotheses about the role of the mitotic apparatus in chromosome movement are based on the idea that the microtubules can generate some sort of mechanical force. In fact there is an equilibrium of forces that is probably at work during metaphase. When this equilibrium is broken at the beginning of anaphase, there are thought to be two processes occurring: motion of chromosomes toward the pole in each half-spindle and separation of the half-spindles by microtubule elongation. The dynamic equilibrium hypothesis and the sliding hypothesis have both been postulated to explain the mechanisms responsible for chromosome movement at anaphase.

Cytokinesis is usually the final event of cell division. Current hypotheses suggest that cleavage by furrowing results from the action of a contractile ring present in the cell cortex.

### 9-2 MEIOSIS

Meiosis occurs in the cells that produce gametes in sexually reproducing organisms, and is the mechanism that has evolved to ensure that the diploid number of chromosomes is recovered upon fertilization in organisms that have biparental inheritance. At a certain moment of the $G_2$ period there is a change that directs the cell toward meiosis, which involves two cell divisions with only one duplication of the chromosomes.

During preleptonema the chromosomes are very thin and difficult to see; during leptonema they become more visible. At this stage the chromosomes have duplicated and contain two chromatids. During zygonema the homologous chromosomes align and pair in a process called synapsis, which involves the formation of a synaptonemal complex. The SC is considered to be the structural basis of pairing. At pachynema pairing has been completed. Each chromosome is now a bivalent or tetrad composed of two homologs (four chromatids). Pachynema is characterized by the exchange of segments between two chromatids belonging to different homologs. At diplonema the paired chromosomes begin to separate but remain attached at the chiasmata. This phase may last for a very long time before diakinesis, when further contraction of the chromosomes takes place. The phases of division I that follow diakinesis are similar to those of mitosis, resulting in the separation of the homologous chromosomes. By the end of telophase I the number of chromosomes is haploid, but each chromosome still has two sister chromatids. At division II the sister chromatids separate. It has been postulated that the differences in chromosome distribution during mitosis and meiosis are the result of differences in kinetochore orientation.

Several special biochemical events occur in meiotic cells. These include a lengthy premeiotic S phase; the persistence of a small amount of late-replicating DNA, which seems to be essential for synapsis and the maintenance of pairing; and a small amount of DNA synthesis during pachynema, which is probably related to the process of recombination.

There are several essential differences between mitosis and meiosis. (1) Mitosis occurs in somatic cells and meiosis in germinal cells. (2) Mitosis results in diploid daughter cells; meiosis results in haploid daughter cells. (3) Meiosis is characterized by a long S phase and a short or nonexistent $G_2$ phase. (4) In mitosis all chromosomes behave independently; in meiosis the homologous chromosomes are paired. (5) Meiosis generally lasts much longer than mitosis. (6) Genetic variability is one of the characteristic consequences of meiosis. The three general features of meiosis that may have genetic consequences are (1) the pairing of the homologs, (2) the process of crossing over and recombination, and (3) the segregation of the homologous chromosomes.

## 9–3 HEREDITY AND CYTOGENETICS

The laws that govern the transmission of hereditary characters were originally discovered by Gregor Mendel, who studied crosses between peas having pairs of different or contrasting characteristics. His findings, arrived at through careful quantitation and logical reasoning, can now be explained in terms of the behavior of chromosomes and genes. The law of segregation of the genes states that genes are distributed independently, without mixing. The law of independent assortment of genes states that genes located in separate chromosomes are independently distributed during meiosis. As a consequence of these laws, the genotypic and phenotypic traits of the parents reappear in statistically predictable proportions in their offspring.

Independent assortment and free segregation are limited to some extent because there is linkage of genes within a given chromosome. Recombination may rupture the linkage between two genes, and the frequency of recombination of two linked genes is a function of the distance that separates them along the chromosome.

A change in the number of chromosomes in a given karyotype may lead to euploidy or aneuploidy. Such changes may occur either spontaneously or as a result of damage caused by radiation or chemical mutagens. Aneuploidy may result from meiotic nondisjunction, leading to the development of a monosomic or trisomic individual; when nondisjunction occurs during mitosis, only some somatic cells may be aneuploid and the individual will be a mosaic. Changes in chromosome structure can be intra- or interchromosomal, and include deficiencies or deletions (which may be terminal or intercalary); duplications; inversions (pericentric or paracentric); translocations (reciprocal or Robertsonian, for example); the formation of isochromosomes; and sister chromatid exchanges.

Comparative study of the karyotypes of different species has revealed many interesting facts about the evolution of plants and animals. In plant species, aneuploidy and polyploidy are the major sources of variation. In animals, however, centric fusion (leading to a decrease in chromosome number) and dissociation or fission (leading to an increase in chromosome number) are the two primary mechanisms by which the chromosome number can change during evolution.

# STUDY QUESTIONS

- Draw diagrams that illustrate the stages of mitosis and meiosis in a diploid cell having four chromosomes.
- Describe the structure and function of the kinetochore.
- What are the dynamic equilibrium and sliding hypotheses?

- Describe the structure and function of the synaptonemal complex.
- How are mitosis and meiosis alike? How are they different?
- Define the following terms: allele, locus, genotype, phenotype, dominant, recessive.
- Given a $P_1$ cross between red, long-stemmed flowers (genotype *RRLL*) and white, short-stemmed flowers (genotype *rrll*), what would be the genotypic and phenotypic characteristics of the $F_1$ dihybrids? the $F_2$ offspring of an $F_1$ dihybrid cross?
- What are the possible consequences of meiotic chromosomal nondisjunction? Why are they likely to be more serious than the consequences of mitotic nondisjunction?
- Give examples of intra- and interchromosomal aberrations, and suggest ways in which they might occur.
- What conclusions can you draw from the fact that human chromosomes are similar in number and banding pattern to those of the hominoid apes?

# SUGGESTED READINGS

**Darlington, G. D.** (1978) A diagram of evolution. *Nature, 276:*447.

**Fuge, H.** (1974) Ultrastructure and function of the spindle apparatus and chromosomes during nuclear division. *Protoplasma, 82:*289.

**Latt, S. A., and Schreck** (1980) Sister chromatid exchange analysis. *Human Genetics, 32:*297.

**Mazia, D.** (1961) How cells divide. *Sci. Am., 205:*100.

**Mendel, G.** Experiments in plant hybridization. Reprinted in J. A. Peters, ed. (1959) *Classical Papers in Genetics.* Prentice-Hall, Englewood Cliffs, New Jersey.

**Moses, M., Counces, D., and Poulson, D.** (1975) Synaptonemal complex complement of man in spreads of spermatocytes. *Science, 187:*363.

**Myers, R. H.** (1980) The chromosome connection (The theory of evolution may be evolving). *The Sciences, 20:*18.

**Stern, C.** (1973) *Principles of Human Genetics.* Freeman, San Francisco.

**Westergaard, M., and von Wettstein, D.** (1972) The synaptonemal complex. *Ann. Rev. Genet., 6:*74.

**Yunis, J. J., ed.** (1977) *Molecular Structure of Human Chromosomes.* Academic, New York.

# Human Cytogenetics

In the last two decades human cytogenetics has become a specialized science in itself. The purpose of including this subject in an elementary textbook of cell and molecular biology is to illustrate one of the many biological and medical applications of a field of knowledge which in the past was considered largely speculative. Studying both normal and abnormal human karyotypes should give clearer insight into the importance of the information presented in previous chapters about chromosomes and heredity.

## 10-1 THE NORMAL HUMAN KARYOTYPE

In 1956 Tjio and Levan established the fact that 46 was the correct diploid number of chromosomes in the human karyotype. The karyotype is composed of 23 pairs of homologs: 22 pairs of autosomes + XY in males, and 22 pairs + XX in females. Three years later Lejeune et al. made the first discovery of an alteration in the normal karyotype, a *trisomy*, in individuals with *Down's syndrome* (formerly called *mongolism*). This was followed by the recognition of other chromosomal aberrations affecting both the sex chromosomes and the autosomes. It was then possible to explain about twelve congenital human syndromes on a chromosomal basis, and to demonstrate that about 0.5 percent of newborns had some type of chromosomal aberration.

A new advance was made by Caspersson et al. in 1968, with the use of the fluorescent dye *quinacrine mustard* (Fig. 10–1). This development was followed by the application of many other banding techniques which revealed the substructure of chromosomes in considerable detail. These techniques allowed the detection of finer chromosomal aberrations, doubling the number of recognizable congenital diseases in the population, and made it possible to di-

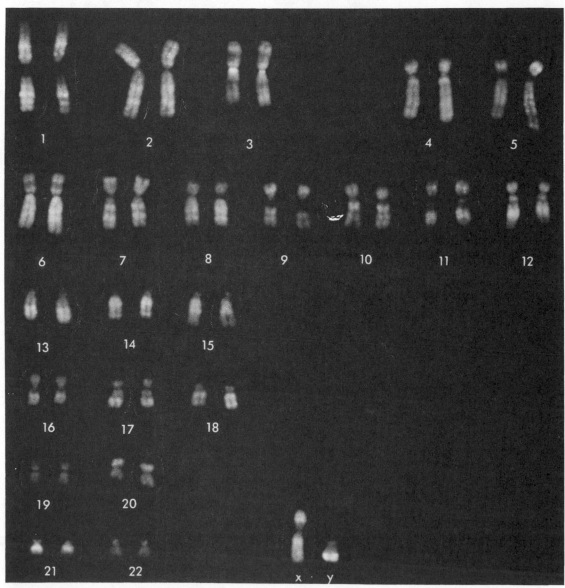

**Figure 10-1**  Human karyotype showing the fluorescent bands (Q bands) produced by staining with quinacrine mustard. (Courtesy of T. Caspersson.)

agnose more than thirty new syndromes, in which almost every chromosome pair was affected. These techniques have also facilitated the genetic mapping of human chromosomes.

## Karyotyping Techniques

Several preparative techniques are used to study the human karyotype. Cultures of fibroblasts, bone marrow, skin, and peripheral blood are used. Mitoses are blocked at metaphase with colchicine, and hypotonic solutions permit better separation of the chromosomes. The karyotype itself is usually obtained from photomicrographs. The individual chromosomes are cut out and then lined up by size with their respective homologs. The technique can be improved by determining the *centromeric index*, which is the ratio of the lengths of the long and short arms of the chromosome. More recently, a system has been introduced that involved a computer-controlled microscope that permits (1) scanning of slides, (2) location of cells in metaphase, (3) counting of chromosomes, and (4) transmission of digitally expressed images for computation and storage.

Table 10–1 and Figure 10–2 show the way in which the chromosome pairs are classified in the human karyotype. The various groups A–G are characterized by size and then by the position of the centromere in metacentric, submetacentric, and acrocentric chromosomes. Observe that five of the pairs (13, 14, 15, 21, and 22) have *satellites* which correspond to the *nucleolar organizers*. The X chromosome is submetacentric and belongs to group C. The Y chromosome belongs to group G of small acrocentrics but has no satellites.

It is possible to obtain cells for karyotyping and diagnosis of chromosomal aberrations even before a child is born. This is done by puncturing the mother's abdomen with a needle and removing a sample of amniotic fluid from the uterus, a procedure called *amni-*

**centromeric index**—ratio of the lengths of the long and short arms of a chromosome

**satellite**—chromosomal region of highly repetitive, constitutive heterochromatin

**amniocentesis**—puncture of the uterine wall with a needle for the purpose of obtaining amniotic fluid, which is analyzed to determine whether the fetus has a genetic abnormality

### TABLE 10-1  CHARACTERISTICS OF THE CHROMOSOMES IN THE HUMAN KARYOTYPE

| Group | Pairs | Description |
|-------|-------|-------------|
| A | 1–3 | Large almost metacentric chromosomes |
| B | 4–5 | Large submetacentric chromosomes |
| C | 6–12 + X | Medium-sized submetacentric chromosomes |
| D | 13–15 | Large acrocentric chromosomes with satellites |
| E | 16–18 | No. 16, metacentric; Nos. 17–18, small submetacentric chromosomes |
| F | 19–20 | Small metacentric chromosomes |
| G | 21–22 + Y | Short acrocentric chromosomes with satellites. (The Y chromosome belongs to this group but has no satellites.) |

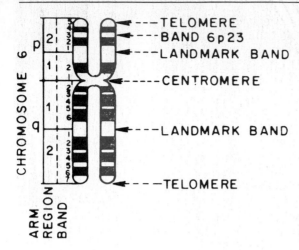

**Figure 10-2** Nomenclature set at the Paris conference (1971) to identify any particular region and band in a chromosome. *p*, short arm; *q*, long arm. Landmark bands divide each arm into regions which are composed of bands. (Courtesy of O. Sanchez and J. J. Yunis. From O. Sanchez and J. J. Yunis, New chromosome techniques and their medical applications. In *New Chromosome Syndromes*, p. 1. Academic, New York, 1977.)

*ocentesis* (Fig. 10–3). The amniotic fluid contains cells from the fetus which can be cultured. This technique is not entirely without risk to the fetus, but is very useful for karyotyping the babies of mothers with a history of chromosomal abnormality, or in whom the probability of conceiving an affected child is high. Amniocentesis also permits the prenatal determination of the fetal sex, which may be of importance in certain sex-linked diseases.

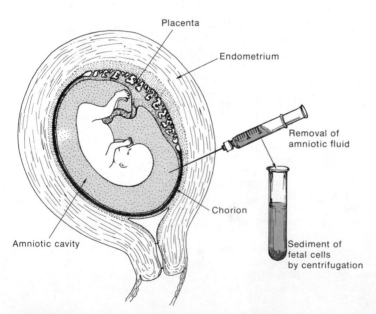

**Figure 10-3** Diagram showing the procedure of amniocentesis, used to remove and culture fetal cells from the amniotic fluid. In the diagram only the wall of the uterus is indicated. To reach the amniotic cavity the needle must first perforate the abdominal wall and then the uterus.

## Banding Techniques

Detailed consideration of the various banding techniques now in use is beyond the scope of this book. We have already mentioned the quinacrine mustard method that stains the so-called Q bands (Fig. 10–1). All the other banding techniques are based on the use of a prestain treatment that partially denatures the DNA, followed by a stain. That used most frequently is *Giemsa* (G banding), a stain used in the past for blood smears.

On the basis of the banding patterns it is possible to identify three major types of chromatin: *centromeric heterochromatin, intercalary heterochromatin,* and *euchromatin.* The staining of chromosome bands provides new morphological features by which to identify a particular chromosome and its finer parts. In addition to the arms, centromere, and telomeres, banding makes it possible to distinguish special landmarks that subdivide the arms into regions. Using nomenclature established at the Paris conference in 1971 (Fig. 10–2), it is now possible to identify the position of any particular band or interband in a given chromosome. The following parameters are used: (1) chromosome number, (2) arm symbol, where p designates the short arm and q the long arm, and (3) region and band number. For example, in Figure 10–2, 6p23 indicates band 3 in region 2 of the short arm of chromosome 6.

As shown in Figure 10–4, with the use of banding techniques some 320 bands per haploid chromosome set can be identified at metaphase (left chromatids). This number can be increased to 1256 if the chromosomes are analyzed at late prophase (right chromatids).

**quinacrine**—dye that combines with AT-rich chromosomal regions to produce specific and reproducible fluorescent banding patterns when viewed under ultraviolet light

**Giemsa**—stain that yields a chromosomal banding pattern similar to that produced by quinacrine

## 10–2 SEX CHROMOSOMES AND SEX DETERMINATION

The well-known fact that male and female individuals are found in approximately equal numbers in the population suggests that sex is determined by a hereditary mechanism. There is physiological and cytological proof that sex is determined as soon as the egg is fertilized, and that it depends on the genetic characteristics of the gametes. Among the physiological evidence is the finding that identical twins, which originate from a single zygote, are always of the same sex. Cytological evidence was first obtained by McClung, who demonstrated that the karyotype of a cell is composed not only of autosomes but also of one or more sex chromosomes that are distinguished from the autosomes by their morphological characteristics and behavior.

Most organisms have a pair of sex chromosomes which, in the course of evolution, have been specialized for sex determination. One of the sexes has a pair of identical sex chromosomes (XX); the other may have a single sex chromosome, which may be unpaired (XO) or paired with a Y chromosome (XY).

Human gametes are not identical with respect to the sex chromosomes. The male is heterozygous (XY) and produces two types of spermatozoa in similar proportions; one type carries an X chro-

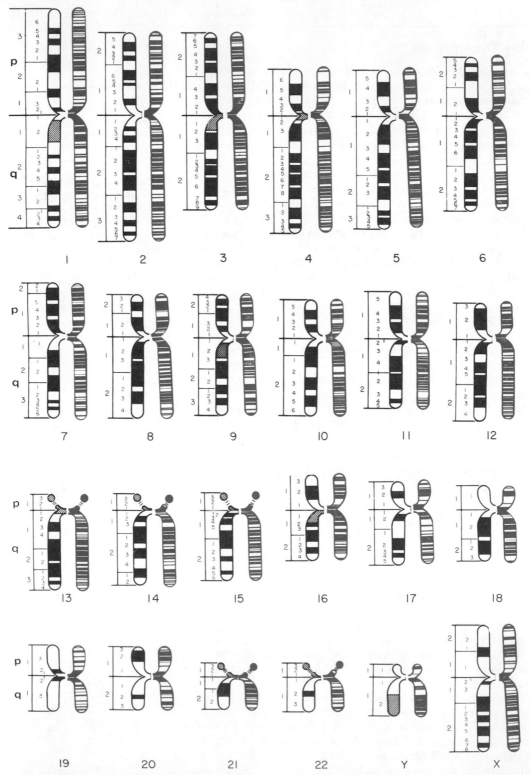

**Figure 10-4** Human karyotype studied with the banding techniques. Chromosomes are disposed following the Paris conference. Left chromatid, in black, represents banding pattern at midmetaphase; right chromatid, in blue, represents G banding pattern at late prophase. (Courtesy of J. J. Yunis.)


276

mosome and one carries a Y. The female, being homozygous, produces only X chromosome-containing gametes. Thus only two combinations of gametes are possible at fertilization, and the result is 50 percent males and 50 percent females.

The human Y chromosome contains genes for an essential testis-determining factor that determines the male sex. The genes involved are situated in a region near the centromere.

We have seen that the X and Y chromosomes differ in size and morphology. In the XY pair there is a *homologous region*, in which recombination may take place during meiosis, and a *differential region* that is unpaired and thus cannot recombine.

## X Chromatin

In 1949 Barr and Bertram made the important discovery that the interphase nucleus of female cells contains a small chromatin body that does not appear in male cells. This was called *sex chromatin* or the *Barr body*, and since the 1971 Paris conference, *X chromatin*.

The X chromatin can be found as a small body in different positions within the nucleus (Fig. 10–5). It can be studied easily by making smears of epithelial cells from the oral mucosa or by studying certain leukocytes in which, in females, it appears as a small rod called the *drumstick* (Fig. 10–5, D).

**X chromatin—an X chromosome randomly selected and inactivated at an early stage of development in the cells of mammalian females**

**Figure 10–5** Sex chromatin in a nerve cell of a female cat. **A,** near the nucleolus; **B,** in the nucleoplasm; **C,** under the nuclear membrane (from M. L. Barr). **D,** normal leukocyte with a drumstick nuclear appendage from a human female. ×1800; **E,** same as **D,** in a male. (Remember that 90 percent of females also lack the drumstick, as in the male.) × 1800. **F,** one sex chromatin corpuscle (*arrow*) in a nucleus from an oral smear. ×2000; **G,** same as **F,** from a male. Notice the lack of sex chromatin. ×1800. **H,** nucleus from the XXX female with two sex chromatin bodies. Vaginal smear. ×2000. **I,** similar, from an XXXX female. The three Barr bodies are indicated by arrows. ×2000. (From M. L. Barr and D. H. Carr. In Hamerton, J. L., ed. *Chromosomes in Medicine*, Medical Advisory Committee of the National Spastics Society in association with Wm. Heinemann, Little Club Clinics in Developmental Medicine, No. 5, 1963.)

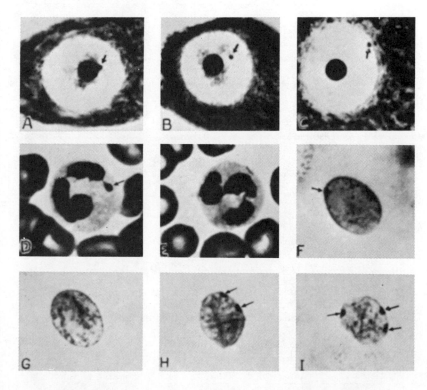

X chromatin is derived from only one of the two X chromosomes present in the cell. The number of corpuscles of X chromatin at interphase is equal to $nX - 1$. This means that there is one Barr body fewer than the number of X chromosomes. This relationship between X chromatin and sex chromosomes is particularly evident in some humans who have an abnormal number of sex chromosomes (see Table 10–2).

We discussed briefly in Chapter 7 that the differential behavior of the two X chromosomes in the female led Lyon to propose the *inactive X hypothesis*. This hypothesis states that (1) only one of the X chromosomes is genetically active; (2) the X that becomes inactive may be of either maternal or paternal origin, and is apparently inactivated at random; and (3) inactivation occurs early in embryonic life and remains fixed. It is now recognized that only a part of the X chromosome condenses into a Barr body; this condensed chromosome is said to contain *facultative heterochromatin*, as opposed to the *constitutive heterochromatin* found in other chromosomes. Inactivation of the X chromosome takes place even in 3X and 4X individuals, in which only one X remains uncondensed (Fig. 10–5, H and I). We have already seen that the inactivated X chromosome(s) remains heterochromatic in the somatic cells, and that during the cell cycle it is characterized by late replication.

inactive X hypothesis—hypothesis that one X chromosome is active in female cells; the other (selected randomly) is permanently inactivated early in embryonic life

## TABLE 10-2  SEX ANEUPLOIDS IN MAN

| Y / X | O | Y | YY | Sex Chromatin |
|---|---|---|---|---|
| **x** | Monosomic XO Turner's syndrome 2X − 1 2n = 45 | Disomic XY Normal 2X 2n = 46 | XYY | 0 |
| **xx** | Disomic XX Normal 2X 2n = 46 | Trisomic XXY Klinefelter's syndrome 2X + 1 2n=47 | Tetrasomic XXYY Klinefelter's syndrome 2X + 2 2n =48 | 1 |
| **xxx** | Trisomic XXX Metafemale 2X + 1 2n = 47 | Tetrasomic XXXY Klinefelter's syndrome 2X + 2 2n = 48 | | 2 |
| **xxxx** | Tetrasomic XXXX Metafemale 2X + 2 2n = 48 | Pentasomic XXXXY Klinefelter's syndrome 2X + 3 2n = 49 | | 3 |
| **Phenotype** | ♀ | ♂ | ♂ | |

# Y Chromatin

Banding studies have shown that a large portion of the Y chromosome is heterochromatic and appears at interphase as a strongly fluorescent body called *Y chromatin*. The number of Y chromatin bodies is identical to the number of Y chromosomes. For example, individuals with XYY have two Y chromatin bodies.

Y chromatin—large, heterochromatic portion of the Y chromosome

The human Y chromosome consists of two segments, one that does not fluoresce and is genetically active, and a second that fluoresces and is genetically inactive. This latter segment may be polymorphic, showing wide variation in size between individuals. Among mammals, only man and the gorilla share this fluorescent region of the Y chromosome, suggesting that it is of recent evolutionary origin. This region corresponds to a highly repetitive (satellite) DNA which is specific to the Y chromosome and probably has no genetic function.

# Sex Determination and Differentiation

Although the primary determination of sex is made at fertilization, the embryo acquires many of its sex characteristics by a more complex mechanism. A hormonal factor may assume control of the genetic determination during development, thereby changing the phenotypic direction of sex at an early developmental stage.

In the human embryo, the gonads and the primordia of the urogenital tract are identical in males and females until the sixth week of development. At this time the gonad has already been invaded by the primary XX or XY germ cells. A gene (or set of genes) present in the Y chromosome then causes the undifferentiated gonad to differentiate into a testis, and the absence of this gene(s) allows the gonad to become an ovary.

**H-Y ANTIGEN.** It is known that the Y chromosome directs the organogenesis of the testis by inducing the production of a protein(s) that becomes localized at the plasma membrane and determines, by cell recognition, the formation of the seminiferous tubule. It has been proposed that a Y-linked determining gene of mammals produces this *H-Y antigen*, which is found in various cell types in males but is absent in females. This antigen is also a histocompatibility factor (H) that occurs, for example, in the skin of male mice and determines the rejection of female skin transplants.

H-Y antigen—cell-surface protein encoded by a Y-linked gene and thought to have a role in sex determination

**ACTION OF TESTOSTERONE.** The male gonad differentiates into a testicle at the seventh week, whereas the female gonad differentiates between the eighth and ninth weeks of development. At the time of differentiation an important hormonal factor is the production of *androgens* by somatic cells in the embryonic male gonad.

In mammals, the administration of *testosterone* (a male hormone) to the mother produces in the fetus a shift in the differentiation of XX genitalia to the male type, producing what is called *masculine pseudohermaphroditism*. This hormone acts locally to accelerate the development of the testis, whereas in the female the absence of the hormone permits the slower development characteristic of the ovary.

## 10-3  ABNORMALITIES IN THE HUMAN KARYOTYPE

reciprocal translocation—
translocation in which small
chromosome segments are
exchanged between
chromosomes of different
pairs

Many of the chromosomal abnormalities discussed in Chapter 9 are found in the karyotypes of individuals who have genetic diseases or syndromes. Changes in the number of chromosomes, such as monosomy or trisomy, or structural aberrations such as translocations, deficiencies, duplications, and so forth have all been observed. One of the aberrations which the banding techniques have been most helpful in detecting is the *reciprocal translocation*, especially of the *balanced* type, in which small segments of chromosomes are exchanged between different pairs. For example, Figure 10-6 shows a

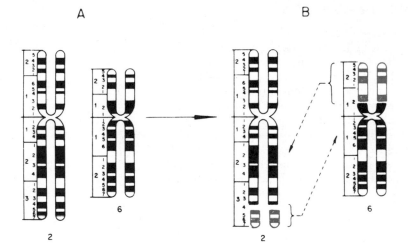

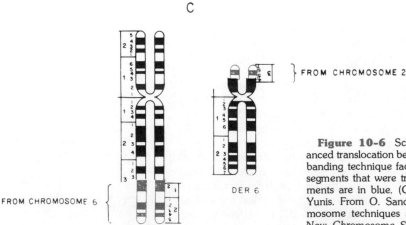

**Figure 10-6** Schematic representation of a balanced translocation between chromosomes 2 and 6. The banding technique facilitates the recognition of the two segments that were translocated. The translocated segments are in blue. (Courtesy of O. Sanchez and J. J. Yunis. From O. Sanchez, and J. J. Yunis, New chromosome techniques and their medical application. In *New Chromosome Syndromes*, p. 1. Academic, New York, 1977.)

balanced translocation between a segment of the long arm of chromosome 2 and a segment of the short arm of chromosome 6 (i.e., a 2/6 translocation). The figure makes it evident that this translocation would be difficult to detect without the use of banding techniques, because the general morphology of both chromosomes has undergone little change.

In aneuploid individuals the genetic message of each chromosome is maintained intact. The alteration is quantitative, residing in the disequilibrium established by an excess (trisomy) or deficiency (monosomy) in the amount of genetic material. In all instances of trisomy the tendency is toward an involution of the nervous system, resulting in a more or less severe mental deficit. Aneuploidy of one of the larger chromosomes is usually lethal, being more severe than an equivalent abnormal dosage of a smaller chromosome. Malformation, mental retardation, sterility, and spontaneous abortion are all selective mechanisms that tend to eliminate from the general population those individuals carrying deleterious genetic imbalances.

We saw in Chapter 9 that aneuploidy originates by *nondisjunction*. The immediate cause of nondisjunction is the lagging, during anaphase, of one sister chromatid, which during telophase remains in one of the cells together with the other sister chromatid. This change gives rise to a cell line that lacks one chromosome or has one chromosome in excess for that pair (Fig. 10–7). Nondisjunction may occur during meiosis, giving rise to an aneuploid ovum. When this ovum is fertilized by a normal spermatozoon, all the cells in the resulting zygote and organism are aneuploid. In some cases the alteration may be in the male gamete, or in both, thus producing more complex types of abnormalities. If nondisjunction occurs during a mitotic division which precedes the formation of germ cells, the effects are similar to those of meiotic nondisjunction. If mitotic nondisjunction occurs during embryonic development, a *mosaic* of different cell lines is produced.

nondisjunction—failure of one or more homologous chromosome pairs to separate properly during cell division

mosaic—tissue or organism whose cells exhibit more than one genotype, for example as a result of mitotic nondisjunction during embryonic development

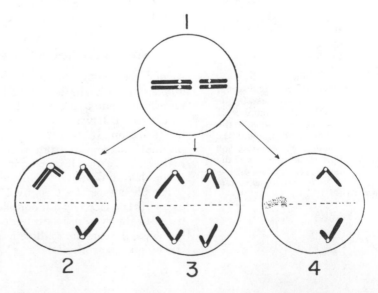

**Figure 10-7** Mitotic nondisjunction and chromosome loss. **1,** normal metaphase. **2,** nondisjunction anaphase giving rise to monosomic and trisomic nuclei, **3,** normal anaphase. **4,** a chromosome loss results in two monosomic nuclei.

## Sex Chromosome Abnormalities

Table 10–2 shows some of the most common sex chromosome abnormalities in which there is an aneuploidy in the XY pair.

**KLINEFELTER'S SYNDROME.**   Most affected individuals are males who show several characteristic phenotypic anomalies. They have small testes, enlarged breasts, a tendency to tallness, obesity, and underdevelopment of secondary sex characteristics. Spermatogenesis does not occur, thereby resulting in complete infertility. These individuals have an X chromatin body and a total of 47 chromosomes (44 autosomes + XXY) in their karyotype.

Males with 48 chromosomes (44 autosomes + XXXY) and two Barr bodies have also been described. They have features of Klinefelter's syndrome and are mentally retarded. Individuals with 49 chromosomes (44 autosomes + XXXXY) have also been reported. They have extensive skeletal anomalies, and are extremely deficient in both sexual and mental development. These persons have three X chromatin bodies.

*Klinefelter's syndrome— genetic disorder characterized by an XXY karyotype, outwardly male phenotype, and various anomalies including sterility*

**XYY "SYNDROME."**   Males having two Y chromosomes have been identified in the past in maximum security institutions. It was proposed that such individuals had a strong tendency toward antisocial behavior and aggression. More recently, XYY individuals have been found in the normal population in 1 out of 650 male infants, suggesting that the correlation with violence may not be as strong or significant as was previously thought.

**TURNER'S SYNDROME.**   Individuals with Turner's syndrome are phenotypically females, usually of short stature. The ovary does not develop, however, and shows a complete absence of germ cells. As a result of this ovarian dysgenesis, menstruation does not occur and secondary sexual characteristics do not develop. The karyotype shows 45 chromosomes (44 autosomes + X), and there is no X chromatin.

*Turner's syndrome—genetic disorder characterized by an XO karyotype, outwardly female phenotype, and various anomalies including sterility and growth retardation*

## Autosomal Abnormalities: Down's Syndrome

Among the most common autosomal aberrations is *Down's syndrome*, which is characterized by multiple malformations, mental retardation, and markedly defective development of the central nervous system. It was discovered that the individual with this syndrome has an extra chromosome: pair 21 is trisomic. This aberration probably originates from the nondisjunction of pair 21 during meiosis. The extra chromosome may in some cases become attached to another autosome by translocation, usually to pair 22.

The phenotype of a trisomy 21 individual is usually recognizable at birth. The face has a moonlike aspect, with increased separation between the eyes and skin folds at their inner corners. The nose is flattened, the ears are malformed, the mouth is constantly open, and the tongue protrudes.

*Down's syndrome— autosomal genetic disorder characterized by the presence of three number 21 chromosomes in the karyotype (trisomy 21), mental retardation, and multiple malformations of the facial features and central nervous system*

Down's syndrome is the most common congenital disease and is present in more than 0.1 percent of births. Its frequency increases as the mother's age exceeds 35. The occurrence of trisomy 21 is sporadic, and there is generally no recurrence in a single family. However, in rare cases when the aberration arises by translocation, the disease may affect siblings and may appear in successive generations. Fortunately this *translocation trisomy* represents only 3 to 4 percent of all cases of the syndrome. In this type there is no increase in frequency with the age of the mother, and when the syndrome is properly diagnosed by karyotype analysis, the parents can be warned of a possible repetition in their family.

**translocation trisomy—rare trisomy arising from translocation rather than nondisjunction**

## Detection of New Syndromes

We mentioned above that the use of banding techniques has permitted the detection of more than 30 chromosomal syndromes which result from a partial trisomy or deletion from each of almost all the chromosomes. Such alterations have been found in children with mental retardation or minor congenital defects, and also in some having normal intelligence and minor anomalies which were not previously suspected of having a chromosomal basis.

Technical advances have permitted the identification of chromosome defects even when there are few clinical symptoms. They have also made it possible to detect balanced translocations (Fig. 10–6), which are important because they may be transmitted to successive generations. In such cases, proper counseling and prenatal diagnosis may prevent the transmission. In Figure 10–8, dotted lines are used to indicate the cases of total or partial trisomy that have been reported in the different chromosome pairs. The cases of total or partial monosomy (deletions) are indicated with solid lines.

Another important advance that has resulted from use of the banding techniques is the demonstration of chromosomal defects in the cells of certain types of tumors. The most characteristic of these is the association of *chronic myelogenous leukemia* with the so-called *Philadelphia chromosome* (Ph 1). This involves a balanced translocation between chromosomes 9 and 22. The Ph 1 chromosome is acquired with the disease and is not found in cultured fibroblasts or in the identical twins of affected individuals. In a *retinoblastoma* (a cancer of the retina) a deletion in the long arm of chromosome 13 has been identified. Cancer cells themselves usually have severe chromosomal abnormalities, such as polyploidy and aneuploidy.

**Philadelphia chromosome—a number 22 chromosome from which the long arm has been deleted and translocated to the end of the long arm of chromosome 9; an aberration associated with the disease chronic myelogenous leukemia**

## 10–4 HUMAN CHROMOSOMES AND THE GENETIC MAP

In recent years important advances have been made in the localization of specific genes to a particular chromosome and even to a chromosomal region (i.e., arm or band). In humans the number of *structural genes*—those that determine the polypeptide sequence of

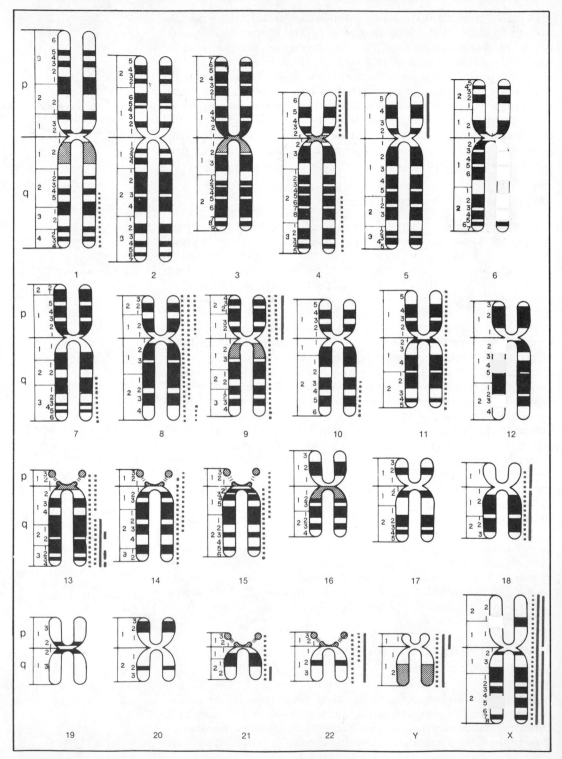

**Figure 10-8**  Human karyotype showing the syndromes of total or partial trisomy of various chromosomes (indicated by dotted lines) and total or partial monosomy (solid lines). (Courtesy of J. J. Yunis.)

the different proteins—could be on the order of 50,000. Before the era of molecular genetics, the presence of a gene was inferred (as in Mendel's experiments) by the existence of two alternative forms of a given trait. More than 1200 genes have been identified by this method and by the occurrence of diseases resulting from mutations.

# Methods of Gene Mapping

**SEX CHROMOSOME LINKAGE.**  As described earlier, genes on the same chromosome behave as though genetically linked. In humans numerous genes have been mapped to the sex chromosomes using standard methods of linkage analysis. The most common approach consists of following a certain variation in a family.

The easiest genes to identify are the *sex-linked genes,* which are present either in the X chromosome or in the nonhomologous region of the Y chromosome. In 1911 Wilson assigned the specific gene for *daltonism,* or red-green color blindness, to the X chromosome. This mutation is expressed in males, 8 percent of whom have daltonism, but it is very rare in females. This is because males have only one X chromosome and therefore only one allele for this gene. As a result, any mutation of this gene will produce a phenotypic change. Females, however, have two copies of the same gene, and in the heterozygous condition color blindness (a recessive trait) is not expressed. Women can nevertheless be genetic *carriers* and can transmit color blindness to their progeny.

*Hemophilia,* a disease involving a defect in blood clotting, is also inherited through an X-linked recessive gene. This disease is transmitted by females but is expressed in males. Figure 10–9 shows the pedigree of the descendents of Queen Victoria and the occurrence of hemophilia among them.

The few cases of daltonism or hemophilia in females are due to a homozygous condition in which both X chromosomes are altered at the same locus. In such cases a female child must receive affected X chromosomes from both parents; the statistical chances of this occurring are very small.

At present about 100 genes have been assigned to the X chromosome. We mentioned earlier that the Y chromosome carries a testis-determining factor that produces the H-Y antigen. A few diseases also follow the male line and pass directly from father to son. There are also other mutations that are partially sex-linked because the genes are localized in the homologous segments of both the X and Y chromosomes.

**SOMATIC CELL AND IN SITU HYBRIDIZATION.**  At present the most widely used technique for gene mapping is *somatic cell hybridization.* The first step of this technique involves the fusion of a cultured human cell with a mouse or hamster cell. Fusion is facilitated by the use of inactivated Sendai virus or by chemical agents. Genetic analysis is possible because the resulting hybrid cells, over subsequent cell divisions, tend to lose the human chro-

sex-linked genes—genes located on the X chromosome or on the nonhomologous portion of the Y chromosome

daltonism—red-green color blindness, a recessive trait known to be X-linked

carrier—female who has an allele for an X-linked trait present in the heterozygous condition; she does not express the trait herself but may transmit it to her offspring

hemophilia—disease involving defective blood clotting, transmitted by an X-linked recessive gene

somatic cell hybridization—technique by which cells from two different species can be united to produce a single cell or heterokaryon

Sendai virus—organism that adheres to the cell surface and facilitates cell fusion

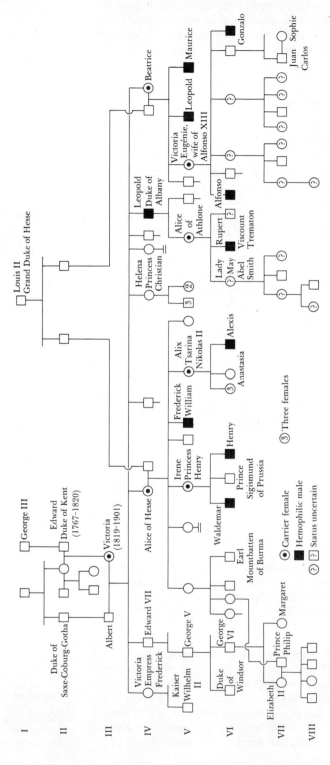

**Figure 10-9** Pedigree of the descendants of Queen Victoria showing carriers and affected males who possessed the X-linked gene conferring the disease hemophilia. (From V. A. McKusick, *Human Genetics.* 2nd ed. Prentice-Hall, Englewood Cliffs, 1969.)

mosomes preferentially. Sometimes the loss of a certain genetic trait can be correlated with the elimination of a particular chromosome, thus allowing the assignment of a gene to the chromosome that was lost. Somatic cell hybridization has so far allowed the mapping of more than 70 genes on autosomes. Furthermore, by using translocated chromosomes, it is even possible to localize genes to a particular region of the chromosome.

Another experimental strategy involves selecting those hybrid cells that retain a specific chromosome. This is accomplished by using mouse cells which are unable to grow in a particular selective medium unless the deficiency is compensated for by a human gene. These mouse cells have an *auxotrophic mutation;* that is, they require a particular nutrient in the medium.

**auxotroph**—strain or organism unable to synthesize a specific nutrient which must therefore be added to the culture medium

The method of *in situ hybridization* makes use of a radioactive complementary RNA to identify a specific chromosomal region. With this method it is easier to localize genes that are present in multiple copies. For example, the genes for the 18S and 28S ribosomal RNA (i.e., the nucleolar organizers) are found in the satellites of chromosomes 13, 14, 15, 21, and 22, and the gene for 5S ribosomal RNA is at the distal end of the long arm of chromosome 1.

**in situ hybridization**—technique in which radioactively labeled RNA is used to identify the specific chromosomal region(s) complementary to it

Structural genes have now been assigned to each one of the 22 autosomes and to the X and Y chromosomes. More than 20 genes have been localized in the large chromosome 1, and about 110 other genes have been assigned to specific autosomes. Among recent localizations are the genes for the ABO blood group, at the distal end of the long arm of chromosome 9; the Rh blood factor, in the short arm of chromosome 1; and the histone genes, in the short arm of chromosome 7. The HLA or haplotype complex, the major histocompatibility system, is present in a segment of chromosome 6 and is inherited in a block. (A defective HLA complex is frequently associated with autoimmune diseases, and careful matching of the HLA is essential for success with many types of human transplants.) Of the 100 genes known to be present in the X chromosome, at least 16 have been localized in a definite arm or region. For example, the loci of the hemophilia and color blindness genes are in the distal portion of the long arm. It is hoped that eventually geneticists will have a bank of cell lines containing each of the human chromosomes, thus making it possible to assign genes to every one.

**HLA complex**—the major histocompatibility system in the human genome, located on the short arm of chromosome 6

# SUMMARY

## 10-1 THE NORMAL HUMAN KARYOTYPE

The normal human karyotype contains 46 chromosomes: 22 pairs of autosomes plus the XX or XY sex chromosomes in females and males, respectively. The karyotype is usually obtained from photomicrographs of late prophase or metaphase cells, from which the individual chromosomes are cut out and then lined up along with their respective homologs. The various human chromosome groups A–G are characterized by size and by the position of the centromere. By a procedure called amniocentesis it is possible to obtain cells for karyotyping and diagnosis of chromosomal aberrations even before a child is born.

Banding techniques have made it possible to distinguish many morphological features by which the chromosomes and their finer parts can be identified.

At the Paris conference in 1971 a system of nomenclature was established that can be used to designate the position of any band or interband in a given chromosome. Two of the most commonly used stains are quinacrine mustard (Q banding) and Giemsa (G banding).

## 10–2  SEX CHROMOSOMES AND SEX DETERMINATION

There is physiological and cytological proof that sex is determined at the moment of fertilization by the sex chromosome constitution of the zygote.

The nuclei of female cells contain a small X chromatin body that is not found in the cells of normal males. The X chromatin is derived from one of the two X chromosomes present in the cell, and represents the X chromosome that was inactivated at an early stage of development. X inactivation takes place even in 3X and 4X individuals, in whose cells only one X chromosome remains uncondensed.

The number of Y chromatin bodies in the nuclei of male cells is equal to the number of Y chromosomes. The Y chromatin body represents only a portion of the chromosome and shows wide variation in size between individuals.

Although sex determination occurs at fertilization, the acquisition of sexual characteristics is a complex process governed to a large extent by hormonal control. The production of H-Y antigen and testosterone induces the differentiation of the testis; without these hormones, in XX fetuses, the developing gonad will become an ovary.

## 10–3  ABNORMALITIES IN THE HUMAN KARYOTYPE

Both numerical and structural abnormalities have been found in the karyotypes of humans who have genetic diseases or syndromes. Numerical abnormalities tend to be the result of nondisjunction, whereas structural aberrations tend to result from various kinds of translocations. The most common sex chromosome abnormalities involve aneuploidy in the XY pair, and include Klinefelter's syndrome (XXY), XYY syndrome, and Turner's syndrome (XO). The most frequent autosomal aberration is Down's syndrome or trisomy 21. In some cases the extra 21 chromosome becomes attached to another autosome (usually a 22 chromosome) by translocation.

Banding techniques have assisted greatly in the discovery and diagnosis of chromosomal aberrations and syndromes. They have made it possible to detect balanced translocations and have been used to demonstrate chromosomal defects in certain types of cancer cells.

## 10–4  HUMAN CHROMOSOMES AND THE GENETIC MAP

In recent years important advances have been made in the localization of specific genes to a particular chromosome and even to a chromosomal region. One method of gene mapping is to study cases of sex chromosome linkage in families whose members carry or manifest genes for such traits as color blindness or hemophilia. Another technique of gene mapping is somatic cell hybridization, in the course of which the loss of a certain genetic trait can be correlated with the elimination of a particular human chromosome. The technique of in situ hybridization makes use of a radioactive complementary RNA to localize a specific chromosomal region. Hundreds of structural genes have been assigned to the 22 autosomes as well as to the X and Y chromosomes by making use of these techniques.

# STUDY QUESTIONS

- If you wanted to examine your own karyotype, what procedures would you need to follow? What would you expect to find?
- What is amniocentesis? Under what circumstances is it performed?
- What does the notation 3q25 refer to?
- How and when is the sex of a human fetus determined?
- What causes some of the common sex chromosome abnormalities? Give examples.
- How could you determine whether an individual with Down's syndrome had a translocation trisomy if nothing was known about his family history?
- If a woman with daltonism and a man with normal color vision have two boys and two girls, what can you predict (in terms of statistical probabilities) about the genotypes and phenotypes of the children?
- What are somatic cell and in situ hybridization?

# SUGGESTED READINGS

**Bühler, E. M.** (1980) A synopsis of the human Y chromosome. *Human Genet., 55:*145.

**Comings, D. E.** (1980) Prenatal diagnosis and the "new genetics." *Human Genet., 32:*453.

**Epstein, C. J., Smith, S., Travis, B., and Tucker, B.** (1978) Both X chromosomes function before visible X chromosome inactivation in female mouse embryos. *Nature, 274:*500.

**Ford, C. E.** (1977) Twenty years of human cytogenetics. *Hereditas, 86:*5.

**Hook, E. B.** (1973) Behavioral implication of the human XYY genotype. *Science, 179:*139.

**Jost, A. D.** (1970) Development of sexual characteristics. *Sci. J., 6:*67.

**Lyon, M. F.** (1972) X-chromosome inactivation and developmental patterns in mammals. *Biol. Rev., 47:*1.

**McKusick, V. A.** (1971) The mapping of human chromosomes. *Sci. Am., 224:*104.

**McKusick, V. A., and Ruddle, F. H.** (1977) The status of the gene map of the human chromosomes. *Science, 390:*405.

**Srivastave, P. K., and Lucas, F. V.** (1976) Evolution of human cytogenetics, an encyclopedic essay. *J. Human Genet., 24:*235 and 337.

**Stern, C.** (1973) *Principles of Human Genetics.* 3rd Ed. Freeman, San Francisco.

# chapter 11

# Molecular Biology of the Gene

<table>
<tr><td>

## 11-1 THE GENETIC CODE

**Three Nucleotides Code for One Amino Acid**
**Synonyms in the Code**
**Initiation and Termination Codons**
**Effect of Mutations on Protein Synthesis**
**Confirmation of the Genetic Code by DNA**
  **Sequencing**
**Overlapping Genes in $\phi$X174**
**The Genetic Code Is Universal, Except for**
  **Mitochondria**

</td><td>

## 11-2 GENETIC ENGINEERING

**Restriction Enzymes**
**Plasmids and Cloning**

## 11-3 TRANSCRIPTION AND PROCESSING OF RNA

**Transcription in Prokaryotes**
**Transcription in Eukaryotes**
**Characteristics of Eukaryotic mRNAs**
**Intervening Sequences in Eukaryotic Genes**

</td></tr>
</table>

DNA carries genetic information in a coded form from cell to cell and from parents to offspring. All the information necessary to produce a new organism is contained in the linear sequence of the four bases, and the faithful replication of this information is ensured by the double-stranded DNA structure in which A pairs only with T, and G only with C.

We have seen that when a gene is *expressed*, its information is first copied into ribonucleic acid (RNA), which in turn directs the synthesis of the ultimate gene products, the specific proteins. We have also seen that in molecular biology *transcription* is used as a synonym for RNA synthesis, and *translation* as a synonym for protein synthesis. In some RNA-containing viruses the RNA can be copied into DNA by the action of the enzyme reverse transcriptase, but the translation of RNA into protein is unidirectional and cannot be reversed.

In this chapter we will analyze the way in which DNA encodes the information for proteins. We will then consider the topics of DNA sequencing and the isolation of eukaryotic genes using "genetic engineering" techniques by which these genes can be grown in *E. coli*. Finally, we will discuss how genes are transcribed, and how RNA is processed within the cell.

## 11-1 THE GENETIC CODE

By 1910, long after Mendel had discovered that genes were inherited as discrete factors, the work of several other investigators suggested that the mechanisms of heredity could be explained in terms of chromosomal behavior and therefore that genes were located in the chromosomes. In 1924 Feulgen developed his now well-known histochemical reaction and showed that chromosomes contain DNA. For many years it was thought that the chromosomal proteins would turn out to be the genetic material, mainly because the DNA composition (four nucleotides) was considered far too simple to code for the great diversity of proteins. Subsequent studies showed that eukaryotic nuclei contain a *constant* amount of DNA and drew attention to the role of DNA in heredity. By then, however, the demonstration that DNA is the genetic material had already been made in experiments with microorganisms.

**transformation—conversion of normal cultured cells into cancerous cells**

In 1928 Griffith found that nonpathogenic strains of *Pneumococcus* would kill mice if injected together with a dead (heated) culture of a pathogenic strain of *Pneumococcus*. A nonpathogenic bacterium had been somehow *transformed* into a virulent strain. In 1944 Avery showed that a preparation of pure *Pneumococcus* DNA had the transformation capacity. This proved that DNA was the genetic material. The virulent strains of *Pneumococcus* have an extracellular polysaccharide capsule which provides resistance to the immunological reaction of the mouse. This capsule is absent in the nonpathogenic strains. In Avery's experiment the gene coding for the enzyme *UDP-glucose dehydrogenase*, required for synthesis of the polysaccharide capsule and contained in the DNA of the virulent strain, was transferred to the DNA of the nonvirulent strains.

## Three Nucleotides Code for One Amino Acid

**codon—triplet of consecutive nucleotides specifying a particular amino acid**

The *codons,* or hereditary units that contain the information coding for one amino acid, consist of three nucleotides (a *triplet*) (Table 11–1). This information resides in the DNA, from which it is transcribed into messenger RNA; the mRNA thus has a sequence of bases complementary to the DNA from which it is copied. DNA and mRNA have only 4 different bases, whereas proteins contain 20 different amino acids. The code is thus read in groups of three bases, three being the minimum number needed to code for 20 amino acids. [The possible permutations of the four bases are $4^3 = 64$. If the genetic code consisted of doublets, the number of codons would be insufficient ($4^2 = 16$), and if groups of four bases were utilized, the possibilities would be many more than necessary ($4^4 = 256$).]

The length of the coding portion of a gene depends on the length of the message to be translated, that is, the number of amino acids in the protein. For example, a sequence of 1500 nucleotides may contain 500 codons that code for a protein having 500 amino acids. The message is read from a fixed starting point signaled by special *initiation codons*, which we will discuss later in this chapter.

The sequence of triplets determines the sequence of amino acids in a protein. Amino acids, however, cannot by themselves recognize a given mRNA triplet; in order to do this, each amino acid is first attached to an adaptor molecule called *transfer* RNA (tRNA). The structure of tRNA is described in detail in Chapter 12; here it is sufficient to note that every tRNA molecule has an *amino acid at-*

**transfer RNA—type of RNA molecule that identifies an amino acid in the cytoplasm and transports it to the ribosome**

**TABLE 11-1 THE GENETIC CODE**

| 1st Base | 2nd Base | | | | | | | | 3rd Base |
|---|---|---|---|---|---|---|---|---|---|
| | U | | C | | A | | G | | |
| U | UUU | Phe | UCU | Ser | UAU | Tyr | UGU | Cys | U |
| | UUC | Phe | UCC | Ser | UAC | Tyr | UGC | Cys | C |
| | UUA | Leu | UCA | Ser | UAA | Termination | UGA | Termination | A |
| | UUG | Leu | UCG | Ser | UAG | Termination | UGG | Trp | G |
| C | CUU | Leu | CCU | Pro | CAU | His | CGU | Arg | U |
| | CUC | Leu | CCC | Pro | CAC | His | CGC | Arg | C |
| | CUA | Leu | CCA | Pro | CAA | Gln | CGA | Arg | A |
| | CUG | Leu | CCG | Pro | CAG | Gln | CGG | Arg | G |
| A | AUU | Ile | ACU | Thr | AAU | Asn | AGU | Ser | U |
| | AUC | Ile | ACC | Thr | AAC | Asn | AGC | Ser | C |
| | AUA | Ile | ACA | Thr | AAA | Lys | AGA | Arg | A |
| | AUG | Met | ACG | Thr | AAG | Lys | AGG | Arg | G |
| | AUG | fMet | | | | | | | |
| G | GUU | Val | GCU | Ala | GAU | Asp | GGU | Gly | U |
| | GUC | Val | GCC | Ala | GAC | Asp | GGC | Gly | C |
| | GUA | Val | GCA | Ala | GAA | Glu | GGA | Gly | A |
| | GUG | Val | GCG | Ala | GAG | Glu | GGG | Gly | G |

*tachment site* and a different site for the recognition of mRNA triplets. The latter site is called an *anticodon* and consists of three nucleotides which can base-pair with the complementary codon of mRNA. The translation of the message into protein occurs on the ribosomes, which ensure the ordered interaction of all the components involved in protein synthesis.

anticodon—triplet of tRNA nucleotides complementary to those of the mRNA codon

The discovery of how cells translate a four-letter code into a series of 20 components was a major scientific achievement. The breakthrough that led to this discovery came in 1961, when Nirenberg and Mattaei discovered that synthetic polyribonucleotides used as artificial mRNAs stimulated the incorporation of amino acids into polypeptides in a cell-free protein-synthesizing system. This system was an extract of *E. coli* which had been broken open and from which the cell walls had been removed by centrifugation. This extract contained the ribosomes, tRNAs, and other factors required for protein synthesis. Radioactive amino acids were incorporated after the addition of ATP, GTP, and mRNAs.

The first RNA used was polyuridylic acid (poly U), and the result was synthesis of polyphenylalanine (a polypeptide made exclusively of phenylalanine). Thus it was deduced that the codon for phenylalanine consisted exclusively of U residues (UUU). Other homopolymers were also tested. Poly A stimulated the uptake of lysine (codon AAA), and poly C that of proline (codon CCC). (Poly G forms complex triple-stranded structures in solution and is therefore unable to function as a synthetic mRNA.)

The base composition of many codons was determined by using polynucleotides consisting of two bases. Thus, a random copolymer of U and G contains eight different triplets (UUU, UUG, UGU, GUU, UGG, GUG, GGU, and GGG) and directs the synthesis of several amino acids, as indicated in Table 11–1. The use of copolymers allowed the determination of the nucleotide *composition* but not the nucleotide *sequence* of the codons. For example, UUG could not be distinguished from UGU or GUU by this method. The determination of codon sequences was later made possible by the use of trinucleotide templates of known base composition.

In 1964 Leder and Nirenberg found that trinucleotides induced the binding of specific *aminoacyl-tRNAs* (AA-tRNAs) to ribosomes. The complexes formed when ribosomes were incubated with the various triplets and $^{14}$C-AA-tRNAs could be detected easily, since they were retained by nitrocellulose filters (unbound AA-tRNA passes through the filter). The short triplets could be made using standard organic chemistry synthesis techniques, and about 50 codons were soon deciphered using the binding assay. For example, UUG induced the binding of leucine-tRNA, UGU bound cysteine-tRNA, and GUU bound valine-tRNA.

aminoacyl-tRNA—tRNA molecule carrying an amino acid

Another method which allowed the determination of codon sequences was developed in the laboratory of Khorana. Using chemical synthetic and enzymatic methods, it was possible to produce long polyribonucleotides with alternating doublets or triplets of known sequence. These synthetic mRNAs were then used to direct protein synthesis in a cell-free system. When an alternating doublet was used, a polypeptide chain consisting of two alternating amino acids was formed. For example, with the alternation of G and U the result was:

GUG UGU GUG UGU GUG
Val Cys Val Cys Val

This showed that GUG and UGU code for valine (Val) and cysteine (Cys) and also confirmed that each codon is a triplet.

When the template consisted of three alternating bases, various polypeptide chains containing only one type of amino acid each were produced. For example, consider the repeating sequence

UUG UUG UUG UUG UUG

Since this message can be read in three "frames," each one out of phase by one base, the three resulting homopolypeptide chains are as follows:

UUG UUG UUG UUG UUG
Leu Leu Leu Leu Leu

U UGU UGU UGU UGU
  Cys Cys Cys Cys

UU GUU GUU GUU GUU
   Val Val Val Val

## Synonyms in the Code

By 1966 all 64 possible codons had been deciphered. As shown in Table 11-1, 61 codons correspond to amino acids and three represent signals for the termination of polypeptide chains. Since there are only 20 amino acids, it is clear that several triplets can code for the same amino acid; that is, some of the triplets are synonyms. Proline, for example, is encoded by CCU, CCA, CCG, and CCC. Note that in most cases the synonymous codons differ only in the base occupying the third position of the triplet and that the first two bases are more inflexible in coding. Consequently, mutations that change the third base frequently go unnoticed *(silent mutations)* since they may not change the amino acid composition of the protein.

DNA sequencing studies have confirmed that all possible codons are utilized in vivo. The fact that all the possible sequences are used probably has selective advantages, since it minimizes the effect of harmful mutations. If any codons had been left without a meaning, mutations giving rise to them would interfere severely with normal protein synthesis.

The use of 61 different codons to code for 20 amino acids posed an additional question: Is each triplet recognized by a special tRNA molecule? It was found that there are fewer than 61 tRNA species and that one tRNA can recognize more than one codon (although always for the same amino acid). This is because the third base of the anticodon (the one that is less important in coding) may have some degree of "wobble," meaning a certain amount of flexibility that allows this base to establish hydrogen bonds with bases other than the normal complementary ones.

degeneracy—property of the genetic code such that several different triplets can specify the same amino acid

silent mutation—alteration in the DNA molecule that does not produce any amino acid change in the protein synthesized

## Initiation and Termination Codons

Figure 11-1 shows that mRNA is translated in the 5' → 3' direction, which is also the direction of protein synthesis. The polypeptide

A (5')    ATGGTTAAC....CTCTAAATGACAAAA....TAT. → (3')

B (3')    TACCAATTG....GAGATTTACTGTTTT....ATA. ← (5')

                                                    ↓ I

C (5')    AUGGUUAAC....CUCUAAAUGACAAAA....UAU. → (3')
           ↑                    ↗    ↖
         (start)          (release) (start)         ↓ II

D (NH₂-end) Met Val Asn....Leu          Met Thr Lys....Tyr. → (COOH-end)
            |                             |
          N-formyl                      N-formyl

**Figure 11-1** Diagram illustrating the transcription and translation steps in the expression of genetic information. **I**, transcription; **II**, translation. **A**, DNA strand 5' → 3'; **B**, DNA strand 3' → 5'; **C**, polycistronic messenger RNA 5' → 3' copied from **B**; **D**, polypeptide chains. The starting and termination codons are underlined. Note that mRNA is translated in the 5' to 3' direction and that proteins are synthesized starting with the amino terminus and ending with the carboxyl end.

initiation signal—the AUG codon that specifies the incorporation of N-formylmethionine at the 5' end of a new protein chain

termination signal—one or more of the nonsense codons UAG, UAA, and UGA that are recognized by protein releasing factors

point mutation—change in a single base of the DNA molecule

missense mutation—change that causes a codon to specify a different amino acid

frameshift mutation—insertion or deletion of a single base(s), causing the message beyond the mutation to be read out of phase

chain termination (nonsense) mutation—conversion of an amino acid-specifying codon to a chain-terminating codon, resulting in premature protein termination

chain is always assembled sequentially, beginning at the end bearing the NH$_2$ terminus. The *initiation signal* for protein synthesis is an AUG codon. This codon has a dual function. When it appears at the beginning of a message, as the *starting codon*, it will code for the incorporation of *N*-formylmethionine; when it appears at any other position it will code for normal methionine. Although natural mRNAs always start translation at AUG, this is not an absolute requirement for artificial mRNAs in in-vitro systems, such as the homopolymers discussed above. (Protein synthesis is discussed in detail in Chapter 12.)

The *termination signal* is provided by the codons UAG, UAA, and UGA (sometimes called *nonsense* codons; see Table 11–1 and Fig. 11–1). When the ribosome reaches a termination codon, the completed polypeptide chain is released. Unlike all other codons, UAG, UAA, and UGA are not recognized by special tRNAs but rather by special proteins, the *releasing factors* of protein synthesis.

## Effect of Mutations on Protein Synthesis

Mutations that change the DNA sequence are reflected in the sequence of the corresponding RNA and protein. Mutations involve several base pairs, as in large deletions of segments of DNA, or may change a single base, resulting in a *point mutation*. The analysis of single base mutations was extremely useful in determining how the code works.

Single base mutations may have a profound effect on protein synthesis. *Missense mutations* change the meaning of a codon from one amino acid to another. For example, a codon that changes from GAA to GUA mediates the incorporation of a valine instead of a glutamic acid residue in hemoglobin, and produces a human genetic disease called *sickle cell anemia*.

*Frameshift mutations* arise from the insertion or deletion of single bases and cause the remainder of the message after the mutation to be read out of phase, producing an incorrect protein. For example, in a hypothetical mRNA the codons would normally be translated as follows:

UAU  CCA  UAU  CCA  UAU
Tyr  Pro  Tyr  Pro  Tyr

The insertion of a single G residue between the third and fourth bases produces a completely different protein from there on:

inserted base
↓
UAU  GCC  AUA  UCC  AUA  U
Tyr  Ala  Ileu  Ser  Ileu

Another type of mutation is the *chain termination* or *nonsense mutation*. As shown in Figure 11–2, these mutations arise when a codon coding for an amino acid changes into a chain termination codon, resulting in the production of a shorter protein. Nonsense

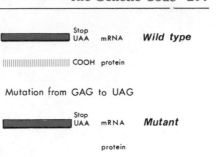

**Figure 11-2** Suppression of a chain termination mutation by a second mutation on a tRNA molecule.

mutations rarely go unnoticed, since the incomplete protein is generally inactive; in the case of missense mutations, however, the change in one amino acid is often compatible with some biological activity.

The phenotypic effects of point mutations can sometimes be reversed by a second mutation in a different gene, which is called a *suppressor mutation*. There are several types of suppressor mutations, but the most interesting are tRNA mutations that act as nonsense (chain termination) suppressors. These mutations substitute a base in the anticodon of a tRNA so that it can incorporate an amino acid at the codon that has been mutated to a chain termination codon. In the example shown in Figure 11-2, a mutation in the tyrosine tRNA anticodon triplet causes it to read a UAG (chain termination) codon and insert tyrosine at its site, thus allowing completion of the polypeptide chain (although with one amino acid change). The tRNA mutations that suppress UAG codons are sometimes called *amber* suppressors for historical reasons (due to the aspect of the plaques of the T$_4$ bacteriophage mutants in which the phenomenon was originally found). UAA suppressors are sometimes called *ochre* suppressors.

A particularly interesting case is that of mutant tRNAs that can suppress frameshift mutations caused by insertions of single bases. Frameshift suppressor tRNAs have an extra base in the anticodon (for example, CCCC instead of CCC) and read a four-letter codon instead of a triplet, thereby restoring the correct reading frame.

**suppressor mutation**—any change that reverses the phenotypic effects of a point mutation, for example by restoring the correct reading frame

## Confirmation of the Genetic Code by DNA Sequencing

The genetic code can also be deciphered by comparing the amino acid sequences of proteins with the nucleotide sequences of the genes that code for them. However, the methods for sequencing nucleic acids had not yet been developed in the 1960s, and, as we have seen, the code was solved using cell-free systems with artificial mRNAs. Methods that allow very rapid sequencing of DNA have recently been developed, and the complete nucleotide sequences of living organisms such as bacteriophage φX174 (a virus that infects *E. coli*) and SV40 (a virus that infects monkey cells and can cause tumors) have been determined. The nucleotide sequences obtained by the new methods confirmed the genetic code and established that all the possible codons are used in vivo. They also provided completely unexpected insights into the way genes are organized.

There are several sequencing methods, all of them based on the generation of DNA fragments of different lengths which start at a fixed point and terminate at specific nucleotides. The DNA fragments are separated by size on *polyacrylamide gels,* and the nucleotide sequence is read directly from the gel. The DNA fragments can be produced by chemical cleavage or by enzymatic copying of single-stranded DNA.

Figure 11–3 outlines the different steps involved in Sanger's chain-termination method. Single-stranded DNA is copied using DNA polymerase I from *E. coli*. This enzyme will not work on single-stranded DNA unless a short primer is annealed to it, producing a stretch of double-stranded DNA. The primer has on its deoxyribose

**chain-termination method—** technique for determining the nucleotide sequence of a DNA molecule; involves the production of DNA chains that start at a unique site and end at specific bases, followed by separation of the chains according to size

(1) Primer (restriction fragment) is annealed to single-stranded DNA template.

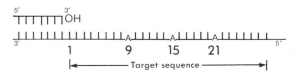

(2) Extension with DNA polymerase and four ³²P nucleoside triphosphates. Termination occurs at specific nucleotides because a small amount of one of the four dideoxynucleotides is added to each reaction mixture. For example, ddTTP stops the copy at T.

(3) Gel

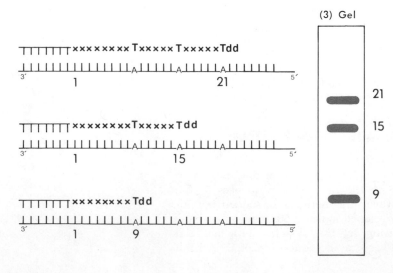

**Figure 11-3** Steps in Sanger's chain-termination sequencing method. (1) Single-stranded DNA is copied by DNA polymerase I from *E. coli*. (2) Extension of DNA copy is terminated, by adding a 2', 3' dideoxynucleotide which lacks the 3' OH group and prevents the attachment of further nucleotides. (3) To read the sequence the labeled DNA is denatured and electrophoresed in a polyacrylamide gel.

a free 3' OH group to which the next nucleotide can be attached. The short primers are generated using restriction endonucleases, which are enzymes that cut DNA at specific sites. The primer is then extended using DNA polymerase and radioactive nucleotides. The chains are terminated at specific bases by adding a 2',3'-dideoxynucleotide. Once a 2',3'-dideoxynucleotide has been incorporated, DNA elongation stops because the nucleotide lacks a 3' OH group

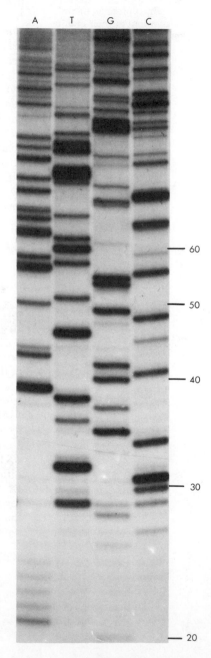

**Figure 11-4** Polyacrylamide gel showing the sequence of a segment of bacteriophage φX174. Each channel represents DNA chains terminated at a specific nucleotide. The sequence can be read clearly (except for weaker G bands at positions 33, 47, 52, and 99). The sequence is:

| | | | |
|---|---|---|---|
| GAAAAAGCGT | CCTGCGTGTA | GCGAACTGCG | ATGGGCATAC |
| 20 | 30 | 40 | 50 |
| TGTAACCATA | AGGCCACGTA | TTTTGCAAGC | TATTTAACTG |
| 60 | 70 | 80 | 90 |

(Courtesy of W. Barnes.)

in the sugar moiety, thus preventing the attachment of the next nucleotide. Four parallel reactions are performed, each containing a small amount of one chain-terminating agent (either ddATP, ddGTP, ddCTP, or ddTTP). These produce DNA chains of various lengths that start at a unique site and end at specific bases.

To read the sequence, the four reactions are electrophoresed on a polyacrylamide gel that separates the fragments according to size. The autoradiograph of one such gel is shown in Figure 11-4. The fragments appear as a series of bands, each differing in length by one nucleotide. Each of the four tracks in the gel indicates chains terminating at one of the four nucleotides. The DNA sequence is read directly from the gel. For example, the band in position 30 in Figure 11-4 is under the track labeled C and therefore represents a cytosine in the DNA sequence. This figure is a good example of a sequencing gel, and the student is advised to read the sequence unassisted and to compare the results with the sequence indicated in the figure legend. The section between positions 30 and 70 can be read rather easily. The number of nucleotides that can be sequenced with these rapid methods is limited only by the resolution of the polyacrylamide gels. In a good experiment, 200–300 nucleotides can be sequenced in only a day's work. These methods, together with new genetic engineering techniques that allow the preparation of large amounts of purified genes, have led to important advances in our understanding of how genes are organized, as exemplified by the work with $\phi$X174.

**polyacrylamide gel—semisolid medium through which proteins migrate according to their charge and/or molecular weight, in response to an electrical current**

## Overlapping Genes in $\phi$X174

**$\phi$X174—small virus that infects bacterial cells and contains a circular molecule of single-stranded DNA**

$\phi$X174 is a small virus having a circular chromosome of single-stranded DNA. After it infects *E. coli*, the complementary strand is synthesized and the double-stranded chromosome then replicates to produce more phage progeny. The sequence of the 5375 nucleotides of this virus, established by Sanger and his colleagues, unexpectedly revealed that sometimes genes can overlap, producing two proteins from the same stretch of DNA.

$\phi$X174 codes for ten proteins, and it was estimated from the molecular weight of these proteins that more than 6000 nucleotides would be required to code for them. However, three of the genes of $\phi$X174 overlap with other genes. In these regions, two proteins with different amino acid sequences are coded for by the same stretch of DNA. This is possible because the proteins are encoded in different reading frames. The most impressive case of coding economy is found in gene K, which overlaps with genes A and C. In this region, four nucleotides are read in the three possible translational frames.

Gene *C*            fMet     Arg

Gene *A*            Lys    Term.

A–A–A–T–G–A–G–G–A

Gene *K*            Asn     Glu

A sequence of 140 base pairs in which two protein-coding genes overlap one nucleotide out of phase has also been found in simian virus 40 (SV40). It is possible that overlapping genes are peculiar to small viruses which can only store a limited amount of DNA within the viral capsid and are therefore under strong selective pressure to use their DNA as efficiently as possible. Cells from most higher organisms contain large amounts of DNA and are probably not under the same constraints as viruses. Whether overlapping genes exist in animal cells will be known only when more gene sequences become available.

SV40—virus that infects monkey cells and contains a circular DNA molecule

## The Genetic Code Is Universal, Except for Mitochondria

Although most of our knowledge about the genetic code comes from experiments with *E. coli* and other microorganisms, essentially similar results have been obtained with other systems, such as amphibian, mammalian, and plant tissue. It may be said that the genetic code is universal, i.e., that there is a single code for all living organisms.

The genetic code developed at the same time as the first bacteria, some three billion years ago, and it has changed very little throughout the evolution of animal and plant species. Presumably once the initial code evolved there were strong selective pressures to maintain it invariant, for the change in a single codon assignment would disrupt too many preexisting proteins and would therefore be lethal.

One important exception to the universality of the code, however, was found in 1979, when the sequence of mitochondrial DNA was compared with the amino acid sequence of certain mitochondrial proteins. The complete sequence of human (16,569 base pairs) and bovine (16,346 base pairs) mitochondrial DNA has been determined by Barrell, Sanger, and their colleagues. These DNAs display a high coding economy without spacer DNA between adjacent genes, but their most surprising feature was that mitochondria have a slightly different genetic code. The most important difference is that in mitochondria the codon UGA codes for tryptophan instead of a termination signal. Other differences are that *AUA* codes for methionine (instead of isoleucine) and that AGA and AGG (which normally code for arginine) are perhaps used as termination codons. This different code in mitochondrial DNA may be a possible cause of why the information coded in mitochondrial DNA has never become incorporated into the nuclear genes: Once a different code was established, the information cannot be translated correctly by the cellular protein synthesis machinery.

## 11–2  GENETIC ENGINEERING

The studies on φX174 and SV40 were possible because their chromosomes are small and can be obtained in pure form. A eukaryotic chromosome, however, contains a DNA molecule that is very large and consequently much more difficult to study. This problem has

genetic engineering—array of techniques that facilitate the manipulation and duplication of pieces of DNA for industrial, medical, and research purposes

now been overcome by the development of techniques (collectively known as *genetic engineering*) which allow short segments of eukaryotic DNA to be excised and ligated into small bacterial chromosomes called *plasmids*. The DNA can then be duplicated in large amounts in *E. coli*. Although genetic engineering is primarily a technological achievement, we discuss it here because it is now very widely used and will have practical applications in industry and medicine in the large-scale production of useful proteins.

## Restriction Enzymes

restriction endonuclease— any of several enzymes capable of recognizing and cutting a specific symmetrical nucleotide sequence in DNA

Genetic engineering is based on the use of *restriction endonucleases*, which are enzymes that recognize and cut specific nucleotide sequences in the DNA. These enzymes provide a precise molecular scalpel and have many uses in DNA research.

The different restriction endonucleases are named according to the microorganisms from which they are isolated (see Table 11-2). For example, Eco RI is the enzyme isolated from *E. coli* containing the drug-resistance plasmid RI. Restriction enzymes usually recognize nucleotide sequences that are four or six nucleotides long. Those that recognize four nucleotides produce shorter fragments (an average of 250 bases long) than those that recognize six nucleotides (an average of 4000 bases long). An important feature of these recognition sequences is that they are symmetrical. For example, the following recognition sequence has an axis of symmetry from which the nucleotides read identically on both strands in the 5′ to 3′ direction:

$$5' \ \ G \ \ A \ \ A \ | \ T \ \ T \ \ C \ \ 3'$$
$$3' \ \ C \ \ T \ \ T \ | \ A \ \ A \ \ G \ \ 5'$$

Some enzymes cut in the middle of the recognition sequence, producing blunt-ended fragments (see the case of Hae III in Table 11-2). Many other enzymes produce breaks in the two strands that are separated by several nucleotides. This results in single-stranded segments at the ends of the cleaved DNA that can base-pair between themselves. These segments are called *sticky ends* and can anneal with any DNA fragments cut by the same enzyme (Table 11-2).

sticky ends—short single-stranded DNA segments, produced by restriction enzyme cleavage, that can base-pair between themselves

Most bacteria contain restriction endonucleases whose function is to provide protection against foreign DNA. If foreign DNA (for example, from a bacteriophage) invades the cells, it will be cleaved by the restriction enzymes. Bacteria protect their own DNA from cleavage by having *modification enzymes* which methylate DNA in the same DNA sequences recognized by their particular restriction enzyme, thus rendering it resistant to cleavage.

## Plasmids and Cloning

plasmid—small piece of autonomous, extrachromosomal circular DNA found in some bacterial cells

Bacteria sometimes contain *plasmids* (small circles of DNA that replicate autonomously) in addition to their chromosomes. Some of the best known plasmids are the resistance factors (R factors) which carry the genes for resistance to various antibiotics. In 1973 it was

**TABLE 11-2 RECOGNITION SEQUENCES OF SOME RESTRICTION ENDONUCLEASES**

| Name | Recognition Sequence | Ends After Cleavage* | | Source |
|------|----------------------|------------|---|--------|
| Eco RI | ↓<br>—G A A T T C—<br>—C T T A A G—<br>↑ | —G<br>—C T T A A | A A T T C—<br>G— | E. coli containing drug-resistant plasmid RI |
| Hind III | ↓<br>—A A G C T T—<br>—T T C G A A—<br>↑ | —A<br>—T T C G A | A G C T T—<br>A— | Hemophilus influenzae, serotype D |
| Bam I | ↓<br>—G G A T C C—<br>—C C T A G G—<br>↑ | —G<br>—C C T A G | G A T C C—<br>G— | Bacillus amyloliquefaciens |
| Hae III | ↓<br>—G G C C—<br>—C C G G—<br>↑ | —G G<br>—C C | C C—<br>G G— | Hemophilus aegyptius |

*Note that some sequences produce "sticky" ends.

shown that fragments of foreign DNA could be introduced in vitro into plasmids with the help of restriction endonucleases, and that these recombinant molecules could replicate when introduced into *E. coli.*

Figure 11-5 shows the different steps involved in the production of a recombinant DNA. The circular plasmid is cut open at a single site with a restriction enzyme. Eukaryotic DNA is also cleaved with the same enzyme, which generates sticky ends. The two DNAs are allowed to reanneal to each other at their sticky ends and are then joined together by the enzyme DNA ligase. The recombinant plasmids are next introduced into *E. coli* made permeable to DNA by incubation in calcium chloride. The cells that have taken up a plasmid molecule are then selected by the use of antibiotics. For example, if the plasmid carries the gene for tetracycline resistance, those cells that do not have such a plasmid can be killed by adding tetracycline to the culture medium. Once an *E. coli* colony is obtained, millions of copies of the eukaryotic DNA segment can be grown. Since all these copies are derived from an *individual* hybrid molecule, this procedure is also called *gene cloning.*

Genetic engineering has already made possible the use of bacteria to produce proteins of medical importance, such as human insulin, interferon, and growth hormone. A future prospect is to introduce nitrogen fixation genes into bacteria that live in the roots of non-leguminous plants to eliminate the need for nitrogen fertilizers. Genetic engineering will also be invaluable in learning how genes are controlled in eukaryotic organisms. Genes cannot only be grown in

cloning—asexual production of a line of cells, organisms, or DNA segments genetically identical to the original

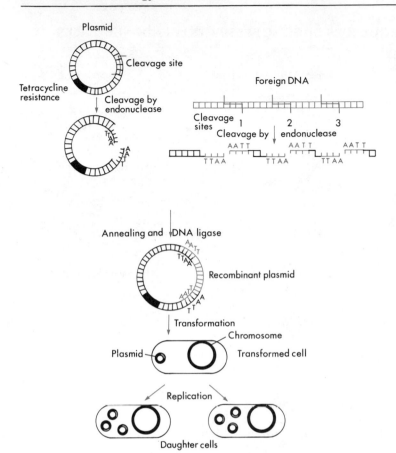

**Figure 11-5** Production of a recombinant DNA molecule by genetic engineering (Courtesy of S. Cohen.)

large quantities and sequenced, but their *expression* can also be studied after their reintroduction into eukaryotic cells, thus allowing a new type of genetics.

In its usual form a genetic experiment begins with the observation of mutant individuals, which in eukaryotic organisms are usually recognized by some phenotypic abnormality. Using the in vitro manipulations that have recently become possible, mutations can be introduced into eukaryotic genes that have been cloned in plasmids (Fig. 11-6). For example, a number of nucleotides can be deleted by treating the DNA with a restriction endonuclease, or a particular change can be induced in a single nucleotide by a chemical mutagen. The mutation can be characterized precisely by the new methods for determining DNA sequences. Then the biological effect of the mutation can be tested by injecting the mutated DNA into oocytes and observing its function. One can determine, for example, which regions of the DNA contain the initiation or termination signals for RNA synthesis (Fig. 11-6). It is now clear that this new kind of genetics, in which purified genes are mutated in a chemically defined way and subsequently tested for biological activity, will considerably widen the possibilities for future experiments.

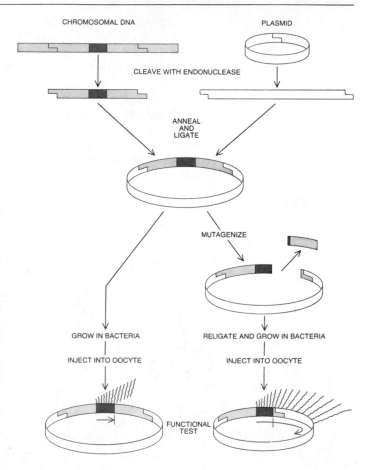

CHROMOSOMAL DNA

PLASMID

CLEAVE WITH ENDONUCLEASE

ANNEAL
AND
LIGATE

MUTAGENIZE

GROW IN BACTERIA

RELIGATE AND GROW IN BACTERIA

INJECT INTO OOCYTE

INJECT INTO OOCYTE

FUNCTIONAL
TEST

**Figure 11-6**   Gene cloning and microinjection can be combined in a new kind of genetic experiment. Chromosomal DNA bearing a gene under study is recombined with a bacterial plasmid by recombinant-DNA techniques: the DNA's are cleaved with an endonuclease (a restriction enzyme) that provides "sticky ends," so that the gene can be inserted into the plasmid. Some of the recombinant plasmids are then mutagenized by deleting a sequence of nucleotides at the end of the gene under study; the precise extent of the deletion can be determined by nucleotide-sequencing methods. The plasmids are cloned in bacteria to provide large amounts of the original gene (*left*) and the mutated gene (*right*), which are injected into oocytes for a functional test. In this case the original gene is transcribed into RNA molecules (*color*) of the proper size. The mutated gene is not: transcription continues beyond the end of gene into plasmid DNA. Presumably deletion eliminated the termination signal recognized by RNA polymerase.

## 11-3 TRANSCRIPTION AND PROCESSING OF RNA

We have seen that the genetic information contained in DNA cannot act directly as a template for protein synthesis but must first be transcribed into messenger RNA. In this section we will analyze the transcription process.

RNA is synthesized by the enzyme *RNA polymerase*. This is an important step in gene regulation since the rate of expression of a particular gene is often controlled by the frequency with which RNA polymerase starts the transcription of that gene. The transcripts synthesized by RNA polymerase, however, are frequently not the final products used by the cell. To become functionally competent, the *primary transcripts* must be modified by a series of chemical alterations known as *processing*. There are four main types of modifications that can be made to the RNA: (1) *cleavage of large precursor RNAs* into smaller RNAs, involving the removal of extra segments from the beginning or end of the precursor; (2) *splicing*, by which the *intervening sequences* (extra segments of nucleotides inserted within the genes themselves) are excised and the molecule is reli-

**RNA polymerase—enzymes involved in the synthesis of RNA from DNA**

**primary transcript—initial, unmodified RNA molecule**

**processing—general term for the various chemical modifications that make the primary eukaryotic RNA transcript functional**

gated; (3) *terminal addition of nucleotides,* involving the addition of poly A residues to the 3′ end and "cap" nucleotides to the 5′ end of eukaryotic mRNAs; and (4) *nucleoside modifications,* such as the methylations that are common in transfer and ribosomal RNAs.

## Transcription in Prokaryotes

Messenger RNAs code for proteins. The average life span of mRNAs in *E. coli* is about two minutes, after which the molecules are broken

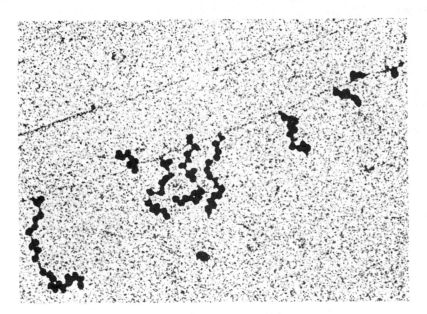

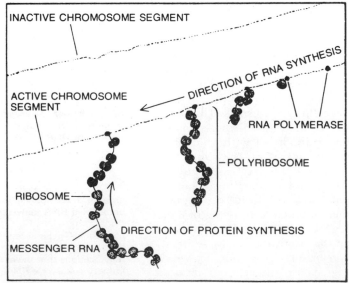

**Figure 11-7** Electron micrograph of the transcription and translation processes in bacteria. Two stretches of DNA—one naked, the other with nascent messenger RNA arranged at right angles—are visible. The bottom diagram facilitates the interpretation of the electron micrograph. This illustration shows clearly that in bacteria, transcription (RNA synthesis) and translation (protein synthesis) are coupled events. In other words, soon after mRNA synthesis begins, ribosomes become attached to the mRNA and initiate protein synthesis. (Courtesy of O. L. Miller, Jr.)

INACTIVE CHROMOSOME SEGMENT

ACTIVE CHROMOSOME SEGMENT

DIRECTION OF RNA SYNTHESIS

RNA POLYMERASE

POLYRIBOSOME

RIBOSOME

DIRECTION OF PROTEIN SYNTHESIS

MESSENGER RNA

down by ribonucleases. Since prokaryotes do not have a nuclear membrane, the ribosomes are free to attach to the mRNA molecules as they are being synthesized, as can be visualized in Figure 11–7. In other words, prokaryotic mRNAs are translated into protein at one end while the other end is still being transcribed. The half-life of these mRNAs is so brief that sometimes degradation starts even before synthesis is completed. In bacteria, therefore, any change in the rate of mRNA synthesis will be followed a few minutes later by a change in the synthesis of the corresponding protein.

The length of mRNA is heterogeneous, reflecting the length of the polypeptide chain for which it will code. The average protein length is between 300 and 500 amino acids, which are encoded by between 900 and 1500 nucleotides. In bacteria, however, mRNAs frequently code for several proteins, and in these cases (called *polycistronic mRNAs*) the molecules are very long. For example, the five enzymes involved in the metabolism of tryptophan are all encoded by a single mRNA molecule.

**polycistronic mRNA—mRNA molecule that codes for more than one protein**

In prokaryotes all types of RNAs are transcribed by the same enzyme. Using DNA as the template, this RNA polymerase catalyzes the formation of RNA from the four ribonucleoside triphosphates. The enzyme copies the base sequence of one of the DNA strands according to the Watson-Crick base-pairing rules. Since only one DNA strand is used as a template, transcription is said to be *asymmetrical*. The polymerization reaction is as follows:

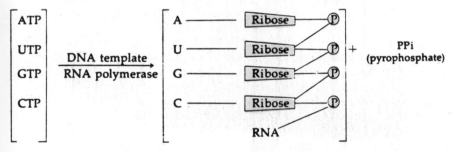

In addition to copying the nucleotide sequence of the DNA template precisely, RNA polymerase is able to recognize a variety of genetic signals on the chromosome, such as the signals for starting and stopping RNA synthesis at precise sites.

There are an estimated 3000 genes in *E. coli*, and they are all transcribed by the same enzyme. Considering these multiple functions, it is not surprising that RNA polymerase is a large and complex enzyme. A complete molecule or *holoenzyme* consists of several polypeptides: two $\alpha$ of 40,000 daltons, one $\beta$ of 155,000, one $\beta'$ of 165,000, and one $\sigma$ of 90,000. The total molecular weight of the enzyme is therefore 490,000 daltons. The sigma ($\sigma$) factor is loosely bound to the rest of the enzyme and can be separated by physical means. The $\beta\beta'\alpha_2$ enzyme (without sigma) is called the *core polymerase*. The sigma factor is required for the enzyme to recognize the correct start signals on the DNA. Figure 11–8 shows that as soon as the RNA chain is started, the sigma factor is released from the core

**core enzyme—unit consisting of four polypeptides (two $\alpha$, $\beta$, and $\beta'$) of prokaryotic RNA polymerase**

**sigma factor—dissociable component of prokaryotic RNA polymerase; required for the enzyme to recognize the correct transcription initiation site**

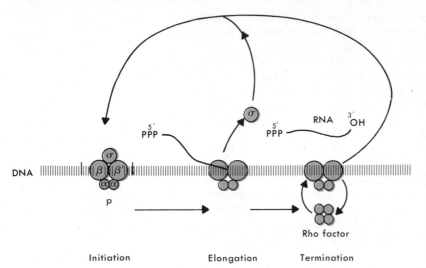

Initiation       Elongation       Termination

**Figure 11-8** Diagram showing the three stages of transcription by RNA polymerase. At *initiation* the enzyme composed of subunits $\beta$, $\beta'$, $2\alpha$ and $\sigma$ is at the site of the promoter ($p$). During *elongation* the enzyme transcribes the gene and subunit $\sigma$ is released. *Termination* occurs at the termination site, and the termination factor *rho* causes the release of the transcribed RNA molecule.

enzyme and can be used in the transcription of other RNA molecules.

The following three stages are commonly distinguished in the transcription by RNA polymerase: (1) binding to promoters and chain initiation, (2) elongation, and (3) termination (see Fig. 11-8).

The start signals on DNA are called *promoters* and represent the initial binding site for the RNA polymerase. This binding is a crucial step in the regulation of gene expression. All prokaryotic promoters have the common sequence TATAATG (or slight variations of it) located some 10 nucleotides before the end of the mRNA. This AT-rich region probably favors local separation of the DNA strands, a step required for RNA polymerase to gain access to the DNA bases. In addition to the TATAATG sequence, the enzyme also recognizes a DNA region located about 35 bases before the start of the mRNA (Fig. 13-3). Single base changes in this region can inactivate promoters, but different promoters have no obvious common sequences in this region. Individual variations between promoters are not surprising, since different genes are expressed with varying efficiencies; that is, some promoters are "stronger" than others. *E. coli* has proteins called *repressors* which can bind to specific sequences of DNA (called *operators*), turning off the expression of a gene or set of genes. Repressors work by binding to a DNA site that overlaps with the promoter, thereby preventing the binding of RNA polymerase (Fig. 13-2).

During elongation, RNA polymerase copies the DNA sequence accurately, progressing at a rate of about 30 nucleotides per second. Only one strand of the DNA is transcribed. Elongation of the RNA proceeds only in the 5' to 3' direction, which as we have seen is the same direction used in DNA replication and protein synthesis.

Termination of transcription occurs when the enzyme arrives at a stop signal in the DNA. A *termination factor* or *rho factor* (a protein with ATPase activity) causes the release of the completed RNA molecules. The termination signal seems to be a GC-rich region followed by a row of four to eight consecutive U residues. Gene expression in

promoter—segment of DNA that represents the initial RNA polymerase binding site

rho factor—protein with ATPase activity that causes the release of completed RNA molecules

**Figure 11-9** Eukaryotic RNA is synthesized in the nucleus and then exported to the cytoplasm. **A,** autoradiograph of a *Tetrahymena* cell incubated in ³H-cytidine for 1.5 to 12 minutes. Notice that all labeled RNA is restricted to the nucleus. **B,** the same as **A,** after 35 minutes. RNA begins to enter the cytoplasm. **C,** the same, incubated for 12 minutes in ³H-cytidine and then for 88 minutes in a nonradioactive medium (a "chase"). Notice that while the nucleus has lost all labeled RNA, the cytoplasm is heavily labeled. (Courtesy of D. M. Prescott, from *Reproduction of Eukaryotic Cells.* Academic, New York, 1976.)

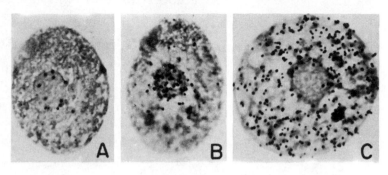

*E. coli* can also be regulated at the level of transcription termination (Chapter 13).

## Transcription in Eukaryotes

Eukaryotic transcription differs in several respects from that of prokaryotes. Figure 11–9 illustrates one of the most obvious differences, which is that in higher cells RNA is synthesized within the nucleus. The nuclear membrane introduces a barrier between transcription and translation, and before the mRNA can reach the ribosomes it must be exported to the cytoplasm. Because of the time factor involved in transport, it is not surprising that eukaryotic mRNAs are more stable than those found in prokaryotes, with half-lives that can be measured in hours or days instead of minutes. In addition, these mRNAs are generally monocistronic, and they often have special types of post-transcriptional modifications, some of which will be described below.

We have seen that in *E. coli* all the genes of the chromosome are transcribed by the same enzyme. In higher organisms three different nuclear RNA polymerases have been identified, each one of which translates different classes of genes. Figure 11–10 shows how the three enzymes can be separated from each other by ion-exchange chromatography. Nuclear extracts of all eukaryotic cells show these three peaks of activity. The enzymes have been classified as I, II,

**Figure 11-10** Separation of RNA polymerases I, II, and III by chromatography in DEAE-Sephadex. The peaks are eluted by increasing the salt concentration. *Solid line,* without α-amanitin; *dotted line,* assays performed with 0.1 μg/ml α-amanitin showing selective inhibition of polymerase II.

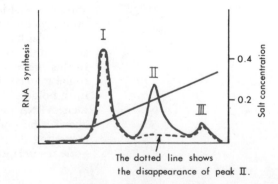

The dotted line shows
the disappearance of peak II.

**TABLE 11–3 PROPERTIES AND FUNCTIONS OF EUKARYOTIC RNA POLYMERASES**

| Enzyme | Localization | Gene Transcripts | Inhibition by α-Amanitin |
|--------|-------------|------------------|--------------------------|
| I | Nucleolus | 18S and 28S rRNAs | Insensitive |
| II | Nucleoplasm | mRNA | Sensitive to low concentration |
| III | Nucleoplasm | tRNA, 5S RNA | Sensitive to high concentration |

α-amanitin—toxin that inhibits the eukaryotic RNA polymerases to different extents and thus is useful for characterizing them

and III according to their order of elution from the column. Another useful criterion for distinguishing the various polymerases is their sensitivity to inhibition by *α-amanitin*, the toxin of the poisonous mushroom *Amanita phalloides*. This toxin is of medical interest because it is the main cause of lethal intoxication by mushrooms in humans.

Table 11–3 summarizes the properties of the nuclear RNA polymerases. RNA polymerase I is not affected by α-amanitin and is localized in the nucleolus. This enzyme transcribes the large ribosomal RNAs (18S and 28S). RNA polymerase II is completely inhibited by low concentrations of α-amanitin and is responsible for the synthesis of mRNA. RNA polymerase III is inhibited by high concentrations of α-amanitin and synthesizes tRNA and 5S RNA. The eukaryotic RNA polymerases are very complex enzymes, each one having between six and ten protein subunits. The protein subunits of each enzyme type are different, except for two small polypeptides of 29,000 and 19,000 daltons that are common to all three RNA polymerases. The different forms of RNA polymerase thus differ in their molecular structure and possibly in their mechanism(s) of regulation.

The DNA in mitochondria and chloroplasts is transcribed by yet another enzyme. In contrast to the nuclear enzymes, mitochondrial RNA polymerase is a very simple enzyme formed by a single peptide. It is not inhibited by α-amanitin but rather by *rifampicin*, an antibiotic that inhibits prokaryotic RNA polymerase. This is another property of mitochondria that suggests their origin was prokaryotic.

rifampicin—antibiotic that inhibits mitochondrial and prokaryotic RNA polymerases

O. Miller has devised a remarkable technique by which transcription complexes can be visualized directly in the electron microscope. In this method, isolated nuclei are lysed gently, and the chromosomal material is centrifuged on top of an electron microscope specimen holder. Transcribing genes are spread together with their attached RNA polymerases and growing RNA chains. Figure 11–11 shows ribosomal RNA genes during transcription, obtained by spreading nucleoli of frog cells. Several conclusions about eukaryotic transcription can be drawn from this type of study:

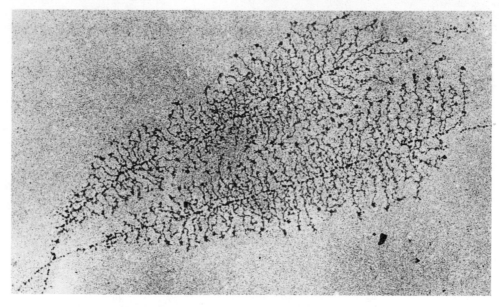

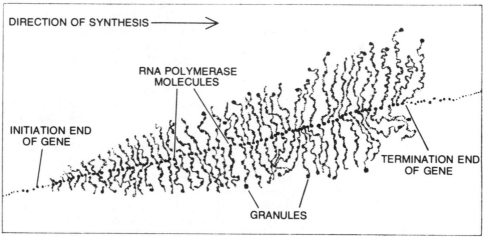

**Figure 11-11    Above,** electron micrograph showing two nucleolar genes in the process of transcription of the ribosomal RNA; **below,** labels are self-explanatory. ×35,000. (Courtesy of O. L. Miller and B. R. Beatty.)

(1) The template for transcription is a complex of DNA and protein that has the characteristic beaded appearance of chromatin nucleosomes (see Fig. 8–12).

(2) Eukaryotic RNA synthesis starts and ends at precise sites on the DNA (the promoter and terminator sequences). Between these two sites the RNA chains gradually increase in length. As a result, genes that are very actively transcribed show the so-called "fern leaf" or "Christmas tree" configuration (shown in Fig. 11–11 in rRNA genes and in Fig. 13–11 in lampbrush chromosomes).

(3) At any one time, only a very small fraction of the chromatin is being transcribed.

(4) In fully transcribing genes the RNA polymerase molecules are packed very close together, with one polymerase more often than every 200 base pairs of DNA. Since the elongation rate of RNA synthesis is 20 to 30 nucleotides per second, it follows that a maximally expressed gene can complete one RNA chain at least every 10 seconds.

(5) The growing or *nascent* RNA becomes associated with protein as it is being transcribed, producing ribonucleoprotein particles (RNP) rather than free RNA. Processing of the initial transcripts into smaller RNAs can also start before the RNA chains are completed.

(6) In eukaryotes the nuclear envelope introduces a barrier between transcription and protein synthesis. The mRNA must be transported into the cytoplasm before it can be utilized.

## Characteristics of Eukaryotic mRNAs

One of the main features of prokaryotic mRNA is its very short half-life, on the order of minutes. In contrast, metabolically stable mRNAs are found in eukaryotes. The best example of a long-lived mRNA is that in the mammalian reticulocyte, an immature red blood cell that has lost its nucleus but still retains the ribosomes and other components of the protein-synthesizing machinery. Reticulocytes contain mRNA that continues to produce hemoglobin for many hours and even days after the nucleus (and therefore the globin genes) has been lost.

Several eukaryotic mRNAs have been purified. This is accomplished by taking advantage of differentiated tissues that synthesize large amounts of a given protein and therefore accumulate considerable amounts of stable, specific mRNAs. For example, 90% of the protein synthesized by reticulocytes is globin, and it is possible to isolate from these cells a globin mRNA which sediments at 9S. The complete nucleotide sequence of globin mRNA is known, and certain features common to all mRNAs have emerged.

The estimated length of the 9S globin mRNA is between 650 and 700 nucleotides. Of these, only 589 are encoded by the DNA. The rest are accounted for by a poly A tail that is attached to the mRNA after transcription is completed. The mRNA molecule is much longer than is required to code for the globin protein, which has 146 amino acids and is therefore encoded by only 438 nucleotides. The location of the extra nucleotides is diagrammed in Figure 11–12.

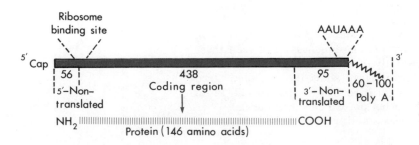

**Figure 11–12** Rabbit globin mRNA. Note that eukaryotic mRNAs have two noncoding regions, a 5' cap and a 3' poly A tail. The number of nucleotides in each section is indicated.

In addition to the sequence coding for protein, all eukaryotic mRNAs have two regions that are not translated into protein. The 5' noncoding region, which is located before the start of the protein-coding region, consists of about 50 nucleotides in most mRNAs and contains the ribosome binding site. There is also a 3' noncoding region, whose function is unknown even though it is about twice the size of the 5' noncoding region. Some mRNAs have a very large 3' noncoding region; in ovalbumin mRNA, for example, it is 600 nucleotides long. All mammalian mRNAs also have the sequence AAUAAA close to the start of the poly A sequence. The significance of this hexanucleotide is not known, but it could be a signal for mRNA processing or for the addition of poly A. Whatever their function, the noncoding regions are thought to be important because their sequence has tended to be conserved throughout evolution (e.g., globin mRNAs from different species have similar noncoding sequences).

**THE 5' CAP.** Most eukaryotic mRNAs have their starting ends blocked by the post-transcriptional addition of a "cap" of 7-methyl-guanosine. Normally the phosphodiester bonds of an RNA molecule connect the 3' position of the ribose of one nucleotide to the 5' position of the ribose of the next one (Fig. 11–13). The linkage of the cap to mRNA is different because the riboses of 7-methyl-G and the terminal nucleotide are linked by a 5' to 5' triphosphate bridge, as shown in Figure 11–13. The cap structure $m^7G^{5'}ppp^{5'}Xp$ has no free phosphates, and this protects mRNA against attack by phosphatases and other nucleases.

The cap is important in mRNA translation. When it is removed, mRNA is unable to form an initiation complex with 30S ribosomal subunits. Similarly, cap analogs such as $m^7G^{5'}$ monophosphate are potent inhibitors of protein synthesis initiation. Bacterial MRNAs lack caps and are translated poorly by eukaryotic ribosomes, but if they are capped artificially in vitro they become very active tem-

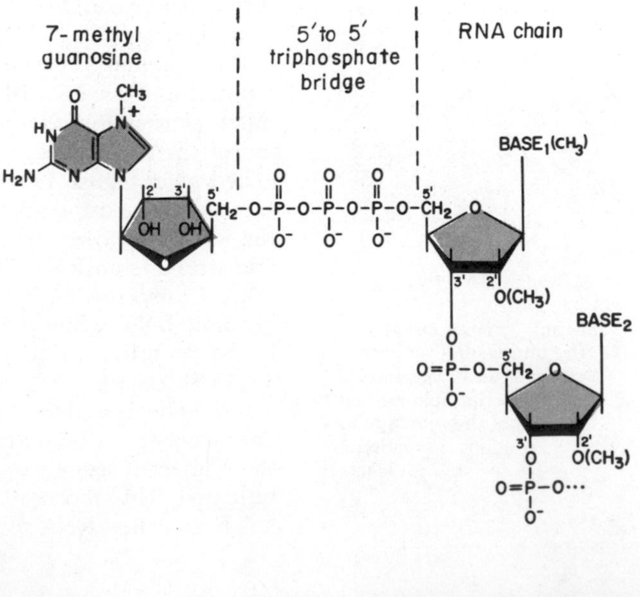

5' cap—7-methyl-guanosine structure added post-transcriptionally to many eukaryotic mRNAs; aids in protein synthesis initiation and may also have a protective function

**Figure 11-13** Structure of the 5' cap of messenger RNA. Observe that the linkage of the 7-methyl-guanosine to the terminal nucleotide is by a 5' to 5' triphosphate bridge.

plates in mammalian cell-free systems. However, the dependence of protein synthesis on caps is not absolute; some mRNAs (such as poliovirus mRNA) do not have caps but can still be translated efficiently.

**THE 3′ POLY A SEGMENT.** Most eukaryotic mRNAs contain a sequence of polyadenylic acid attached to their 3′ end. This poly A segment is 100 to 200 nucleotides long, is not coded in the DNA, and is added sequentially to the mRNA after transcription is completed. Addition is catalyzed by a poly A synthetase that uses ATP as substrate. Bacterial mRNAs do not have poly A.

3′ poly A region—long polyadenylic acid segment added post-transcriptionally to many eukaryotic mRNAs; its function is not yet established

Not all eukaryotic mRNAs are polyadenylated; up to 30 percent of mRNAs in polysomes lack poly A. Among these the most prominent are the histone mRNAs. The function of poly A is still unclear. It has been suggested that it could be involved in nuclear processing of mRNA precursors and in transport between the nucleus and the cytoplasm.

## Intervening Sequences in Eukaryotic Genes

Segments of eukaryotic DNA can be grown in *E. coli* plasmids using genetic engineering technology. Those plasmids containing a given eukaryotic gene can be identified by using radioactive hybridization probes prepared from purified mRNA. When a variety of such cloned eukaryotic genes became available during 1977, molecular biologists were in for quite a surprise.

Unexpectedly, it was found that in eukaryotes the information for covalently contiguous mRNA is frequently located in noncontiguous DNA segments. In other words, genes are interrupted by insertions of noncoding DNA. These inserted DNA sequences, which are absent in the mature mRNA, are called *intervening sequences* or *introns*. Introns have been found in globin, ovalbumin, immunoglobulin, tRNA, and many other genes. Not all eukaryotic genes are interrupted; those coding for histones and some tRNAs, for example, are continuous.

intron—intervening sequence of DNA, located within a gene, that does not code for any part of the final protein (a coding portion of the gene is referred to as an exon

The β-globin gene is interrupted by a DNA fragment 600 base pairs long inserted within the protein-coding sequence. A second, much shorter intervening sequence is found closer to the starting end of the coding sequence. The larger globin intron can be visualized as in Figure 11–14. The electron micrograph shows a cloned segment of mouse DNA that was hybridized to globin mRNA. When the RNA hybridizes to the DNA, a "bubble" can be seen in which one strand is an RNA-DNA hybrid and the other is single-stranded DNA. Globin mRNA hybridizes to two discontinuous regions of the genomic DNA, while the intervening sequence, which is not present in the 9S mRNA, remains as a loop of double-stranded DNA (Fig. 11–14**B**).

splicing—precise excision of the intervening sequences from an RNA primary transcript, followed by ligation of the message to produce a functional molecule

Intervening sequences are transcribed into precursor RNAs which are larger than the mature mRNAs. To obtain a functional mRNA these internal sequences must be excised precisely and the molecule religated. This mechanism, shown schematically in Figure 11–15, has been called RNA *splicing*.

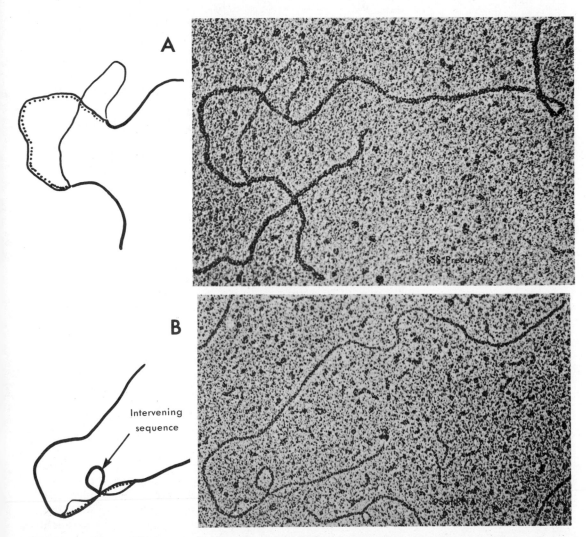

**Figure 11-14**   Visualization of the intervening sequence of the globin gene in the electron microscope. A cloned segment of DNA containing the mouse β-globin gene was hybridized with 15S globin mRNA precursor **(A)** and mature 9S globin mRNA **(B).** The hybridized RNA is represented by a dotted line in the diagram. The 15S precursor hybridizes in a continuous way, showing that the intervening sequence is transcribed into RNA. Note in **B** that the mature globin mRNA hybridizes to two discontinuous regions of the DNA, and the intervening sequence remains as a loop of double-stranded DNA. It was from electron micrographs such as these that intervening sequences were first discovered. (Courtesy of Philip Leder.)

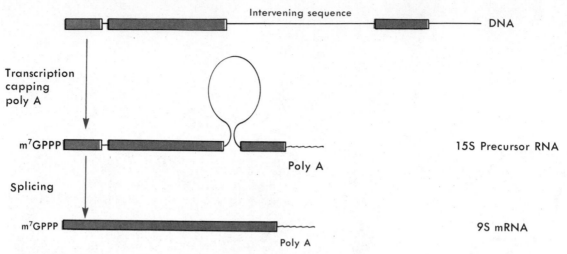

**Figure 11-15** Diagram showing the processing of globin mRNA transcripts. The globin gene has two intervening sequences (one is very short) which are transcribed into the 15S precursor RNA. The intervening sequences are excised and the molecule religated (a process called *splicing*), producing the mature 9S globin mRNA.

The β-globin gene is not transcribed initially as a 9S molecule, but rather as a precursor that sediments at 15S in sucrose gradients. This sedimentation value represents a molecule two to three times longer than globin mRNA, and this extra length is due to the presence of the intervening sequence. Mature 9S mRNA results from removal of the intervening sequences, a process that takes place rapidly, since the 15S precursor has a half-life of less than two minutes.

While it is clear that several genes are interrupted at the DNA sequence level (the ovalbumin gene has eight introns), we do not know why this would be beneficial to eukaryotic cells. It has been suggested that some evolutionary advantage could be gained by the deletion or addition of whole functional units of amino acids (for example, from a neighboring gene). In this way novel proteins could evolve more rapidly than by single nucleotide changes. The results for many protein-coding genes suggest that protein functional domains are separated by intervening sequences in the DNA and are consistent with the view that in evolution, new proteins might be constructed from parts of old ones brought together by the splicing mechanism. The splicing mechanism has not yet been detected in prokaryotes.

**heterogeneous nuclear RNAs (hnRNAs)—large nuclear RNAs of variable length, most of which are degraded within the nucleus and many of which represent unspliced mRNA precursors**

When eukaryotic cells are treated with a radioactive RNA precursor for short periods, most of the labeled molecules are incorporated into a population of large nuclear RNAs of variable length, the so-called *heterogeneous nuclear* RNAs (hnRNAs). These rapidly labeled nuclear RNAs are not related to rRNA or to tRNA; they contain 3' poly A tails and 5' caps; and they are larger than mature mRNAs. The average length of eukaryotic mRNA is 1800 nucleotides, and that of hnRNA is 4000, but some molecules can be 20,000 nucleotides long. Most of the hnRNA is never released into the cytoplasm and is degraded within the nucleus.

Nucleic acid hybridization studies have shown that hnRNA contains more sequences than mRNA. It is estimated that 6 to 20 percent of the information available in the genome is represented in hnRNA, while only 1 percent is represented in cytoplasmic mRNA. Most of the hnRNA consists of mRNA precursor molecules from which the intervening sequences have not yet been removed, as in the case of the globin 15S precursor. Once the intervening sequences are removed the mature mRNAs can exit into the cytoplasm.

# SUMMARY

## 11-1 THE GENETIC CODE

The hereditary units that contain the information to specify one amino acid are called codons, and consist of three nucleotides. The DNA is transcribed into a complementary mRNA on which a protein molecule is synthesized. The length of the coding portion of a gene depends on the number of amino acids in the protein, and the sequence of triplets determines the sequence of amino acids. The actual translation of the message takes place on the ribosomes.

The discovery of how cells translate a four-letter code into the 20 amino acids was a major scientific achievement, and was accomplished primarily by the use of homo- and copolymers of known base content in a cell-free protein-synthesizing system. The determination of codon sequences was later made possible by the use of trinucleotide templates of known base composition.

By 1966 all 64 codons had been deciphered; 61 represent amino acids and three are signals for chain termination. Some of the triplets are synonyms and can code for the same amino acid; in most cases the synonymous codons differ only in the base occupying the third position of the triplet. DNA sequencing studies have confirmed that all possible codons are utilized in vivo. It has also been found that there are fewer than 61 tRNA species and that one tRNA can recognize more than one codon (although always for the same amino acid).

mRNA is always translated sequentially in the 5' → 3' direction, and the polypeptide chain is always assembled sequentially beginning at the NH$_2$ terminus. The AUG codon, when it appears at the beginning of a message, is the signal to start translation (elsewhere in the sequence it codes for normal methionine). Chain termination is signalled by UAG, UAA, or UGA, which are recognized not by tRNAs but by protein releasing factors.

Types of mutations that affect the synthesis of mRNA and proteins are point mutations (changes in a single base), missense mutations (changes in the meaning of a codon), frameshift mutations (causing sections of the message to be read out of phase), and nonsense or chain termination mutations. Suppressor mutations can sometimes cancel the effect of these alterations in the DNA.

The nucleotide sequences obtained by DNA sequencing methods have confirmed the genetic code and established that all the possible codons are used in vivo. These sequencing methods are based on the generation of DNA fragments of different lengths which start at a fixed point and end at specific nucleotides. The fragments are separated by size on gels, from which the sequence can be read directly.

Sequencing methods led to the discovery that the bacteriophage φX174 contains overlapping genes in its DNA. It is not yet known whether overlapping genes exist in higher organisms.

With one exception, the genetic code is the same for all living organisms, and once the code evolved there were strong selective pressures to maintain it in invariant form. The mitochondrial genetic code, however, has a few differences, which have important biological consequences.

## 11-2 GENETIC ENGINEERING

The techniques collectively referred to as genetic engineering allow short segments of eukaryotic DNA to be excised from the genome, ligated into small bacterial chromosomes called plasmids, and cloned in large amounts. These techniques rely on the use of restriction endonucleases, enzymes which recognize and cut specific, symmetrical nucleotide sequences in the DNA.

## 11-3 TRANSCRIPTION AND PROCESSING OF RNA

RNA is synthesized by RNA polymerase. The primary transcripts thus produced must usually be processed further before they can be used by the cell. Types of processing include (1) cleavage of large precursor RNAs, (2) splicing, (3) terminal addition of nucleotides, and (4) nucleoside modifications.

Prokaryotic mRNAs are very short-lived, and are often translated into protein at one end while the other end is still being transcribed. The length of these mRNAs is heterogeneous; some mRNAs (polycistronic mRNAs) code for several proteins. All prokaryotic RNAs are transcribed by the same enzyme, which is a large and complex multi-subunit molecule consisting of several polypeptides. One of these peptides, the sigma factor, is loosely bound to the core polymerase and is required for the enzyme to recognize the correct start signals on the DNA. The transcription process involves three stages: (1) binding to promoters and chain initiation, (2) elongation, and (3) chain termination.

Eukaryotic transcription differs in many respects from that of prokaryotes. Transcription takes place in the nucleus; eukaryotic mRNAs are considerably more stable than those of prokaryotes, and are generally monocistronic. Three different nuclear RNA polymerases have been identified, each one of which translates a different class of genes. These polymerases can be distinguished by their order of elution from ion-exchange columns and by their sensitivity to inhibition by $\alpha$-amanitin. A maximally expressed gene can complete one RNA chain at least every 10 seconds, and the growing chain becomes associated with protein as it is being transcribed. The mRNAs must be transported into the cytoplasm before they can be utilized.

In addition to the protein-coding sequence, all eukaryotic mRNAs have 5′ and 3′ regions that are not transcribed into protein. The function of these noncoding sequences is unknown, but they are thought to be important because their sequences have tended to be conserved throughout evolution. Most eukaryotic mRNAs have their starting ends blocked by a 7-methyl-guanosine cap. This structure protects the mRNA against attack by nucleases, and is necessary for the formation of the initiation complex with 30S ribosomal subunits. Most eukaryotic mRNAs also contain a poly A tail, the function of which is still unclear.

It has recently been learned that some eukaryotic genes are interrupted by intervening sequences of DNA which do not code for protein. Intervening sequences are transcribed along with the rest of the gene to form precursor mRNAs which are larger than the mature RNAs; these introns must then be excised and the molecule religated. It is not known why this mechanism exists or what advantage it confers on eukaryotic cells. It has not been detected in prokaryotes.

# STUDY QUESTIONS

- How did the experiments of Griffith and Avery show that DNA was the genetic material?
- Describe the experiments that led to the elucidation of the genetic code.
- What is meant by synonyms in the code? What does the "wobble" of the third nucleotide in a codon refer to?
- Given the sequence

| 1 | 4 | 7 | 10 | 13 | 16 | 19 |
|---|---|---|----|----|----|----|
| AUG | GUU | GCU | UCG | ACG | CCC | UAA |

- Refer to Table 11–1 and explain what kind of mutation has occurred and what the results will be if:
  (1) U → C at position 6
  (2) A → G at position 13
  (3) a C residue is inserted between positions 5 and 6
  (4) C → A at position 11
  (5) an A residue is inserted between positions 7 and 8 and residue 15 is deleted.
- Explain how Sanger's chain termination method works.
- What steps would be necessary if you wanted to clone the human globin gene in an E. coli plasmid?
- What kinds of modifications are involved in mRNA processing?
- What are the major differences between RNA transcription in prokaryotes and eukaryotes?
- What are intervening sequences? How and at what stage of processing are they removed from RNA?

# SUGGESTED READINGS

**Abelson, J.** (1979) RNA processing and the intervening sequence problem. *Annu. Rev. Biochem.,* *48*:1035.

**Barrell, B., et al.** (1980) Different patterns of codon recognition by mammalian mitochondrial tRNA. *Proc. Natl. Acad. Sci. USA, 77*:3164.

**Bonitz, S. G., et al.** (1980) Codon recognition rules in yeast mitochondria. *Proc. Natl. Acad. Sci. USA, 77*:3167.

**Cohen, S. N.** (1975) The manipulation of genes. *Sci. Am., 233*:24.

**Crick, F. H. C.** (1979) Split genes and RNA splicing. *Science, 204*:264.

**De Robertis, E. M., and Gurdon, J. B.** (1979) Gene transplantation and the analysis of development. *Sci. Am., 241*:74.

**Fiddes, J. C.** (1977) The nucleotide sequence of a viral DNA. *Sci. Am., 237*:54.

**Gilbert, W., and Villa-Komaroff, L.** (1980) Useful proteins from recombinant bacteria. *Sci. Am., 242*:74.

**Itakura, K.** (1980) Synthesis of genes. *Trends Biochem. Sci., 5*:114.

**Kourilsky, P.** (1980) Genetic engineering. *LaRecherche, 11*:390.

**Miller, O. L., Jr.** (1973) The visualization of genes in action. *Sci. Am., 228*:34.

**Roberts, R. J.** (1980) Restriction and modification enzymes and their recognition sequences (a review). *Gene, 8*:329.

# The Machinery for Protein Synthesis

In this chapter we will discuss protein synthesis, the last step in the flow of genetic information. The translation of the 4-base code into a 20-amino acid protein sequence involves many cellular components; in addition to mRNA, these include the ribosomes, about 60 tRNA molecules, and many soluble enzymes. We will examine these components individually and then study how they interact to produce specific proteins.

## 12-1 RIBOSOMES

Ribosomes were first observed by Palade in the electron microscope as dense particles or granules. Upon isolation they were shown to contain approximately equal amounts of RNA and protein. Ribosomes are found in all cells and provide a scaffold for the ordered interaction of all the molecules involved in protein synthesis. Cells devote considerable effort to the production of these essential organelles. An *E. coli* cell contains 15,000 ribosomes, each one with a molecular weight of about 3 million daltons. Ribosomes therefore represent 25 percent of the total mass of these bacterial cells.

ribosome—cytoplasmic granule composed of RNA and protein, at which protein synthesis takes place

### Subunit Structure

The ribosome is a spheroidal particle of 23 nm and is composed of a large and a small subunit (Fig. 12–1). Eukaryotic ribosomes sed-

Polysome

_50nm_

**Figure 12-1**   Diagram of the subunit structure of the ribosome and the influence of Mg. A polyribosome formed by five ribosomes is shown. The filament uniting the ribosomes represents messenger RNA. The sedimentation constants (S) of the different particles are indicated.

sedimentation coefficient (S)—measure of the rate at which a substance or particle sediments (usually through some type of gradient) in a centrifugal field

iment in sucrose gradients with a *sedimentation coefficient* of 80S, and in the absence of $Mg^{2+}$ these ribosomes dissociate reversibly into subunits of 40S and 60S. Prokaryotic ribosomes are smaller and sediment at 70S; they have subunits of 30S and 50S (Table 12-1). (Note that the values of the sedimentation coefficients are not additive.) Ribosomes are also found in the mitochondria and chloroplasts of eukaryotic cells. They are always smaller than the 80S cytoplasmic ribosomes and are comparable to prokaryotic ribosomes in both size and sensitivity to antibiotics, although the sedimentation values vary somewhat in different phyla.

During protein synthesis several ribosomes become attached to one mRNA molecule (Fig. 12-1), forming a *polyribosome* or *polysome*. In this way a single mRNA molecule can be translated by several ribosomes at the same time. The mRNA is located in the gap between the two ribosomal subunits (Fig. 12-2). It is possible that the nascent peptide chain grows through a channel or groove in the large ribosomal subunit; this is thought to be the case because ribosomes protect a segment of 30 to 40 amino acids from degradation by proteolytic enzymes.

## Ribosomal RNAs

The major constituents of ribosomes are RNA and proteins present in approximately equal amounts (there is little or no lipid material). The positive protein charges are not sufficient to balance the many negative charges in the phosphates of the RNA, and for this reason ribosomes are strongly negative and bind cations and basic dyes. Ribosomal RNA generally represents more than 80 percent of the total RNA present in cells.

Prokaryotic ribosomes contain three RNA molecules: 16S RNA in the small subunit, and 23S and 5S in the large subunit (Table 12-1). In eukaryotes these RNAs are larger: 18S in the small subunit, and 28S and 5S in the large subunit. Eukaryotic ribosomes also have an extra RNA species that went unnoticed for some time. When 28S RNA is heated briefly or otherwise denatured, a small, noncovalently attached component is released. This small RNA is called

**TABLE 12-1  SIZES OF VARIOUS RIBOSOMES**

| Ribosomes | Size | Subunits | | RNAs | | Number of Proteins | |
|---|---|---|---|---|---|---|---|
| Eukaryotes | 80S | 60S | 40S | 28S + 5.8S<br>5S | 18S | 40 | 30 |
| Bacteria | 70S | 50S | 30S | 23S<br>5S | 16S | 34 | 21 |
| Mitochondria<br>(in mammals) | 55S | 35S | 25S | 21S<br>3S | 12S | – | – |

5.8S RNA (Table 12-1) and is transcribed in the nucleolus as a single unit along with the 18S and 28S RNAs; 5S RNA is synthesized outside the nucleolus (as discussed below).

Ribosomal RNA has a high degree of secondary structure; about 70 percent of it is double-stranded and helical as a result of base-pairing. The double-stranded regions are formed by *hairpin loops* between complementary regions of the same linear RNA molecule. The ribosomal subunits thus contain highly folded RNA to which various proteins adhere.

hairpin loop—double-stranded region formed by base pairing between complementary regions of the same linear RNA molecule

## Ribosomal Proteins

The *E. coli* ribosome contains 21 proteins in the small subunit and 34 in the large one (Table 12-1). All the proteins are different with the exception of one that is present in both subunits. Ribosomal proteins are small, ranging between 7000 and 32,000 daltons in molecular weight, and are rich in basic amino acids.

Ribosomal proteins can be dissociated from the ribosome and then added back to reconstitute active ribosomes. For example, ri-

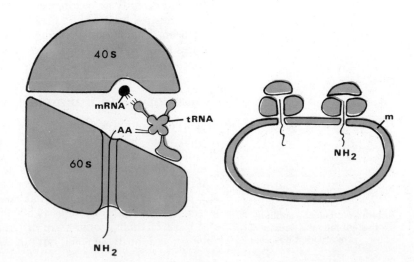

**Figure 12-2  Left,** diagram of a ribosome showing the two subunits and the probable position of the messenger RNA and the transfer RNA. The nascent polypeptide chain passes through a kind of tunnel within the large subunit. **Right,** diagram of the relationship between the ribosomes and the membrane of the endoplasmic reticulum and the entrance of the polypeptide chain into the cavity. *m*, membrane of endoplasmic reticulum (Courtesy of D. D. Sabatini and G. Blobel.)

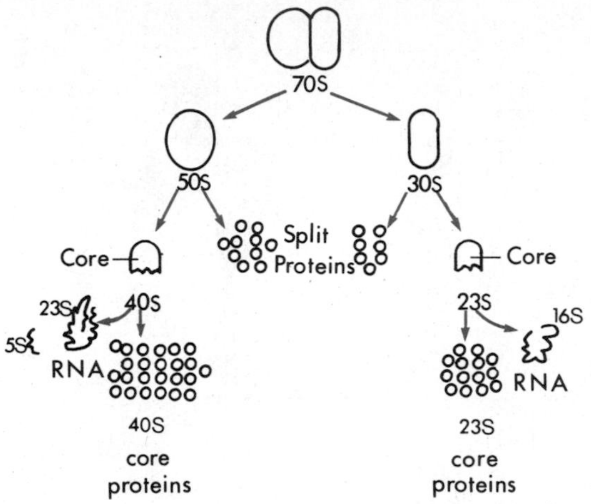

**Figure 12–3** Diagram representing the stepwise dismantling of the two subunits of a 70S ribosome. Note that the proteins may be separated into split and core proteins. The 50S subunit contains 23S and 5S RNAs, and the 30S subunit has 16S RNA.

split proteins—proteins that can be dissociated from the ribosome core by centrifugation in CsCl

bosomal subunits centrifuged in 5 M cesium chloride lose 30 to 40 percent of their proteins, the so-called *split proteins* (Fig. 12–3). When the split proteins are returned to medium containing the inactive ribosome *cores* and incubated at 37°C, active ribosomes can be reconstituted. The total reconstitution of ribosomes from naked RNA and proteins can also be achieved. It is interesting that such a complex organelle can assemble spontaneously by simple physicochemical interactions; this is another good illustration of the principle of self-assembly of biological structures, which as we have seen also applies to viruses and biological membranes.

Reconstitution experiments are useful because they make it possible to identify the function of individual ribosomal proteins. This is managed in experiments in which one ribosomal protein is omitted (or modified) at a time. These studies have also shown that certain ribosomal proteins require a prior attachment of other proteins in order to become incorporated in a stepwise manner; some proteins bind directly to specific regions of the RNAs, and then in turn enable the attachment of additional proteins. All ribosomal proteins have been isolated, and various immunological and chemical procedures have made it possible to construct maps of the topographical positions they occupy within the subunits (Fig. 12–4).

## Prokaryotic and Eukaryotic Ribosomes

chloramphenicol—antibiotic that inhibits the activity of prokaryotic ribosomes

Prokaryotic and eukaryotic ribosomes do not differ in any fundamental way; both perform the same functions by the same set of chemical reactions. We have seen that the genetic code is the same in all living organisms, and it has been demonstrated that eukaryotic ribosomes are able to translate bacterial mRNAs correctly. However, we have also seen that eukaryotic ribosomes are much larger than prokaryotic ones (4,500,000 versus 2,700,000 daltons), and most of their proteins are different. Antibiotics such as *chloramphenicol* inhibit bacterial but not eukaryotic ribosomes (account-

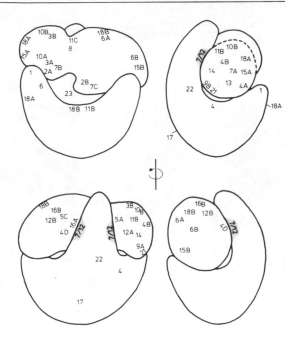

**Figure 12-4**   Four views of a model of a ribosome from *E. coli*. The numbers indicate the topography of individual ribosomal proteins as determined by immunoelectron microscopy. (From H. G. Wittman, *Eur. J. Biochem.*, 61:1, 1976.)

ing for the use of such antibiotics in medical treatment). Protein synthesis by eukaryotic ribosomes is inhibited by *cycloheximide*.

Mitochondrial and chloroplast ribosomes resemble those in bacteria. They are inhibited by chloramphenicol, and hybrid ribosomes containing one bacterial and one chloroplast ribosome subunit, for example, are fully active in protein synthesis. Hybrid eukaryotic ribosomes containing subunits from both plants and mammals are also active in protein synthesis, but are inactive if one of the subunits is derived from bacteria. Some structural resemblance must exist, however, since reconstitution experiments have shown that two proteins from the large subunit of *E. coli* can replace the homologous proteins in mammalian ribosomes. In summary, there is little structural but considerable functional homology between prokaryotic and eukaryotic ribosomes.

**cycloheximide**—compound that inhibits the activity of eukaryotic ribosomes

## 12-2  THE NUCLEOLUS

Most cells have within the nucleus at least one RNA-rich body called the *nucleolus*. The nucleolus is the cellular factory where ribosomes are produced. It is especially large in cells in which protein synthesis is a prominent feature, such as secretory cells or oocytes.

The nucleolus is formed around the *nucleolar organizer*, a segment of DNA that contains many repeats of the genes coding for 18S, 5.8S, and 28S rRNAs. This ribosomal DNA becomes uncoiled and is actively transcribed. The nucleolar organizer is usually located in a *secondary constriction* on the chromosome, i.e., in a chromosomal site that becomes less condensed during mitosis.

**nucleolus**—structure within the nucleus, consisting of chromatin and large amounts of RNA; site of rRNA synthesis and ribosome formation

**nucleolar organizer**—chromosomal region that contains the genes for ribosomal RNAs and induces formation of the nucleolus

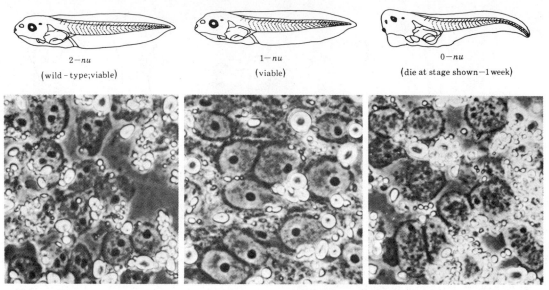

2−nu                          1−nu                          0−nu
(wild−type;viable)            (viable)                      (die at stage shown—1week)

**Figure 12-5**   Number of nucleoli in mutants of *Xenopus laevis*. The wild-type tadpole has two nucleoli *(2-nu)*; the heterozygous mutant has one nucleolus *(1-nu)*, and the homozygous anucleolate mutant has none *(0-nu)*. The 0-*nu* tadpoles do not synthesize ribosomal RNA. (Courtesy of J. B. Gurdon.)

The synthesis of a eukaryotic ribosome is a complex phenomenon in which several regions of the cell are involved. The 18S, 5.8S, and 28S RNAs are synthesized as part of a much longer precursor RNA molecule in the nucleolus; 5S RNA is synthesized on the chromosomes outside the nucleolus; and the 70 ribosomal proteins are synthesized in the cytoplasm. All these components collect in the nucleolus, where they are assembled into ribosomes and transported to the cytoplasm. Ribosome biogenesis is thus a striking example of coordination at the cellular and molecular levels.

Direct evidence that the nucleolus is responsible for the synthesis of rRNA was obtained when it was discovered that a mutant of the South African frog *Xenopus laevis* that lacks nucleoli was unable to synthesize rRNA. Figure 12–5 shows that diploid cells of wild-type *Xenopus* have two nucleoli (2-*nu*); the heterozygous mutant has only one nucleolus (1-*nu*); and the homozygous mutant is anucleolate (0-*nu*). When two 1-*nu* heterozygotes are crossed, 25 percent of the progeny are 0-*nu*. This condition is lethal and the tadpoles die after one week. Up until this stage of development, the embryos rely on maternal ribosomes inherited from the egg cytoplasm (one *Xenopus* egg contains $10^{12}$ ribosomes). The 0-*nu* embryos do not synthesize 18S, 28S, or 5.8S rRNA, and DNA-RNA hybridization studies have shown that they do not contain the genes for rRNA.

*Xenopus laevis*—species of South African frog well suited for many kinds of biological studies

## rRNA Genes

Table 12–2 shows that all organisms have multiple copies of rRNA genes. Few copies are present in bacteria, but in eukaryotes the rDNA is highly repetitive. In the case of *Xenopus*, each nucleolar

**TABLE 12-2  REPETITIVE rDNA CISTRONS IN VARIOUS ORGANISMS**

| Organism | % rDNA | rDNA Cistrons per Haploid Genome |
|---|---|---|
| E. coli | 0.42–0.65 | 8–22 |
| B. subtilis | 0.38 | 9–10 |
| HeLa cells | 0.005–0.02 | 160–640 |
| Drosophila (wild-type) | 0.27 | 130 |
| Xenopus (wild-type) | 0.2 | 450 |

organizer contains 450 rRNA genes. These genes are *tandemly repeated* along the DNA molecule (i.e., head-to-tail) and are separated from each other by stretches of *spacer* DNA which is not transcribed. These linear repeats of genes can be visualized clearly in Figure 12–6, which shows a nucleolar organizer spread for electron microscopy. These rRNA genes are being actively transcribed, and the nascent RNA chains are spread perpendicularly to the DNA axis. Each gene is transcribed into a long RNA molecule (which varies in size from 40S to 45S, depending on the species) which will eventually be processed, giving rise to 18S, 28S, and 5.8S RNA. Since each gene has a fixed initiation site (promoter) and a fixed termination site, the transcripts adopt the characteristic "Christmas tree" or "fern leaf" configuration mentioned in Chapter 11. Nucleolar rRNA genes are transcribed by RNA polymerase I, and these polymerase molecules (about 100 per gene) can be visualized at the origin of each nascent RNA chain (Fig. 12–6).

The genes coding for 5S RNA are not located in the nucleolus, and therefore are transcribed normally in the *Xenopus* anucleolate mutant. The 5S genes are also present in multiple copies; *Xenopus* has 24,000 5S genes which are located in the tips (telomeres) of most chromosomes, as shown by in situ hybridization of $^3$H-5S RNA to chromosome preparations. As with rDNA in the nucleolus, the 5S genes are arranged in tandem repeats separated by segments of spacer DNA. Tandem repeats of genes separated by spacers seem to be a general organizational feature of the eukaryotic genome. The 5S RNA is transcribed by RNA polymerase III and is then transported to the nucleolus, where it is incorporated into the immature large ribosomal subunits.

## rDNA Amplification

Gene amplification is the process by which one set of genes is replicated selectively while the rest of the genome remains constant. The clearest example of gene amplification is seen in the rDNA of amphibian oocytes, such as those of *Xenopus*. *Xenopus* eggs are extremely large cells (1.2 mm in diameter) that accumulate large numbers of ribosomes ($10^{12}$) for use during early development. Oocytes must synthesize all these ribosomes using the genes contained in a single nucleus, and they achieve this by amplifying the number of rDNA genes about 1000-fold (Table 12–3).

gene amplification—process by which a gene(s) is replicated selectively, for example in the rDNA of amphibian oocytes, while the rest of the genome remains constant

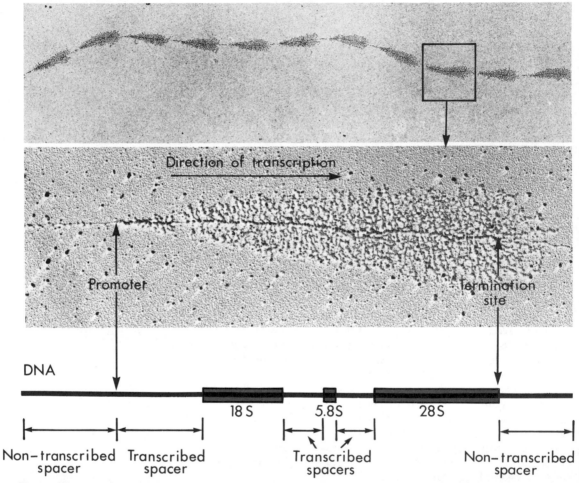

**Figure 12-6** Ribosomal RNA genes are tandemly repeated (head-to-tail) and separated from each other by nontranscribed spacers. A nucleolar organizer was spread for electron microscopy and 11 consecutive rRNA genes can be seen. Note the "Christmas tree" or "fern leaf" configuration of the nascent RNA precursor molecules. The long precursor molecule contains 18S, 5.8S and 28S rRNAs and is associated with proteins, forming a nucleoprotein complex. **Bottom,** the genetic map of an rDNA unit (Courtesy of U. Scheer, M.F. Trendelenburg and W. W. Franke).

**TABLE 12-3 RIBOSOMAL GENES OF *XENOPUS LAEVIS* IN SOMATIC CELLS AND AFTER AMPLIFICATION IN OOCYTES***

| Cell Type | Nuclear DNA (pg) | Number of Chromosome Sets per Nucleus | Chromosomal DNA (pg per Nucleus) | Ribosomal DNA (pg per Nucleus) | Number of rDNA Genes | Number of Nucleoli |
|---|---|---|---|---|---|---|
| Somatic | 6 | 2 | 6 | 0.012 | 900 | 2 |
| Oocyte (after amplification) | 37 | 4 | 12 | 25 | 2,000,000 | 1000–1500 |

*Oocytes are in meiotic prophase and therefore contain a tetraploid amount of DNA; ribosomal DNA is amplified to 1000 times more than expected for that amount of chromosomal DNA. (Modified from J. B. Gurdon, *The Control of Gene Expression in Animal Development.* Oxford University Press, Oxford, 1974.)

Amplification takes place in very small oocytes which are in the early stages of meiotic prophase. During pachynema, excess DNA begins to accumulate on one side of the nucleus, forming a cap that can be stained as shown in Figure 12–7. This cap incorporates ³H-thymidine intensely (while the chromosomes do not), and by the end of the amplification process it contains 25 pg of DNA, an amount equivalent to 2,000,000 rRNA genes (Table 12–3). As the oocytes grow, the extra DNA is accommodated in 1000–1500 extrachromosomal nucleoli. The amplified DNA is not inherited by the embryo, but is lost in the course of development.

The newly synthesized DNA is indeed rDNA. This can be shown by an in situ hybridization procedure in which ³H-ribosomal RNA is hybridized to cytological preparations of *Xenopus* oocytes that have been previously treated with alkali to denature the DNA. The hybridized material can then be located by autoradiography. Figure 12–7 shows that rRNA hybridizes to the cap of amplified DNA.

Unlike the genes coding for the large ribosomal RNAs, the 5S genes are not amplified during oogenesis. Oocytes meet the in-

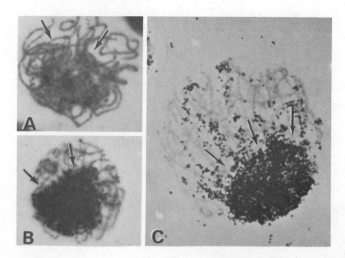

**Figure 12-7** *Xenopus* oocytes during the period of nucleolar DNA synthesis. **A,** at late pachynema, the excess DNA begins to accumulate as granules around the nucleolus (arrows). **B,** later on, the excess DNA appears as a dense mass (arrows). **A** and **B,** Feulgen reaction, ×1700. **C,** large pachynema oocyte prepared with the ³H-rRNA hybridization technique. Observe that the silver grains are deposited mainly on the mass of excess DNA. ×1200. (Courtesy of J. G. Gall.)

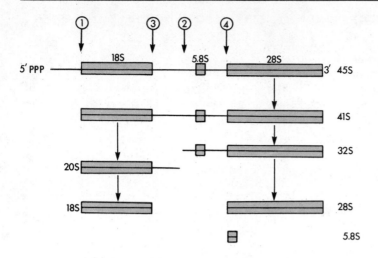

**Figure 12-8** Diagram showing the processing of ribosomal RNA in HeLa cell nucleoli. The orientation of the various rRNAs in the 45S precursor and the four main sites of processing are indicated.

creased demand for ribosomes by activating all 24,000 5S genes. The amplification of genes at specific developmental stages is not a common event. The only instance in which it is known to occur other than in oocyte rDNA is in the DNA puffs of insects (Chapter 13). In both cases, the amplified material is not passed on to future generations.

## rRNA Processing

The biogenesis of rRNA provides an important and clear example of RNA processing. As shown in Figure 12–8, the rRNA genes are transcribed into a long precursor RNA which must be cleaved into 18S, 28S, and 5.8S RNA. In this cleavage process about 50 percent of the precursor RNA is degraded within the nucleus. Several events can be distinguished in the course of processing.

1. Within the 45S or 40S primary transcript the rRNAs are separated by stretches of spacer RNA, and the order of transcription is 5' end – 18S – 5.8S – 28S – 3' end (Fig. 12–8). On a fully active gene about 100 RNA polymerases are transcribed simultaneously on the ribosomal DNA cistron (i.e., gene).

2. 45S RNA is methylated rapidly, even before transcription is completed. Methylations occur primarily on the ribose moiety, and occur only in the 18S and 28S sequences that will be conserved; those segments that will be degraded remain nonmethylated.

3. The 45S RNA has a lifetime of about 15 minutes, after which it is cleaved into smaller components in the nucleolus, according to the following general pattern:

$$45S \rightarrow 41S \begin{array}{l} \nearrow 32S \rightarrow 28S + 5.8S \\ \searrow 20S \rightarrow 18S \end{array}$$

4. The processing of the small ribosomal subunits is faster, and

these appear in the cytoplasm before the large ribosomal subunits have been assembled.

The 45S RNA becomes associated with proteins at the same time as it is being transcribed, and all the subsequent processing events occur within nucleolar ribonucleoprotein particles. The end products are eukaryotic ribosomes, which are assembled in the nucleolus. The other ingredients (5S RNA and ribosomal proteins) are imported from other parts of the cell into the nucleolus, where they become bound to the ribosomal RNAs in an orderly fashion.

## Nucleolar Structure

Under the light microscope nucleoli appear as strongly refringent bodies, due to high concentrations of solid materials (both RNA and proteins). This high RNA content can be demonstrated by staining with basic dyes, both with and without digestion with ribonuclease.

The electron microscope makes it possible to distinguish two characteristic components in most nucleoli: the *granular zone* and the *fibrillar zone* (both are illustrated in Fig. 12–9). The granular zone consists of granules 15 to 20 nm in diameter and frequently occupies the peripheral region of the nucleolus. The fibrillar zone consists of fine fibers 5 to 10 nm in diameter and is located in a more central region of the nucleolus.

When cells are labeled with ³H-uridine for 5 minutes followed by a chase with nonradioactive uridine and examined by autoradiography at the electron microscope level, it is evident that ³H-uridine is incorporated first into the fibrillar zone, and appears only later in the granular area. This suggests the relationship:

$$\text{nucleolar DNA} \rightarrow \text{fibrillar zone} \rightarrow \text{granular zone}$$

The fibrillar area contains the rDNA that uncoils from the chromosome and penetrates the nucleolus, together with the 45S RNA molecules attached to it. When spread for observation in the electron microscope, it has the appearance shown in Figure 12–6. The granular area represents ribosome precursor particles at various stages of assembly and processing. Once the ribosomal subunits are mature, they are released from the nucleolus and exit into the cytoplasm through the nuclear pores (Fig. 12–9).

During mitosis the nucleoli undergo cyclic changes. Nucleoli are formed around the DNA loop that extends from the nucleolar organizer. There may be several nucleoli per cell, but frequently they tend to fuse into one or a few nucleoli at this stage. During late prophase the DNA loop containing the rRNA genes gradually retracts and coils into the nucleolar organizer of the corresponding chromosome. Since this DNA is highly extended as a consequence of intense RNA synthesis, the nucleolar organizer region is one of the last to undergo condensation, thus producing a secondary constriction on the chromosome. The fibrillar and granular components are gradually dispersed into the nucleoplasm. After the cell divides,

granular zone—peripheral region of the nucleolus, containing ribosome precursor particles at various stages of assembly

fibrillar zone—central region of the nucleolus, containing the rDNA that uncoils from the chromosome and the nascent rRNAs

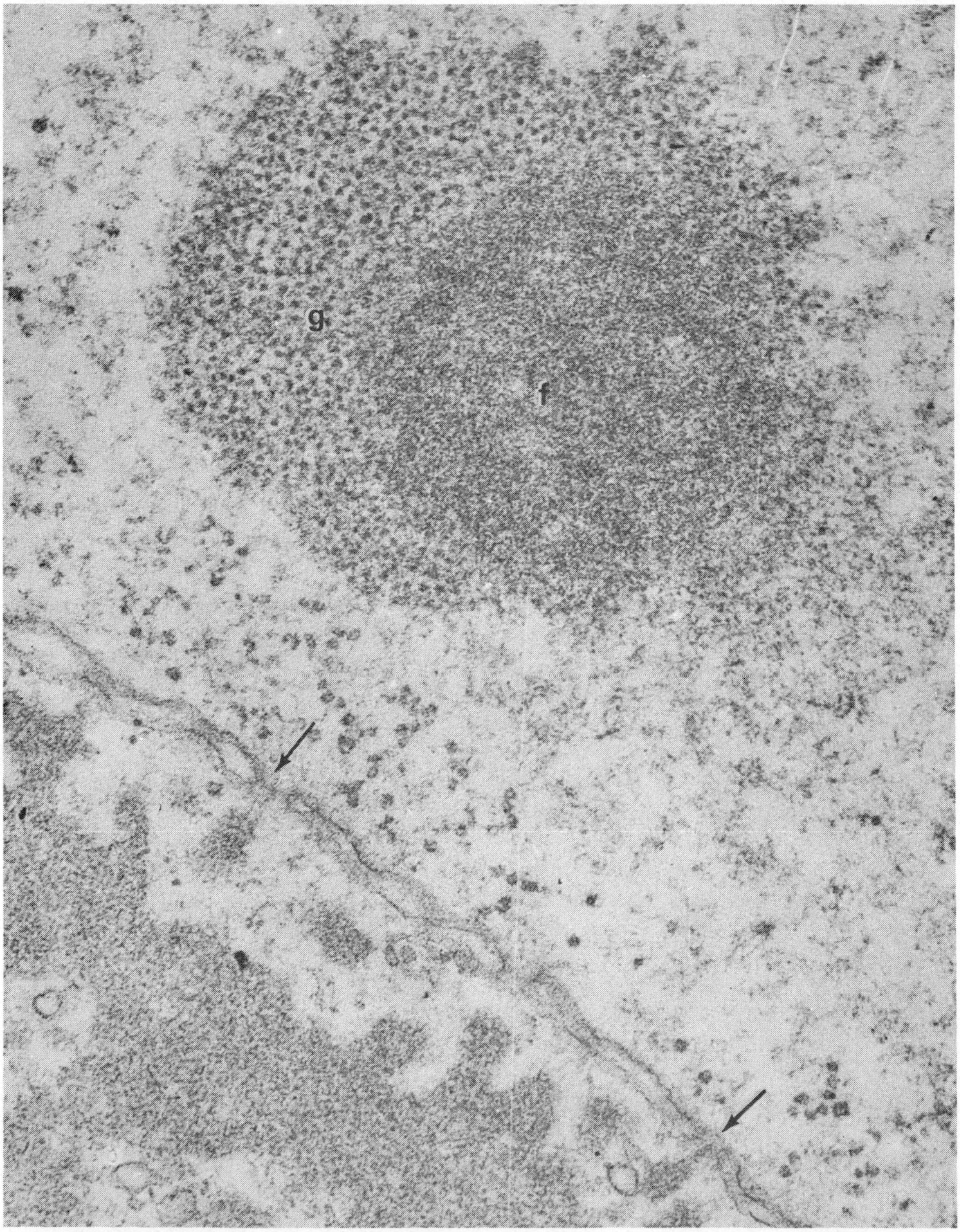

**Figure 12-9** Electron micrograph of an oocyte of *Rana clamitans* showing one of the peripheral nucleoli (containing amplified DNA) with a fibrillar (*f*) central portion and a granular (*g*) peripheral portion. Arrows indicate material entering the cytoplasm through the nuclear pores. ×70,000. (Courtesy of O.L. Miller.)

during telophase, the nucleolar organizer DNA uncoils and the nucleolus is reassembled.

## 12–3  TRANSFER RNA

Transfer RNAs are a group of small RNAs (between 75 and 85 nucleotides in length) that have the important role of serving as molecular adaptors during protein synthesis. Since the 20 amino acids have a shape that is not complementary in any way to the mRNA nucleotide triplets, they cannot recognize the codons by themselves. Transfer RNA has a triplet of nucleotides called an *anticodon* that can establish hydrogen bonds with the codon in mRNA; it also can have the amino acid corresponding to that particular codon attached to one of its ends. This entire structure is called an *aminoacyl-tRNA*, and it enables a particular amino acid to be brought to the ribosome in response to the appropriate codon.

It will be seen below that accurate protein synthesis depends on the correct amino acid being attached to the tRNA molecule. This important activity is performed by specific enzymes called *aminoacyl-tRNA synthetases* or *activating* enzymes. If a tRNA is mischarged it will cause the incorporation of an incorrect amino acid into the protein chain.

In *E. coli* there are about 50 different tRNAs that represent 10 to 15 percent of the total RNA in the cell. These tRNAs utilize 61 codons from among the 64 possible triplets of the genetic code. Some tRNAs are able to recognize more than one codon because the third base of the anticodon (which is less important in coding) may have a certain degree of flexibility or "wobble" that enables it to base-pair with more than one nucleotide (see Chapter 11). The three codons for termination of protein synthesis are exceptional in that they are recognized by protein factors called *releasing factors*, instead of by tRNA. There are 20 *isoacceptor tRNA families*, one for each amino acid, and for each one of these families there is one aminoacyl-tRNA synthetase.

### Cloverleaf Secondary Structure

All tRNAs share common features, of which the most notable is the folding into a "cloverleaf" structure as shown in Figure 12–10. Other common features include the sequence CCA at the 3′ end (which attaches covalently to the corresponding amino acid), the conserved nucleotides indicated by circles in Figure 12–10, and unusual bases such as pseudouridine (Ψ), inosinic acid, methylcytosine, methylguanine, ribothymidine (T), and others. These unusual bases are modified post-transcriptionally (for example, by the action of methylating enzymes).

The cloverleaf structure consists of a series of loops separated by short stems of helical double-stranded regions between four and seven complementary bases long. Figure 12–10 shows the principal elements of this structure, including (1) the *acceptor end* (CCA), at which the specific amino acid becomes attached; (2) the *anticodon*

transfer RNA—type of RNA molecule that identifies an amino acid in the cytoplasm and transports it to the ribosome

aminoacyl-tRNA—tRNA molecule carrying an amino acid

aminoacyl-tRNA synthetase—enzyme responsible for attaching the correct amino acid to the tRNA molecule (i.e., that amino acid specified by the codon with which the tRNA anticodon will pair)

isoacceptor tRNA family—all tRNA molecules specific for a particular amino acid

cloverleaf structure—secondary structure of tRNA, having essentially four stems and three loops occupied by characteristic combinations of nucleotides

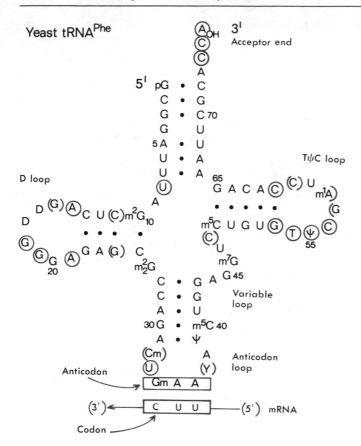

**Figure 12-10** Diagram of the cloverleaf model of transfer RNA showing the amino acid acceptor end (CCA); the anticodon loop with the three bases that read the genetic message; the D loop, which contains several dihydrouridines (D); the TψC loop, which is thought to interact with the 5S ribosomal RNA; and a variable loop. Bases that are invariant in all tRNAs are indicated by circles; bases that tend to be moderately conserved (e.g., always a purine or pyrimidine) are indicated by parentheses. These constant bases interact with each other via hydrogen bonds, producing the three-dimensional folding of the molecule.

*loop,* directly opposite from the acceptor end and having the three bases that recognize and form hydrogen bonds with the mRNA codon; (3) the *D loop,* which contains dihydrouridine; and (4) the *T loop,* which has the conserved sequence TΨCG (for which it is named) and is thought to interact with a complementary region of 5S ribosomal RNA during protein synthesis. There is also a highly variable region (variable loop) which differs greatly in length in the different tRNAs.

The tertiary structure of tRNA has been determined by x-ray diffraction studies. As shown in Figure 12–11, tRNA has a compact configuration in which the side arms are folded together, rather than the open configuration (shown for clarity) in Figure 12–10. The molecule has the shape of an L, with the anticodon at one end of the L and the acceptor at the other. The D and T loops are hydrogen-bonded together at the corner of the structure, and the nucleotides involved in these crucial hydrogen bonds tend to be highly conserved in different tRNAs. All tRNAs share these characteristics of folding, with the exception of mitochondrial tRNAs, which sometimes do not follow these rules.

## tRNA Transcription and Processing

As in the case of rRNA, tRNA is not transcribed directly into its final form but rather into a longer precursor. For example, *E. coli* tyrosine tRNA is synthesized as a precursor having 39 extra nucleotides at the 5′ end and 3 extra nucleotides at the 3′ end. These nucleotides must be removed before the tRNA molecule becomes active. Other *E. coli* tRNAs are synthesized as very long precursors that contain several tRNAs (as many as five to seven) in a single RNA chain.

Other post-transcriptional modifications important for tRNA function are the base modifications (such as methylations) which lead to the unusual bases found in these molecules, and the terminal addition of CCA. The 3′ end of tRNA always terminates with CCA, and its integrity is essential because, as we have seen, the amino acid is attached at this end. The 3′ end can be degraded in vivo and this has shown that cells have an enzyme capable of terminal addition of CCA, which regenerates the active end. In fact, some tRNA genes do not contain in their DNA the information for the CCA end, which must be added post-transcriptionally.

An even more complex form of tRNA processing has been found in eukaryotes. Some (but not all) yeast tRNA genes contain in their DNA sequence an insertion of between 10 and 40 base pairs close to the anticodon site (Fig. 12–12). We have seen (Chapter 11) that many other eukaryotic genes are also interrupted by intervening sequences or introns. The extra nucleotides are transcribed into precursor RNAs and are excised, and the rest of the molecule is then rejoined by splicing enzymes, as shown in Figure 12–12.

## Importance of Aminoacylation

The tRNA molecule is directly involved in two events that determine the fidelity of protein synthesis. The first event is the selection of the correct amino acid by the aminoacyl-tRNA synthetase for incorpo-

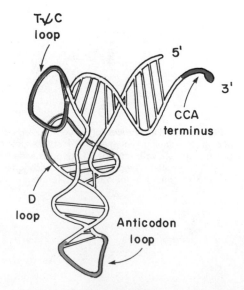

**Figure 12–11**  Tertiary structure of yeast phenylalanine transfer RNA showing a compact configuration with an L form. The position of the various parts shown in Figure 12–10 are indicated. Note that the D and TψC loops are hydrogen-bonded together (Redrawn from A. Rich, 1978.)

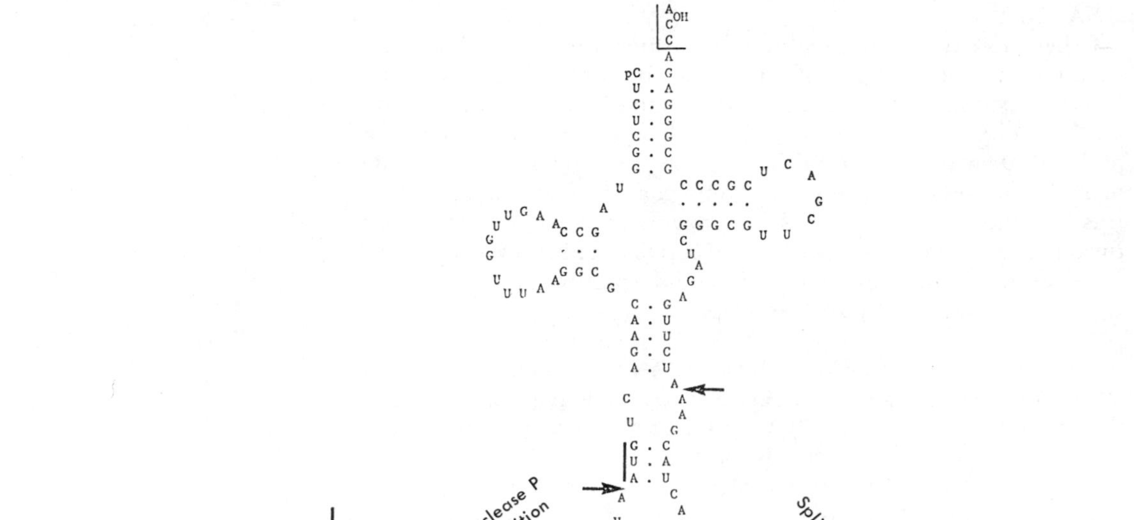

**Figure 12-12** Processing of yeast tyrosine-tRNA. *A*, primary transcript with a 5′ leader sequence of 19 nucleotides and an intervening sequence of 14 nucleotides. *B*, the 5′ leader sequence is cleaved by ribonuclease P (in a two-step process; not shown), and the CCA end is added post-transcriptionally. *C*, the intervening sequence is spliced and mature tRNA<sup>Tyr</sup> is formed. The position of the anticodon is underlined. The primary transcript was detected by microinjecting cloned yeast tRNA genes into the nucleus of *Xenopus* oocytes. (From E. M. De Robertis, and M. V. Olson, *Nature, 278*:134, 1979.)

ration into an aminoacyl-tRNA. The second event occurs in the ribosome and involves the selection of the correct tRNA in response to an mRNA codon.

For each amino acid there is a specific aminoacyl-tRNA synthetase which activates the carboxyl group of that amino acid for covalent bonding with the adenylic acid residue of the CCA end of the tRNA The first step of this reaction requires ATP:

$$AA + ATP \xrightleftharpoons[\text{synthetase}]{\text{aminoacyl}} AA{\sim}AMP + P{\sim}P$$

The AA~AMP intermediate formed remains bound to the enzyme until the proper tRNA arrives, and at that moment the formation of the AA~tRNA complex occurs:

$$AA{\sim}AMP + tRNA \xrightleftharpoons[\text{synthetase}]{\text{aminoacyl}} AA{\sim}tRNA + AMP$$

To perform these two functions, the activating enzyme has two sites: one to recognize the amino acid and another to recognize the specific tRNA.

The bond between the amino acid and the CCA end of the tRNA is a high-energy one which stores the energy from the ATP used in the first reaction. The energy thus stored is used in the ribosome to drive the formation of the peptide bond.

The considerable accuracy of the tRNA synthetases depends on this two-step reaction. The AA~AMP must remain attached to the enzyme until the tRNA arrives; if an incorrect amino acid were selected initially, during this period it would have a higher chance of falling off since it would have a decreased affinity for the enzyme. This results in a double check of the amino acid, first when it initially becomes bound to the enzyme, and second during the period when it must remain attached. This "proofreading" process is an essential feature in the translation of the genetic code.

# 12–4 PROTEIN SYNTHESIS

When not engaged in protein synthesis, most ribosomal subunits exist as a cytoplasmic pool of free subunits rather than as complete ribosomes (Fig. 12–13). The ribosomes are separated into subunits at the end of protein synthesis by a *dissociation factor* that binds to the 30S subunit. This dissociation factor has been identified as an initiation factor (discussed below) which is also required for the binding of mRNA to the 30S subunit. Thus, the same event that dissociates ribosomes enables the 30S subunit to start a new round of protein synthesis.

## Initiation of Synthesis

The first step in protein synthesis is the binding of the small ribosomal subunit to the *ribosome binding site* on the mRNA molecule;

**dissociation factor**—protein that binds to the 30S subunit after protein synthesis is completed and causes separation of the ribosomal subunits

**ribosome binding site**—series of bases on the mRNA molecule, including the AUG codon, to which the small ribosomal subunit binds

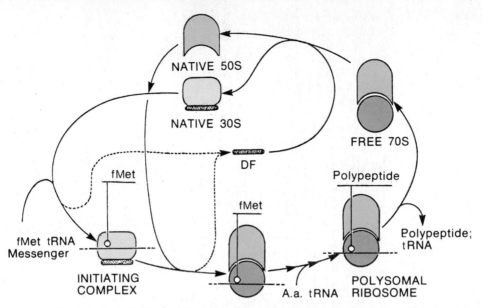

**Figure 12–13** A model of the ribosome-polysome cycle in which the action of a dissociation factor (*DF*) is indicated. DF is identical to initiation factor 3. (From A. R. Subramanian, E. Z. Ron, and B. D. Davis, *Proc. Natl. Acad. Sci. USA, 61*:761, 1968.)

this site contains the AUG codon at which protein synthesis starts. Ribosome binding sites in prokaryotic mRNAs have between three and eight bases that are complementary to the sequence at the 3' end (tail) of the 16S RNA of the small ribosomal subunit. This complementary region occurs about ten nucleotides before the start of the protein, and has been found in all prokaryotic mRNAs studied to date. It is thought that the end of the 16S RNA hybridizes to the mRNA sequences, thus aligning the initiation codon in the correct position to start protein synthesis.

The mRNA start signal is complex and involves other sequences in addition to the AUG codon. These additional sequences differentiate the starting AUG codon from the other AUG codons located within the message. The secondary structure of the mRNA (hairpin loops) may be important in the correct selection of the starting AUG.

**F-Met-tRNA—first aminoacyl-tRNA to initiate protein synthesis in bacteria**

In bacteria, the first AA-tRNA to initiate protein synthesis is always a formyl (—CHO) derivative of methionine, called *F-Met-tRNA*. Since in protein synthesis the peptide chain always grows sequentially from the free terminal —NH₂ group toward the —COOH end, the role of formylmethionine tRNA is to ensure that proteins are synthesized in this direction. In F-Met-tRNA the —NH₂ group is blocked by the formyl group, leaving only the —COOH available to react with the —NH₂ of the second amino acid. In this way synthesis proceeds in the correct sequence. The first amino acid is later separated from the protein by a hydrolytic enzyme.

The ribosome is able to discriminate between an AUG codon within as opposed to at the beginning of the mRNA. When this codon appears at the start of the mRNA, F-Met is incorporated, but when an AUG appears in the middle of a message, normal methionine is incorporated instead. It has been learned that there are two different tRNAs that respond to the codon AUG, an initiator F-Met-tRNA and one for normal methionine. Methionine is formylated only after it has been bound to this special initiator tRNA.

In eukaryotic cells synthesis is initiated by a special Met-tRNA, but in this case the methionine is not formylated. It remains true, however, that the initial methionine is generally split off from the finished polypeptide.

The small ribosomal subunits in *E. coli* also require three protein factors to start protein synthesis. These *initiation factors* are loosely associated with the 30S subunits and can be isolated by washing ribosomes with salt solutions. All three factors are essential for initiation when natural mRNAs are used as templates, but they are not required with artificial mRNAs such as poly U.

IF$_3$ binds to the 30S ribosomal subunit and is required for its binding to the mRNA starting site. IF$_3$ also functions as a ribosome dissociation factor, as diagrammed in Figure 12–13. IF$_2$ binds the initiator F-Met-tRNA and carries it to the ribosome (together with GTP) in response to the first AUG codon. IF$_1$ participates in the interaction between IF$_2$ and the initiator tRNA. The upper part of Figure 12–14 shows the main steps of protein synthesis initiation in

**initiation factors—three proteins (IF$_1$, IF$_2$, IF$_3$) necessary for protein synthesis initiation in prokaryotes, loosely associated with the small subunit and released once it binds to the large subunit**

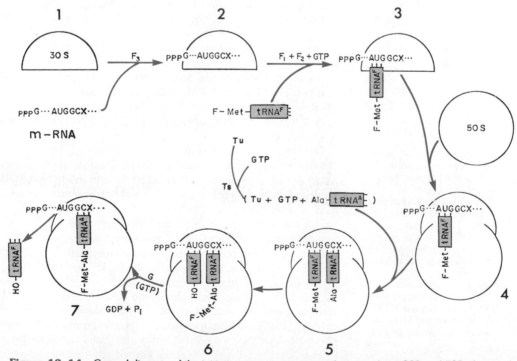

**Figure 12–14** General diagram of the initiation steps in protein synthesis involving 30S and 50S ribosomal subunits, messenger RNA, the initiation factors F$_3$ and F$_1$ + F$_2$, GTP, the elongation factors T and G, formylmethionine-tRNA (F-Met-tRNA) and alanine-tRNA. **1,** isolated 30S; **2,** binding of mRNA and F$_3$ to the 30S; **3,** binding of the F$_1$ + F$_2$ + GTP and F-Met-tRNA to make the initiation complex; **4,** binding of the 50S subunit to make the complete 70S ribosome; **5,** binding of the second aminoacyl-tRNA; **6,** synthesis of the first peptide bond; **7,** liberation of the free tRNA after translocation.

*E. coli*. Once the large ribosomal subunit binds to the small subunit, the initiation factors are released and can be reutilized.

## Elongation Factors

Once protein synthesis has been initiated, additional factors are required for the elongation of the peptide chain. The elongation factors Tu, Ts, and G are soluble proteins that can be isolated from bacterial cellular supernatants.

EFTu (Tu stands for temperature-unstable factor) forms a complex with an aminoacyl-tRNA and GTP and brings them to the ribosome (Fig. 12–14). In contrast with IF₂, which is used only to transport F-Met-tRNA to the ribosome, EFTu is used to carry all other aminoacyl-tRNAs to the site of chain elongation. Once the AA-tRNA is in place, the GTP is hydrolyzed and EFTu is released from the ribosome.

EFTs (Ts stands for temperature-stable factor) catalyzes the formation of the complex between EFTu, AA-tRNA, and GTP (Fig. 12–14).

The EFG factor, also called *translocase,* is involved in the translocation process that occurs when the ribosome moves from one codon to the next along the mRNA. G factor binds GTP (hence its name) and carries it to the ribosome, where it is hydrolyzed to GDP and inorganic phosphate. The energy released is used for the translocation process and for the release of the deacylated tRNA used to translate the previous codon.

The energy for peptide bond formation is stored in the AA~tRNA and comes from an ATP molecule used during the aminoacylation of tRNA. Thus, protein synthesis is expensive in terms of energy: for each amino acid incorporated, one ATP and two GTP molecules are required.

The large subunit contains the enzyme *peptidyltransferase,* or *peptide synthetase,* which catalyzes the formation of the peptide bond. Another function of the 50S subunit is to provide two binding sites for two tRNA molecules: the *aminoacyl* or *acceptor site* and the *peptidyl* or *donor site* (Fig. 12–15).

The stepwise growth of the polypeptide chain involves (1) the entrance of an aminoacyl-tRNA into the aminoacyl site; (2) the formation of a peptide bond and consequent ejection of the tRNA that was occupying the peptidyl site; and (3) the translocation of the tRNA (now carrying the peptide chain) from the aminoacyl to the peptidyl site. This process should be coupled with the simultaneous movement of the mRNA to place the following codon in position (Fig. 12–15). This translocation, in which the ribosome moves along the mRNA in the 5′ to 3′ direction, requires the G factor and GTP.

The velocity with which these coordinated processes occur can be illustrated by the fact that it takes only about 1 minute to construct a hemoglobin chain consisting of 150 amino acids.

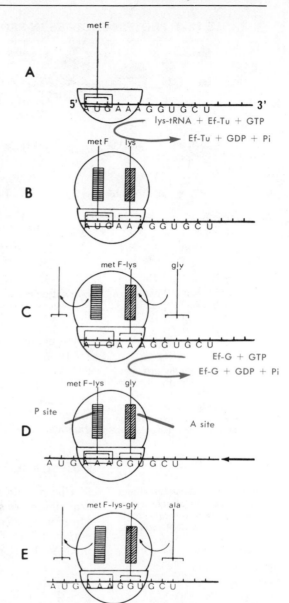

**Figure 12-15** Diagram representing the early stages of translation of messenger RNA (5'→3'). The *initiation site* in the 30S subunit is indicated by a white rectangle. The *aminoacyl* site is indicated by oblique stripes and the *peptidyl* site by horizontal stripes. **A,** initiation complex in which formylmethionine-tRNA (F-Met-tRNA) binds to the first codon in mRNA (AUG). **B,** the 70S ribosome has been formed, and the second aminoacyl-tRNA (Lys-tRNA) binds the second codon (AAA). The function of EFTu and GTP is indicated. **C,** the tRNA is eliminated from the peptidyl site, and the first peptide bond is formed; the function of EFG and GTP is indicated. **D,** translocation of the mRNA and of the peptidyl-tRNA has occurred, and a new aminoacyl-tRNA (Gly-tRNA) binds to the third codon (GGU). **E,** the molecular events of **C** are now repeated. (Adapted from S. Ochoa.)

## Chain Termination

The termination of the polypeptide chain occurs when the 70S ribosome carrying the peptidyl-tRNA reaches the termination codon located at the end of each cistron. Chain termination leads to the release of the free polypeptide and tRNA, and to the dissociation of the 70S ribosome into 30S and 50S subunits.

## TABLE 12-4  PROTEIN FACTORS IN PROTEIN SYNTHESIS

| Phase | Factor | Source | Function |
|---|---|---|---|
| Initiation | $IF_3$ | high-salt 30S | dissociation of ribosomal subunits |
| | $IF_1$ $IF_2$ $IF_3$ $\Big\}$ GTP | ribosomal wash | binding of mRNA and initiator tRNA to 30S subunit |
| Elongation | Tu Ts $\Big\}$ T + GTP | supernatant fraction | binding of aminoacyl-tRNA Tu-GTP complex to ribosome |
| | Peptidyl transferase | 50S ribosomal subunit | peptidyl transfer from peptidyl-tRNA to aminoacyl-tRNA |
| | G + GTP | supernatant fraction | translocation of peptidyl-tRNA; release of free tRNA |
| Termination | $R_1$ $R_2$ | supernatant fraction | release of protein at UAA, UAG, or UGA codons |

*From F. Lipmann, What do we know about protein synthesis? In F. T. Kenney, B. Hamkalo, G. Favelukes, and J. T. August, eds. *Gene Expression and Its Regulation.* Plenum, New York, 1973.

**releasing factors—two proteins ($R_1$, $R_2$) that recognize termination codons in the mRNA, leading to the release of the new protein and ribosome dissociation**

The chain termination codons are UAA, UGA, and UAG (Table 11–1). Unlike all other triplets, these codons are not recognized by tRNAs but rather by two specific proteins, the *releasing factors* $R_1$ and $R_2$. $R_1$ is specific for UAG and UAA, and $R_2$ is specific for UAA and UGA.

Table 12–4 summarizes the protein factors that are used in the various stages of polypeptide synthesis.

## Polysomes

**polysome—translational unit formed by the attachment of several ribosomes to a single mRNA molecule**

During protein synthesis each mRNA molecule is translated by several ribosomes simultaneously. The number of ribosomes in a *polysome* or *polyribosome* depends on the length of the mRNA. A fully active mRNA has a ribosome every 80 nucleotides, and longer mRNAs therefore have larger polysomes.

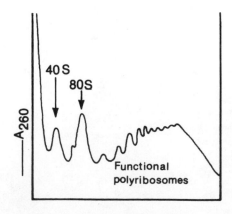

**Figure 12-16** Distribution of ribosomes and polysomes. (From R. P. Perry, J. R. Greenberg, D. E. Kelley, J. La Torre, and G. Schochetman, Messenger RNA: Its origin and fate in mammalian cells. In F. T. Kenney, et al., eds. *Gene Expression and Its Regulation.* Plenum, New York, 1973, p. 149.)

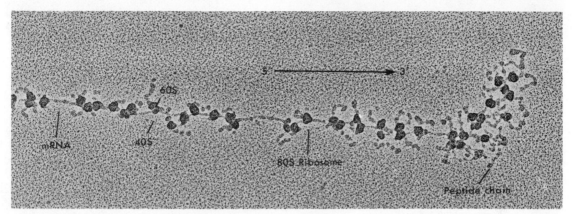

**Figure 12-17** Silk fibroin mRNA visualized during protein synthesis. A nascent peptide can be observed emerging from each ribosome and increasing in length in the 5′ to 3′ direction. The nascent peptide may be visualized in this case because silk fibroin is a fibrous protein and remains extended, whereas most other proteins adopt a globular conformation as they are being synthesized. (Courtesy of O. L. Miller, Jr.)

When eukaryotic cells are lysed and centrifuged on a sucrose gradient, the absorbance at 260 nm (which reflects mainly the location of the ribosomal RNAs, the most abundant nucleic acids in the cell) has the distribution shown in Figure 12-16. The polysomes appear as a series of peaks which are larger than the 80S monomer ribosomes, and therefore sediment closer to the bottom of the centrifuge tube.

The best way to visualize polysomes is to spread them for electron microscopy. Figure 12-17 shows a molecule of silk moth fibroin mRNA with its associated ribosomes. Each ribosome has a nascent protein chain attached to it which lengthens in the direction of protein synthesis (5′ to 3′). Fibroin mRNA is very long, and each polysome may contain up to 80 ribosomes; this means that there can be up to 80 polypeptides translated simultaneously from the same RNA molecule.

We have seen that in prokaryotes protein synthesis starts even before mRNA synthesis is completed, since there is no nuclear en-

| TABLE 12-5 MODE OF ACTION OF SOME ANTIBIOTICS USED IN MEDICAL PRACTICE | |
| --- | --- |
| **Antibiotic** | **Action on Bacterial Cells** |
| Streptomycin | misreading of mRNA |
| Tetracycline | binds to small ribosomal subunit and inhibits binding of AA-tRNA |
| Chloramphenicol | inhibits peptidyl transferase in large ribosomal subunit |
| Rifampicin | inhibits bacterial RNA polymerase |
| Penicillin and ampicillin | inhibit cell wall formation |

velope interposed between the genome and the ribosomes. Bacterial polysomes are therefore visualized with the nascent mRNA still attached to the DNA via the RNA polymerase (Fig. 11–7).

## Antibiotics Inhibit Protein Synthesis

antibiotic—substance whose medical usefulness is based on the fact that it inhibits prokaryotic cells (usually by acting on the protein synthesis machinery) but not eukaryotic cells

Many microorganisms (such as bacteria and fungi) produce substances that inhibit the growth of other microbes in the surrounding medium. Several of these substances, called *antibiotics*, have medical uses because they inhibit bacteria but do not affect eukaryotic cells. Many antibiotics act on the protein synthesis machinery, and their selective action on prokaryotic cells is based on the differences between prokaryotic and eukaryotic ribosomes. The mechanisms of action of some antibiotics are indicated in Table 12–5.

## Segregation of Newly Made Proteins

In several of the previous chapters we have discussed mechanisms by which the cell delivers newly synthesized proteins to the correct cellular compartments. It will be useful to review these briefly.

**SECRETORY PROTEINS.** Secretory proteins and some membrane proteins are translated on ribosomes attached to the endoplasmic reticulum membranes. These proteins contain at their amino end a *signal peptide* of between 15 and 30 amino acids, most of which are hydrophobic. When the signal peptide emerges from the large ribosomal subunit, it is recognized by a receptor protein in the ER membrane. The initial transport of the protein through the membrane is facilitated by the hydrophobic nature of the signal peptide. Once inside the lumen of the ER, the signal peptide is removed by a specific peptidase. This segregation of exported or membrane proteins occurs simultaneously with protein synthesis; that is, it is a *cotranslational* event (Fig. 12–18).

Bacteria do not have an endoplasmic reticulum but are nevertheless able to secrete proteins to the outside of the cell; for example, bacteria resistant to penicillin secrete the enzyme penicillinase into the surrounding medium. Bacterial extracellular proteins also start with a hydrophobic signal peptide which facilitates the movement of the nascent peptide through the plasma membrane. The direct demonstration that the signal peptide is indeed required for traversing the membrane came from genetic experiments with *E. coli* which showed that single amino acid changes in the signal peptide can abolish transport through the membrane. As expected, transport was most affected when hydrophobic amino acids in the signal peptide were replaced by hydrophilic amino acids.

**MITOCHONDRIAL AND CHLOROPLAST PROTEINS.** Many mitochondrial and chloroplast proteins are synthesized on cytoplasmic ribosomes and must gain access to the proper organelles.

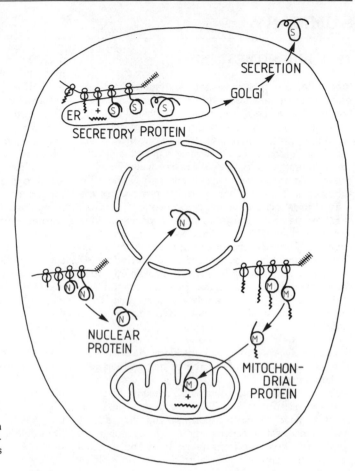

**Figure 12-18** Mechanisms of protein segregation. The correct segregation of secretory, mitochondrial, and nuclear proteins is achieved by different mechanisms.

These proteins are synthesized on free polysomes; the completed polypeptides are released into the cytoplasm and then traverse the organelle membrane (Fig. 12–18). Mitochondrial and chloroplast proteins are initially synthesized as long precursors with extra amino acids, which are removed when they enter the organelle. We do not yet know whether these extra amino acids are hydrophobic, but it is already clear that this mechanism differs from that of secreted proteins in that protein segregation can be achieved after protein synthesis is completed; secretory proteins can cross the ER membrane only as they are being synthesized.

**NUCLEAR PROTEINS.** Nuclear proteins can accumulate selectively in the correct compartment independently of protein synthesis. In this case the segregation property resides in the mature protein, and transient precursor proteins with extra amino acids are not involved (Fig. 12–18).

# SUMMARY

## 12-1 RIBOSOMES

Ribosomes are spheroidal particles that contain approximately equal amounts of RNA and protein. They are found in all cells, and serve as a scaffold for the ordered interaction of the numerous molecules involved in protein synthesis. Eukaryotic ribosomes have a sedimentation coefficient of 80S and large and small subunits of 60S and 40S, respectively. The 70S prokaryotic ribosomes have subunits of 50S and 30S. Ribosomes similar to those found in prokaryotes are present in mitochondria and chloroplasts.

Ribosomes are strongly negative, and bind cations and basic dyes. This is attributable to the many negative charges in the phosphates of the RNA molecules they contain; rRNA usually represents more than 80 percent of the total RNA in a cell. Prokaryotic ribosomes have a 16S RNA in the small subunit and 23S and 5S RNAs in the large subunit; eukaryotic RNAs are 18S in the small subunit and 28S and 5S in the large one, and also contain a 5.8S RNA that is transcribed in the nucleolus along with the 18S and 28S RNAs. 5S RNA is synthesized outside of the nucleolus. The rRNAs have a high degree of hairpin loop secondary structure.

An *E. coli* ribosome contains 55 small proteins, 21 in the small subunit and 34 in the large. These can be dissociated from the ribosome and then added back to reconstitute active particles. The complete assembly of ribosomes from naked RNA and proteins can also be demonstrated, illustrating the principle of biological self-assembly. All ribosomal proteins have been isolated and their function identified, and maps have been constructed of the positions they occupy within the subunits.

Prokaryotic and eukaryotic ribosomes are made up of different RNA and protein molecules, although they are functionally very similar, performing the same operations by the same series of chemical reactions.

## 12-2 THE NUCLEOLUS

The nucleolus is the site of ribosome synthesis in eukaryotic cells. It is formed around the nucleolar organizer, a region of DNA that contains many genes for the 18S, 5.8S, and 28S rRNAs. The nucleolar organizer is usually located in a secondary constriction on the chromosome. The various ribosomal components all collect in the nucleolus, where they are assembled into ribosomes and transported to the cytoplasm. The rRNA-synthesizing function of the

nucleolus has been confirmed in anucleolate *X. laevis* mutants, which were unable to produce 18S, 28S, or 5.8S RNA.

All eukaryotic organisms have multiple copies of the rRNA genes. Within the nucleolar organizer the 18S, 28S, and 5.8S genes are tandemly repeated and separated from each other by stretches of spacer DNA which is not transcribed. Nucleolar rRNA genes are transcribed by RNA polymerase I. The 5S rRNA genes are also present in multiple copies and are also arranged in a tandem repeat configuration. 5S RNA is transcribed by RNA polymerase III at non-nucleolar chromosomal sites, and subsequently it is transported to the nucleolus.

The nucleolar rRNA genes are amplified or replicated selectively in amphibian oocytes for use during early development. The amplified DNA is not inherited by the embryo. The 5S genes are not amplified, although all 24,000 of them are activated at this stage. Gene amplification is not a common event.

Nucleolar rRNA genes are transcribed into a long 45S or 40S precursor which must be cleaved into 18S, 28S, and 5.8S RNA. Within this primary transcript the RNAs appear in the order 5' – 18S – 5.8S – 28S – 3' and are separated by spacer RNA. Methylations of the 45S RNA occur rapidly and primarily on the ribose moiety, and only on those segments of the molecule that will be conserved. After 15 minutes the primary transcript is cleaved into its smaller components, having become associated with proteins while it was being transcribed.

Most nucleoli have a characteristic fibrillar zone, containing the rDNA that uncoils from the chromosome and penetrates the nucleolus along with the 45S RNA molecules attached to it, and a granular zone, representing ribosome precursor particles at various stages of assembly and processing. Once the ribosomal subunits are mature, they are released from the nucleolus and exit into the cytoplasm through the nuclear pores. The nucleoli undergo regular changes during the cell cycle that are related to their rRNA-synthesizing function.

## 12-3 TRANSFER RNA

Transfer RNAs are small RNAs that serve as molecular adaptors during protein synthesis. The tRNA molecule has an anticodon that can hydrogen-bond with the mRNA codon, and it has the amino acid corresponding to that codon attached to one of its ends. Accurate protein synthesis depends on the correct charging of these so-called aminoacyl-tRNAs by specific amino-acyl-tRNA synthetases. Some tRNAs recognize more than one codon third-base "wobble" in the anticodon. There is an isoacceptor tRNA family for each amino acid, and an aminoacyl-tRNA synthetase for each family.

All tRNAs share common features including a "cloverleaf" secondary structure, a CCA acceptor end for amino acid attachment, an anticodon loop, and D and T loops that have been conserved throughout evolution. The tertiary structure resembles an L, with the anticodon at one end and the amino acid attachment site at the other.

tRNA, like rRNA, must be processed from a longer precursor into its final form. Other types of post-transcriptional processing include the methylation of some bases, the terminal addition of CCA, and the excision of intervening sequences in some eukaryotic tRNAs.

Proper aminoacylation of the tRNA molecule is essential for accurate protein synthesis. The first step of this reaction requires ATP, which is later used in the ribosome to provide energy for peptide bond formation. The AA~tRNA complex is formed during the second step.

## 12–4 PROTEIN SYNTHESIS

The first step in protein synthesis is the binding of the small ribosomal subunit to the mRNA binding site, which contains an AUG codon to signal the start of translation, as well as additional sequences that differentiate the starting AUG codon from others within the message. In prokaryotes, the first AA-tRNA is always F-Met-tRNA. In eukaryotes, synthesis is initiated by an unformylated Met-tRNA. In both cases the initial methionine is removed from the finished polypeptide. Three protein initiation factors associated with the 30S subunit are also required for synthesis initiation in *E. coli*. Once the large ribosomal subunit binds to the small subunit, the initiation factors are released.

Other protein factors are required for polypeptide chain elongation. These factors are free in the cytoplasm and involved in bringing the correct AA-tRNAs to the ribosome and in the translocation process that occurs when the ribosome moves from one codon to the next. The large ribosomal subunit contains the enzyme that catalyzes the formation of the peptide bond, and provides aminoacyl and peptidyl sites that can accommodate the two tRNA molecules whose amino acids are being added to the protein chain.

Chain termination occurs when the 70S ribosome reaches a termination codon, which is recognized by one of two protein releasing factors.

During protein synthesis each mRNA molecule is translated by several ribosomes simultaneously, forming a structure called a polysome or polyribosome. This greatly increases the speed with which a long polypeptide can be transcribed. A fully active mRNA molecule has a ribosome every 80 nucleotides.

## STUDY QUESTIONS

- How are eukaryotic and prokaryotic ribosomes alike? How are they different?
- What is a polysome?
- Describe the steps involved in ribosome synthesis in a eukaryotic cell, beginning with the processing of the subunit components.
- How are the rRNA genes organized in the eukaryotic genome?
- How is the nucleolus organized?
- Describe the secondary and tertiary structure of a typical tRNA molecule.
- What is an aminoacyl-tRNA, and how is it produced?
- How is protein synthesis initiated?
- Explain how protein chain elongation takes place.
- Compare the mechanisms by which secretory, mitochondrial, and nuclear proteins are transported to the appropriate locations within the cell.

## SUGGESTED READINGS

**Blobel, G.** (1980) Intracellular protein topogenesis. *Proc. Natl. Acad. Sci. USA,* 77:1496.
**Clark, B.** (1980) The elongation step of protein biosynthesis. *Trends Biochem. Sci.,* 5:207.

**Davis, B. D., and Tai, P. C.** (1980) The mechanism of protein secretion through membranes. *Nature,* *283*:433.

**Hunt, T.** (1980) The initiation of protein synthesis. *Trends Biochem. Sci. 5:*178.

**Kozak, M.** (1980) Evaluation of a "scanning model" for initiation of protein synthesis in eukaryotes. *Cell, 22:*3.

**Maden, B. E. H.** (1977) Ribosomal precursor RNA and ribosome formation in eukaryotes. *Trends Biochem. Sci., 1:*196.

**Nomura, M.** (1969) Ribosomes. *Sci. Am., 221:*28.

**Ochoa, S.** (1979) Regulation of protein synthesis. *CRC Crit. Rev. Biochem., 7:*7.

**Watson, J. D.** (1975) *Molecular Biology of the Gene.* Benjamin, Menlo Park, California.

**Wickner, W.** (1980) Assembly of proteins into membranes. *Science, 210:*861.

**Wittmann, H. G.,** (1976) Structure, function, and evolution of ribosomes. *Eur. J. Biochem., 61:*1.

# chapter 13

# Regulation of Gene Expression

| 13–1 GENE REGULATION IN PROKARYOTES | 13–2 GENE REGULATION IN EUKARYOTES |
|---|---|
| **Inducible Systems: The *lac* Operon** **Repressible Systems: The *trp* Operon** **Enzymatic Control of Metabolism** | **Repetitive DNA** **Polytene Chromosomes** **Lampbrush Chromosomes** |

In previous chapters we have seen that the information encoded in DNA is transcribed into RNA and then translated into protein. In this chapter we will analyze the ways in which a cell regulates the expression of its genes. The ability to switch genes on and off is of fundamental importance, since it enables cells to respond to a changing environment and is the basis for the process of cell differentiation.

The rate of expression of bacterial genes is controlled at the level of mRNA synthesis (transcription). Regulation can occur at both the initiation and the termination of mRNA transcription. Since bacteria obtain their food from the medium that immediately surrounds them, their gene regulation mechanisms are designed to adapt quickly to changes in the environment.

Eukaryotes, however, have a much larger and more complex genome than prokaryotes; furthermore, the cells of higher organisms are surrounded by a constant internal milieu. The ability of such cells to respond to hormones and to impulses from the nervous system, for example, is thus comparatively more important than the ability to respond rapidly to the presence of certain nutrients. The complexity of the genome partly explains why our knowledge of how genes are controlled in eukaryotes is still incomplete. However, valuable information has been obtained from studies of polytene and lampbrush chromosomes, in which the transcription process can be seen directly in the microscope.

## 13–1 GENE REGULATION IN PROKARYOTES

### Inducible Systems: The *lac* Operon

We mentioned earlier that the cell is an energy-efficient unit that makes only the proteins it needs. *E. coli* has about 3000 protein-coding genes, but only a subset of them is expressed at any one time. The best illustration of this efficiency is provided by the enzymes involved in the utilization of the disaccharide lactose: the synthesis of these enzymes may be induced up to 1000-fold in response to the addition of lactose to the culture medium. Regulation by *enzyme induction* is also found in many other *catabolic systems* that degrade sugars, amino acids, and lipids. In these systems the availability of a substrate stimulates the production of the enzymes involved in its degradation.

The three enzymes required for the utilization of lactose are *β-galactosidase*, *lac permease*, and *transacetylase*, and the synthesis of all of them is regulated coordinately by a unit called an *operon* (Fig. 13–1). An operon is a group of genes that are next to each other in the DNA, and that can be controlled (i.e., turned on and off) in a unified manner. The *lac* operon consists of the genes *z, y,* and *a*, which code for the three enzymes, and two elements involved in their regulation, the *promoter (p)* and the *operator (o)*. The genes for the three enzymes are always transcribed together into a single *polycistronic* mRNA, which explains why they are always expressed together.

The expression of the *lac* operon is regulated by the *lac repressor,* a protein having four identical subunits of 40,000 daltons each (Fig.

**enzyme induction**—type of metabolic regulation in which the availability of a substrate stimulates production of the enzymes involved in its degradation

**operon**—group of genes (and regions governing their transcription) that occur sequentially in the DNA and can be regulated in a unified manner

**repressor**—protein that can bind strongly and specifically to a short region of DNA preceding the structural gene(s), thus preventing transcription

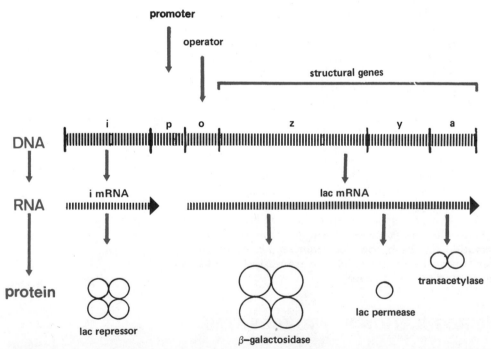

**Figure 13–1**   Diagram representing the *lac* operon. *i,* the regulatory gene that produces the *i* mRNA that codes for the *lac* repressor, a protein with four subunits. The promoter (*p*) is the region of attachment of RNA polymerase; the operator (*o*) is the region where the repressor binds. The *i* gene, the promoter, and the operator are regulatory elements. The structural genes *z, y,* and *a* produce a polycistronic *lac* mRNA, and the three proteins indicated below.

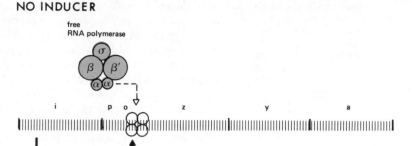

NO INDUCER

free
RNA polymerase

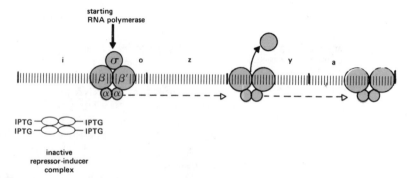

INDUCTION

starting
RNA polymerase

IPTG ⎯ IPTG
IPTG ⎯ IPTG

inactive
repressor-inducer
complex

**Figure 13-2** Diagram representing the regulation of the lactose operon in the absence and in the presence of the inducer. In the absence of the inducer, the *lac* repressor binds tightly to the operator, interfering with the transcription of the structural genes. The binding sites for the repressor and for RNA polymerase overlap, and RNA polymerase is therefore unable to bind when the repressor occupies the operator. The repressor is a tetramer with subunits of 40,000 daltons. When the inducer (IPTG) is present, it binds to the repressor, eliciting a conformational change that prevents its binding to the operator; as a consequence, RNA polymerase is free to transcribe the structural genes. Observe that *E. coli* RNA polymerase has five subunits of different sizes: $\sigma$ (95,000 daltons); $\beta/\beta'$ (about 160,000); and two $\alpha$ (40,000). Note that after transcription has started the sigma ($\sigma$) subunit is released and only the core enzyme remains bound to the DNA.

13–2). In an *E. coli* cell there are about ten *lac* repressor molecules, which are coded by the regulatory gene *i*. The repressor binds strongly and specifically to a short DNA segment called the *operator*, which is located very close to the start of the $\beta$-galactosidase gene. As shown in Figure 13–2, a repressor molecule bound at the operator site prevents the transcription of *lac* mRNA.

operator—short DNA region to which the repressor binds

The lac repressor binds to the following 21 base-pair sequence of operator DNA:

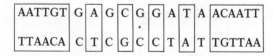

This sequence contains regions of twofold symmetry. Some sequences (boxed areas) on the left side of the operator are also present on the right side but on the opposite strand. Similar symmetrical regions have been found in other operators and in DNA-binding proteins (such as restriction enzymes), and probably reflect the fact that most proteins that recognize specific DNA sequences are multimeric. The two symmetrical sites perhaps represent recognition sites for different subunits of the repressor.

The affinity of repressor binding to the operator is regulated by inducer—molecule which, when bound to the repressor, causes a conformational change that prevents the repressor from binding to the operator

the *inducer,* a small molecule that can bind to the repressor. The natural inducer of the lactose operon is *allolactose,* a metabolite of lactose; however, the analog IPTG (isopropylthiogalactoside) is a more powerful inducer that is preferred in laboratory experiments. Each subunit of the repressor has one binding site for the inducer and, upon binding it, undergoes a conformational change by which it becomes unable to bind to the operator. In this way the presence of the inducer permits the transcription of the *lac* operon, which is no longer blocked by the repressor (Fig. 13–2).

The effect of this conformational change is dramatic. While in the absence of lactose *E. coli* cells have an average of only three molecules of β-galactosidase enzyme per cell, after induction of the lac operon 3000 molecules of β-galactosidase are present in each cell, representing 3 percent of the total protein. Furthermore, these adaptations occur very rapidly; most bacterial mRNAs have a half-life of only a few minutes, and therefore any increase or decrease in the rate of mRNA synthesis is rapidly reflected in the rate of protein synthesis.

promoter—DNA region to which RNA polymerase binds when initiating transcription

The *promoter (p)* is the DNA segment to which RNA polymerase binds when initiating transcription. Figure 13–3 shows the nucleotide sequence of the *lac* promoter region. Two sections are particularly important for RNA polymerase binding: (1) a TATGTTG sequence located 6 to 12 bases before the transcription starting site, which is conserved in most prokaryotic promoters, and (2) a region located 25 to 35 bases before the beginning of the mRNA, which is known to be important because mutations in it severely inhibit *lac* expression.

RNA polymerase binds to a region of about 80 nucleotides of DNA, and Figure 13–3 shows that the RNA polymerase binding site in fact overlaps with the region covered by the repressor (that is, the operator). In vitro studies have shown that a repressor bound to the operator blocks the binding of RNA polymerase. The way in which repressors work is therefore very simple: they bind within the promoter sites and prevent the attachment of RNA polymerase.

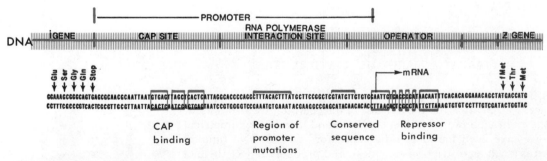

**Figure 13–3** Nucleotide sequence of the *lac* promoter, showing its relation to the overlapping operator sequences. Note that the repressor and CAP protein bind at regions of symmetry. The sequence TATGTTG, which is conserved in prokaryotic promoters, and the region where several promoter mutations occur are indicated. (Data from R. C. Dickson, J. Abelson, W. M. Barnes, and W. S. Reznikoff, *Science, 187:*27, 1975. Copyright 1975 by the American Association for the Advancement of Science.)

Cyclic AMP (cAMP) has a regulatory role in bacteria, activating inducible operons at the level of transcription. *E. coli* has a cyclic AMP receptor protein that is able to bind cAMP with high affinity. This protein is also known as "*c*atabolite gene *a*ctivator *p*rotein." CAP is a dimeric protein which, when complexed with cAMP, is able to bind to a specific site within the *lac* promoter region (Fig. 13–3). RNA polymerase will recognize the *lac* promoter only if the CAP-cAMP complex is already bound to it. Therefore, in the *lac* operon, in addition to the *negative control* provided by the repressor there is also *positive control* provided by the CAP-cAMP complex.

The positive control mechanism is also required for the expression of many other inducible operons, such as those involved in the utilization of maltose, galactose, and arabinose. However, cAMP is not necessary for the synthesis of those enzymes required for the utilization of glucose as an energy source. This mechanism is highly beneficial in *E. coli* because its cAMP levels vary according to the available food source. Bacteria growing in the presence of glucose have a lower cAMP content than those grown in a poorer energy source, such as lactose. When the intracellular cAMP level is low (i.e., when glucose is available) the CAP protein does not bind to the promoter region, and the *lac* operon will not be turned on even in the presence of lactose. As a result, if *E. coli* is grown in the presence of both glucose and lactose, it will utilize only the glucose. This makes good sense because glucose is the richest and most efficient energy source. It is clear that this mechanism of positive control enables *E. coli* to adapt more efficiently to the changing environment of its natural habitat, the human intestine.

## Repressible Systems: The *trp* Operon

The *tryptophan operon* codes for the five enzymes that are required for the synthesis of tryptophan (Fig. 13–4). It has been known for many years that the expression of this operon is regulated by the availability of tryptophan in the culture medium, and that the presence of tryptophan *represses* the synthesis of *trp* enzymes. Regulation by *enzyme repression* is also found in many other *anabolic systems* involved in the synthesis of amino acids or nucleic acid precursors, in which the synthesis of the enzymes of a metabolic pathway is selectively inhibited by the end product of that metabolic chain. This is another mechanism that enables *E. coli* to synthesize enzymes only when they are required.

Enzyme repression can also be explained on the basis of a model similar to that depicted for enzyme induction in the *lac* operon. In enzyme repression the regulatory gene produces a repressor protein which is normally in the inactive form. The repressor, upon binding with a metabolite called a *co-repressor* (in this case the amino acid tryptophan), undergoes a conformational change that enables it to bind to the operator and thereby inhibit the binding of RNA polymerase to the *trp* promoter (Fig. 13–4). The affinity of the repressor for binding to the operator is normally low, but it is increased by the action of the co-repressor. (This is in contrast to the situation in the

catabolite gene activator protein (CAP)—dimeric protein which, when complexed with cAMP, is able to bind within the promoters of some operons in order to facilitate recognition by RNA polymerase

enzyme repression—type of metabolic regulation in which the synthesis of the enzymes required for a metabolic pathway is selectively inhibited by the end-product of that pathway

co-repressor—metabolite which, when bound to the repressor, produces a conformational change that causes the repressor to bind to the operator with greater affinity

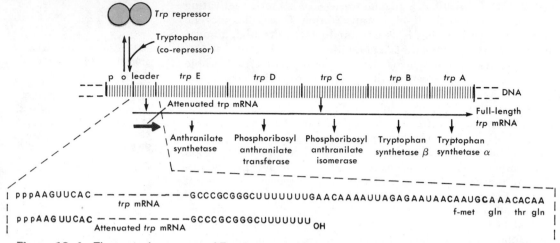

**Figure 13–4** The tryptophan operon of *E. coli* is regulated at transcription initiation (by the *trp* repressor) and at transcription termination. The leader RNA has a termination sequence called the attenuator, and two types of mRNA can be produced: a short leader RNA or a full-length mRNA that codes for the five enzymes of the operon. The proportion of the two RNAs is regulated by the tryptophan concentration in the cell. Note that as with many other terminators, the attenuated mRNA ends in a row of Us. (Data from Bertrand et al., *Science, 189*:22, 1975.)

*lac* operon, in which the repressor is active on its own and loses affinity for the operator when bound to the inducer.)

The tryptophan operon also has a second mechanism of regulation, which operates at the level of *termination of RNA synthesis.* Normally 90 percent of the RNA polymerase molecules that start transcription at the promoter terminate at a nearby *attenuator* site, yielding a short RNA (140 nucleotides) that does not code for *trp* enzymes. Only 10 percent of the molecules continue on to produce a full-length *trp* mRNA. Termination at the attenuator site is regulated by the level of tryptophan in the cells. When the tryptophan level is low, the proportion of RNA molecules yielding full-length *trp* mRNA increases. When it is high, most of the mRNAs terminate prematurely. Other repressible operons, such as those for histidine *(his)* and phenylalanine *(phe)* biosynthesis, also show similar regulation at the level of RNA termination.

The mechanism by which the metabolic end products regulate transcription termination became apparent when the nucleotide sequence of the short "leader" RNAs was analyzed, The three leader RNAs of the *trp*, *his*, and *phe* operons code for a short peptide between 14 and 16 amino acids long. Their striking feature is that many of these amino acids are the same as the one whose synthesis is directed by the corresponding operon; for example, the *his* operon leader has seven histidine codons in a row.

In these operons, transcription termination is regulated by the ribosomes, which in prokaryotes translate nascent RNAs simultaneously with transcription. The natural tendency of RNA polymerase is to terminate transcription prematurely at the attenuator site, but when the cells are starved for that amino acid, this termination signal is ignored. This is because a ribosome remains attached to

**attenuator site**—site of premature mRNA termination occurring in some operons; the frequency with which the site is used is regulated by the cellular level of the end product synthesized by the operon the leader RNA (instead of completing the synthesis of the short peptide) due to the insufficient concentration of Trp (or His, or Phe). The presence of an attached ribosome interferes with transcription termination by preventing the formation of a hairpin loop in the nascent RNA (required for termination) and the long mRNA that codes for the biosynthetic enzymes is produced.

In summary, *E. coli* uses two distinct mechanisms to ensure that the *trp* enzymes are not synthesized unless tryptophan is required. The use of two mechanisms which operate at different levels of tryptophan probably allows a finer control of *trp* enzyme synthesis over a wider range of tryptophan concentrations than would be possible with the repression mechanism alone.

## Enzymatic Control of Metabolism

Enzyme induction and repression provide a coarse control of metabolism by switching off the synthesis of enzymes when they are not required. However, as we discussed in Chapter 2, cells also have fine controls that operate directly at the level of enzyme activity. The most common mechanism is feedback inhibition, in which the end product of a metabolic pathway acts as an allosteric inhibitor of the first enzyme of its metabolic chain. This process is diagrammed in Figure 2–16, and should be reviewed in detail. Whereas gene induction and repression save valuable energy by preventing the synthesis of unnecessary enzymes, control of enzyme activity by allosteric regulation allows an almost instantaneous fine tuning of catalytic activity.

feedback inhibition—type of enzyme regulation in which the end-product of a metabolic pathway acts as an allosteric inhibitor of the first enzyme in that pathway

## 13–2  GENE REGULATION IN EUKARYOTES

## Repetitive DNA

The DNA of *E. coli* is about 1 mm long and contains approximately $4.5 \times 10^6$ base pairs. Table 13–1 shows that the genomes of eukaryotic cells contain much more DNA than this. For example, a bull

## TABLE 13-1  GENOME SIZE AND REPETITIVE DNA IN VARIOUS ORGANISMS

|  | Species | Base Pairs in Haploid Genome | Single Copy (%) | Moderately Repetitive (%) | Highly Repetitive (%) |
|---|---|---|---|---|---|
| Bacterium | *E. coli* | $4.5 \times 10^6$ | 100 | — | — |
| Fruit fly | *Drosophila melanogaster* | $1.4 \times 10^8$ | 74 | 13 | 13 |
| Frog | *Xenopus laevis* | $3.1 \times 10^9$ | 54 | 41 | 6 |
| Mouse | *M. musculus* | $2.7 \times 10^9$ | 70 | 20 | 10 |
| Cow | *B. domesticus* | $3.2 \times 10^9$ | 60 | 38 | 2 |

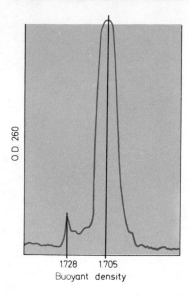

O.D. 260

1728   1705
Buoyant density

**Figure 13-5**   Density gradient in cesium chloride of the DNA extracted from cells of the salamander. Observe the main peak of DNA with a buoyant density of 1.705 and the small satellite peak of higher density (1.728). (From H. C. MacGregor, and J. Kezer, *Chromosoma, 33*:167, 1971.)

spermatozoon (haploid) contains $3.2 \times 10^9$ base pairs, which corresponds to a 700-fold increase in information relative to the size of the *E. coli* genome. This quantitative difference could be reflected, at least in principle, in an increased number of mRNAs and proteins. However, cell DNA content is not necessarily an accurate indicator of genome complexity: much of the DNA in eukaryotic cells is repeated.

The discovery of repeated sequences in the DNA of higher cells was a consequence of the ability of separated complementary DNA strands to recognize each other and to reassociate. The greater the number of repeated sequences, the faster the reassociation of such strands will occur. As an illustration of the amount of repeated DNA in a higher organism, the following percentages have been measured in mouse cells: 10 percent of the total DNA was found as short nucleotide sequences that were repeated in about 1 million copies; another 20 percent of the genome consisted of repeated sequences of 1000 to 100,000 copies; and the remaining 70 percent was non-repetitious DNA, present in only one copy.

The most highly repetitive DNAs in eukaryotes are often called *satellite* DNAs because they can frequently be separated from the bulk of the DNA as a *satellite peak* by equilibrium centrifugation in cesium chloride (Fig. 13-5). Cesium chloride gradients separate DNAs according to their buoyant densities, which in turn depend on their base composition: DNA rich in AT is less dense than GC-rich DNA.

**satellite DNA—highly repetitive DNA consisting of short nucleotide sequences arranged tandemly in the genome**

Satellite DNAs are frequently located in heterochromatic regions of the chromosome. The blocks of constitutive heterochromatin located near the centromeres have been shown to contain satellite DNA in a variety of organisms. This was established by hybridizing labeled nucleic acids in situ to chromosome preparations, as shown in Figure 13-6.

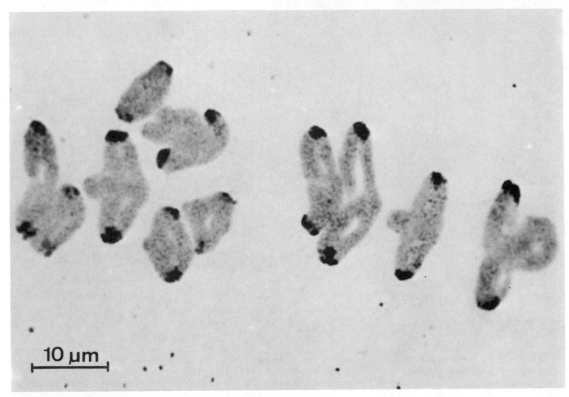

**Figure 13–6**  Meiotic metaphase I chromosomes of salamander processed for *in situ* hybridization. Incubation with tritiated RNA complementary to the heavy satellite DNA shown in Figure 13–5. Observe that the labeling has occurred in the centromeric region of the chromosomes. (Courtesy of H.C. MacGregor.)

Satellite DNAs are formed by short nucleotide sequences that are tandemly repeated many times in the genome. Their function is not known, and most of them are not transcribed into RNA. From an evolutionary standpoint, satellite DNAs tend to change rapidly; there is little selective pressure to keep the sequences constant, and as a result satellites vary widely between species.

Most protein-coding genes are present only once per haploid genome, although there are exceptions. The genes coding for the five histone proteins are repeated about 400 times in the sea urchin and between 10 and 20 times in humans. The genes for all five of these proteins are arranged in clusters at a few chromosome loci, and are found in a tandem (head-to-tail) arrangement with the clusters separated by segments of AT-rich spacer DNA. In Figure 13–7 one such set of genes has been partially denatured to reveal the AT-rich spacers and the five genes.

Tandem repeats of genes alternating with stretches of spacer DNA seem to be a central feature in the organization of the eukaryotic genome.

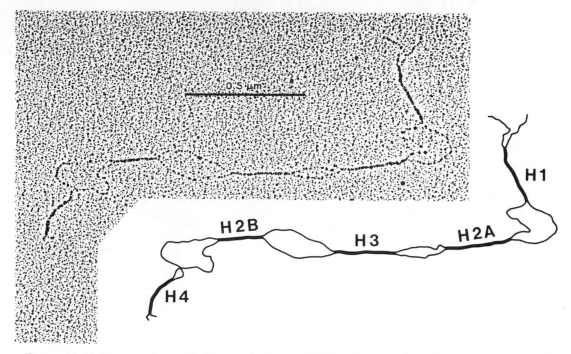

**Figure 13-7** Electron micrograph of a partially denatured DNA molecule containing the five histone genes. The genes are separated by spacers that are AT-rich and therefore melted at a lower temperature than the gene regions. This molecule was cloned in *E. coli* and cut with a restriction enzyme. In the cell, many units of these five genes are tandemly repeated. (Courtesy of R. Portmann and M. L. Birnstiel.)

## Polytene Chromosomes

polytene chromosome— chromosome (usually studied in the cells of Dipteran larvae) in which the DNA replicates in such a way that the daughter chromatids do not separate, producing instead a characteristic morphology of bands and interbands of variable gene activity

Polytene chromosomes have provided the best evidence to date that eukaryotic gene activity is regulated at the level of RNA synthesis. They constitute a valuable material for the study of gene regulation because their gene transcription can be seen directly in the microscope.

Some cells of the larvae of Diptera (flies, mosquitoes, and midges) are very large and have a high DNA content (Fig. 13–8). When the chromosomes in one of these cells become *polytenic*, the DNA replicates by *endomitosis* and the resulting daughter chromatids do not separate; instead they remain aligned side by side. A giant chromosome of a *Drosophila* salivary gland, for example, contains about 1000 DNA fibers and arises from ten rounds of DNA replication (i.e., $2^{10} = 1024$). Polytene cells are unable to undergo mitosis and are destined for cell death. Not all the cells in a Dipteran larva have polytene chromosomes, however; those destined to produce the adult structures after metamorphosis (called *imaginal discs*) remain diploid.

The chromosomes of polytenic cells are visible during interphase, and have a characteristic morphology in which a series of dark *bands* alternate with clear zones called *interbands* (Fig. 13–9). The bands represent regions where the DNA is more tightly coiled, while

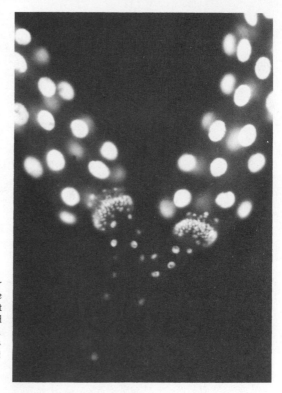

**Figure 13-8**  Salivary glands of a *Drosophila melanogaster* larva showing variable degrees of polyteny. The cells were stained with ethidium bromide, a fluorescent compound that binds to DNA. Note that the secretory cells (top) are larger and contain more DNA than those of the excretory duct (bottom). The rings of small nuclei located between the excretory ducts and the secretory cells are the *imaginal discs*, which are diploid and destined to produce the salivary glands of the adult fly (Courtesy of M. Jamrich.)

in the interbands the DNA fibers are folded more loosely. There are 5000 bands in the four *Drosophila* chromosomes, and it is possible that each one represents a single genetic unit (studies have shown that there are about 5000 genes in *Drosophila*). However, this now seems improbable in the light of experiments that have involved the cloning of *Drosophila* DNA segments in bacteria. These experiments have shown that at least some of the bands contain multiple genes.

Polytene chromosomes have become even more important with the advent of recombinant DNA techniques, because they make it possible to map any DNA segment to specific chromosomal loci by in situ hybridization. Polytene chromosomes are suitable for in situ hybridization because their 1000 DNA molecules are aligned side by side, thereby greatly increasing the possibility of detecting single-copy genes.

One of the most remarkable characteristics of polytene chromosomes is that it is possible to visualize in them the genetic activity of specific chromosomal sites at local enlargements called *puffs* (Fig. 13-9). A puff can be considered a band in which the DNA unfolds into open loops as a consequence of intense gene transcription.

In 1952 W. Beerman compared the polytene chromosomes of different tissues of *Chironomus* larvae and showed that although the pattern of bands and interbands was similar in all tissues, the distribution of puffs differed from one tissue to another. For example, salivary glands have characteristic puffs which are now known to

**puff—local enlargement of a polytene chromosome in which active gene transcription can be visualized**

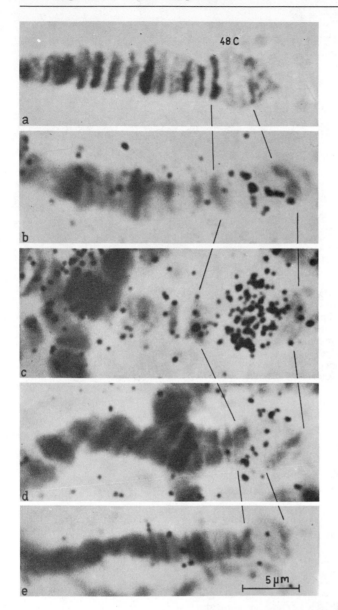

**Figure 13-9** Experiment to demonstrate how ³H-uridine is incorporated into puff 48 C of *Drosophila hydei*. **a**, control; **b**, after 3 minutes of temperature shock (a few silver grains have been deposited); **c**, 15 minutes after temperature shock. Observe the greatly increased incorporation. **d**, 15 minutes after injection of actinomycin D in larvae having puffs as in **c**. **e**, 30 minutes after injection of the antibiotic in animals with puffs as shown in **c**. These results indicate that in 15 to 30 minutes most of the synthesized RNA has been eliminated from the puff. (Courtesy of H. D. Berendes.)

contain genes coding for proteins of the larval saliva; these genes are expressed abundantly in this tissue. Beerman correctly interpreted the puffs as sites of intense RNA synthesis and concluded that different tissues transcribe different genes preferentially. To this day puffs remain the best evidence indicating that cell specialization results from variable gene transcription.

Puffing is a cyclic and reversible phenomenon; at definite times, and in different tissues of the larvae, puffs may appear, grow, and disappear. Puff formation can be studied experimentally using factors that will induce their formation. The steroid hormone *ecdysone*, which induces molting in insects, will induce the formation of spe-

ecdysone—steroid hormone that can be used to induce chromosomal puffing

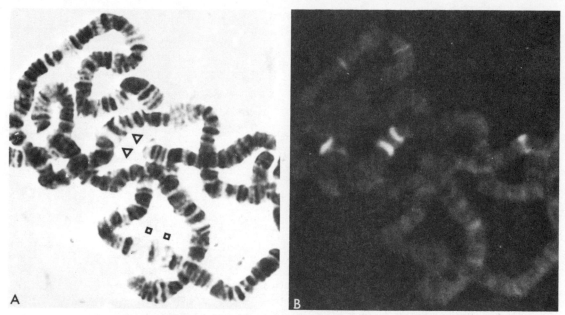

**Figure 13-10** Localization of RNA polymerase II in heat-shock puffs. Larvae were placed at 37° C for 45 minutes before preparation of chromosomes. **A,** orcein stain; **B,** immunofluorescence. Triangles indicate the largest heat-shock puffs, 87 A and 87 C. The squares indicate preexisting puffs. RNA polymerase II accumulates in the heat-shock puffs while it disappears from other regions of the chromosome (even from the preexisting puffs). (Courtesy of E. K. F. Bautz.)

cific puffs when injected into larvae or when added to salivary glands in culture. Puffing can also be induced by *temperature shock*. When *Drosophila* larvae normally grown at 25°C are exposed to a temperature of 37°C, a series of specific genes is activated while most other genes are repressed. Five minutes after heat shock, several new puffs are visible on the giant chromosomes of the salivary glands (Fig. 13–10). These puffs contain RNA polymerase and are very active in RNA synthesis, as can be demonstrated by $^3$H-uridine labeling and autoradiography. The newly made RNAs code for eight specific heat shock proteins, and in this experimental situation it is clear that the induction of these proteins is due to an increased rate of transcription of individual genes.

**temperature shock—** exposure of *Drosophila* larvae to high temperatures in order to induce chromosomal puffing and the production of so-called heat shock proteins

## Lampbrush Chromosomes

Lampbrush chromosomes occur at the diplonema stage of meiotic prophase in the oocytes of all animal species, but are best visualized in those species (such as salamanders) that have a high DNA content and large chromosomes. Lampbrush chromosomes are so named because their many lateral loops of DNA give them a characteristic appearance (Fig. 13–11) reminiscent of the brushes used to clean the chimneys of oil lamps in pre-electricity days.

The axis of a lampbrush chromosome has a row of granules or *chromomeres*, regions in which the DNA is tightly folded and transcriptionally inactive. Each chromomere has two lateral loops (one

**lampbrush chromosome—** chromosome with highly extended lateral loops of DNA; found in all animal oocytes at the diplonema stage of meiosis

**chromomere—**granule found along the axis of a decondensed chromosome, in which the DNA is tightly folded and transcriptionally inactive

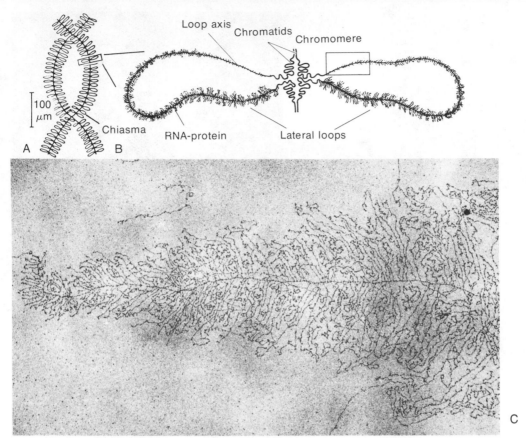

**Figure 13-11** A diagram of the lampbrush chromosomes of *Triturus* oocytes. **A,** low magnification showing two chiasmata. **B,** higher magnification showing two lateral loops and the folding of the chromatids in the chromomere. **C,** electron micrograph of a region corresponding to the inset in **B.** Note the DNA axis upon which fine fibrils of ribonucleoprotein are inserted perpendicularly. Observe that they increase in length toward the right. ×15,000. (Courtesy of O. L. Miller, Jr.)

for each chromatid) in which the DNA is extended as a result of intense RNA synthesis. Each loop in turn has an axis formed by a single DNA molecule, which is coated by a matrix of nascent RNA with proteins attached to it. The matrix is asymmetrical, being thicker at one end of the loop. RNA synthesis starts at the thinner end and progresses toward the thicker end, as can be shown in preparations spread for electron microscopy. The appearance of such molecules is very similar to the "fern leaf" or "Christmas tree" configuration that can be observed during ribosomal RNA synthesis.

The study of lampbrush chromosomes has shown that the units of transcription in eukaryotes are very large, involving DNA segments between 10 and 30 $\mu$m long. It has also been shown that the processing of these huge RNAs starts even before transcription is terminated. This can be inferred from the experiment shown in Figure 13-12, in which lampbrush chromosomes were hybridized in

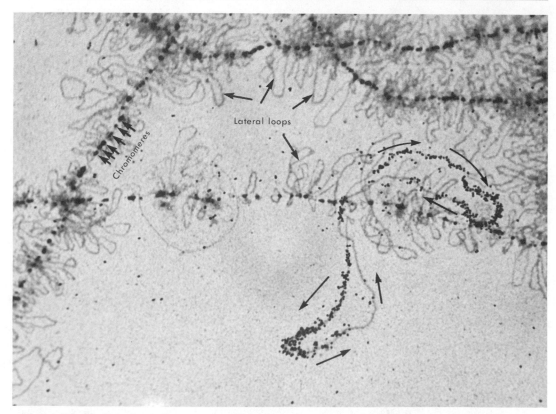

**Figure 13-12**  Lampbrush chromosomes of *Triturus cristatus* oocytes hybridized with labeled cloned genes under conditions in which hybridization to nascent RNA chains could be detected. One pair of loops is labeled. Lateral loops are always in pairs, one for each sister chromatid. The arrows indicate the direction of transcription. The hybridization ceases abruptly before the end of the loop. This mechanism is interpreted in Figure 13–13. ×1000. (Courtesy of H. G. Callan.)

situ to a labeled segment of cloned DNA under conditions that allow hybridization to the nascent RNA chains only. Hybridization did occur, but the labeling did not continue throughout the loop; rather, it ceased suddenly before the end of the loop. That is, the hybridizing sequence was removed from the nascent RNA chain before the transcription of the loop was completed, as indicated in Figure 13–13.

The diplonema stage of meiotic prophase lasts for about 200 days in salamander oocytes, and this is a period of great synthetic activity during which a single nucleus produces all the RNA required to build an egg cell more than 1 mm in diameter. Furthermore, there is evidence that some of the mRNAs synthesized on these chromosomes are stored in the cytoplasm during oogenesis and used later during embryonic development. Lampbrush loops may therefore be considered an adaptation that allows intense transcription of prophase chromosomes by producing localized uncoiling of the chromatin fibers.

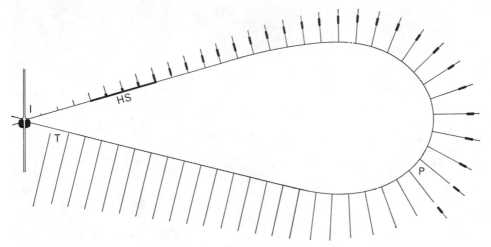

**Figure 13-13** Diagram interpreting the experiment described in Figure 13–12. The DNA in the lamp-brush loop is drawn as a single line that is thicker in the region corresponding to the hybridizing sequences (*HS*). The RNA transcripts project perpendicularly to the loop and hybridize with the labeled DNA, indicated by a heavier tracing. *I,* initiation site; *T,* termination site; *P,* site where processing of the mRNA occurs. At this site the mRNA is cleaved off before the transcription of the loop is completed. (Courtesy of H. G. Callan, from R. W. Old, H. G. Callan, and K. W. Gross, *J. Cell Sci., 27*:57, 1977.)

# SUMMARY

## 13–1 GENE REGULATION IN PROKARYOTES

Regulation by enzyme induction occurs in many catabolic systems, in which the availability of a substrate stimulates the production of the enzymes involved in its degradation. The synthesis of the three enzymes required for lactose utilization, for example, is coordinately regulated by the presence of lactose in the culture medium. *Lac* operon expression is regulated by the *lac* repressor, which binds to a DNA segment called the operator and thus prevents RNA polymerase from binding to the promoter. The affinity of repressor binding to the operator is regulated by the inducer (a derivative of lactose). When bound to the repressor, the inducer causes a conformational change in the repressor that prevents it from binding to the operator. When this occurs, RNA polymerase has access to the promoter site on the DNA, and transcription can take place. cAMP also has a regulatory role in bacteria. *E. coli* also contains a protein, CAP, which when complexed with cAMP is able to bind to a specific site within the *lac* promoter region. RNA polymerase recognizes the *lac* promoter only if the CAP-cAMP complex is already bound to it. This positive control mechanism is also required for the expression of many other inducible operons, although cAMP is not necessary for the synthesis of those enzymes required for the utilization of glucose as an energy source.

Regulation by enzyme repression occurs in many anabolic systems, in which the production of the enzymes of a synthetic pathway is selectively inhibited by the end product of that metabolic chain. The *trp* operon, for example, is repressed by the presence of tryptophan in the medium. In enzyme repression the regulatory gene produces a repressor protein that is normally inactive. Upon binding with a co-repressor, the repressor undergoes a conformational change that enables it to bind to the operator. The *trp* operon is also regulated at the level of termination of RNA synthesis. When the Trp level in the cell is low, the proportion of RNA molecules yielding *trp* mRNA is high; when the Trp level in the cell is high, most of the mRNAs terminate prematurely at an attenuator site.

Enzyme induction and repression provide a coarse control of metabolism by switching enzyme synthesis on or off as necessary. A finer and instantaneous control mechanism is provided by feedback inhibition, in which the end product of a metabolic pathway acts as an allosteric inhibitor of the first enzyme of that pathway.

## 13–2 GENE REGULATION IN EUKARYOTES

Eukaryotes contain far more DNA than prokaryotes, although much of the DNA in eukaryotic cells is repeated. The discovery of repeated sequences was a consequence of the fact that separated complementary strands can recognize each other and reassociate; the greater the number of repeated sequences, the faster strand reassociation will occur. The most highly repetitive eukaryotic DNAs, or satellite DNAs, can be separated from the rest of the DNA by centrifugation in CsCl gradients. Satellite DNAs are frequently located in heterochromatic regions of the chromosome, and are formed by short nucleotide sequences that are tandemly repeated many times. Their function is not known and their sequences vary widely between species. Tandemly repeated genes alternating with stretches of spacer DNA are a characteristic feature of the eukaryotic genome.

Polytene chromosomes have provided the best evidence to date that eukaryotic gene activity is regulated at the level of RNA synthesis. When Dipteran chromosomes become polytenic, the DNA replicates by endomitosis, and the resulting daughter chromatids remain aligned side by side. These chromatids are visible during interphase, and have a characteristic morphology of dark bands and alternating interbands. Within these chromosomes it is possible to observe the genetic activity of specific loci at local enlargements called puffs, which represent DNA undergoing intense gene transcription. Puff distribution varies from one tissue to another and can be induced experimentally, indicating that cell specialization results from variable gene transcription.

Lampbrush chromosomes occur at the diplonema stage of meiotic prophase in the oocytes of all animal species. Study of these chromosomes has shown that eukaryotic transcription units are very long, and that the processing of their mRNAs starts even before transcription is terminated. Lampbrush loops allow the intense transcription of prophase chromosomes by producing localized uncoiling of the chromatin fibers.

# STUDY QUESTIONS

- In what ways is gene regulation in eukaryotes different from that in prokaryotes?
- Why are all the enzymes of an operon expressed coordinately?
- What is the *lac* operon? Describe two mechanisms by which it is regulated.
- Explain how the *trp* operon is controlled and regulated.
- What is repetitive DNA? Where is it found, and how is it organized?
- What are polytene chromosomes? Describe their morphology.
- Describe the phenomenon of puffing. How is it related to the process of cell differentiation?
- What are lampbrush chromosomes?

# SUGGESTED READINGS

**Ashburner, M., and Bonner, J. J.** (1979) The induction of gene activity in Drosophila by heat shock. *Cell, 17:*241.

**Britten, R. J., and Kohne, D. E.** (1968) Repeated sequences in DNA. *Science, 161:*529.

**Fedoroff, N. V.,** (1979) On spacers. *Cell, 16:*997.

**Jacob, F., and Monod, J.** (1961) Genetic regulatory mechanisms in the synthesis of proteins. *J. Mol. Biol., 3:*318.

**Johnston, H., Barnes, W. M., Chumley, F. G., et al.** (1980) Model for regulation of the histidine operon of Salmonella. *Proc. Natl. Acad. Sci. USA, 77:*508.

**MacGregor, H. C.** (1980) Recent developments in the study of lampbrush chromosomes. *Heredity, 44:*3.

**Miller, J. H., and Reznikoff, W. S., eds.** (1978) *The Operon.* Cold Spring Harbor Laboratory, New York.

**Pastan, I., and Perlman, R.** (1970) Cyclic adenosine monophosphate in bacteria. *Science, 169:*339.

**Smith, G. P.** (1978) What is the origin of repetitive DNAs? *Trends Biochem. Sci., 3:*N34.

# chapter 14

# Cell Differentiation

| 14–1 NUCLEOCYTOPLASMIC INTERACTIONS | 14–2 MECHANISMS OF CELL DIFFERENTIATION |
|---|---|
| Nucleocytoplasmic Interactions in *Acetabularia* <br> Reactivation of Erythrocyte Nuclei by Cell Fusion <br> Transplanted Nuclei in *Xenopus* Oocytes | Constancy of the Genome during Cell Differentiation <br> Control at the Level of Transcription <br> Importance of Cytoplasmic Localizations in Development |

Cell differentiation is the process by which stable differences arise between cells. All higher organisms develop from a single cell, the fertilized ovum, which gives rise to the various tissues and organs. The question of how an apparently structureless egg converts itself into a complex and highly organized embryo has interested scientists since the time of Aristotle 2000 years ago, and still remains one of the major unanswered questions of biology.

In most animal species females produce large unfertilized eggs which contain most of the materials and nutrients required to form an embryo. Development is triggered by *fertilization*. The sperm contributes a small, condensed nucleus (the *male pronucleus*), which rapidly enlarges in the egg cytoplasm, fuses with the *female pronucleus,* and finally divides. The fertilized egg then undergoes a series of very rapid cycles consisting of DNA synthesis followed by cell divisions. These divisions are called *cleavage,* because unlike normal cell division, the cytoplasm is partitioned without growth. Then the cells form a hollow sphere *(blastula)* in which tissues are not yet evident (Fig. 14–1). Some of the cells then invaginate in a series of cell movements known as *gastrulation,* and the first signs of morphological differentiation appear (Fig. 14–1). These complex changes take place in a comparatively short time. For example, in the South African frog *Xenopus laevis*, a swimming tadpole containing most differentiated tissues (such as blood, nerve, eye, muscle, and so forth) hatches only 72 hours after fertilization.

In molecular terms, cell differentiation means *variable gene activity* in different cells of the same organism. Cell specialization involves the preferential synthesis of some specific proteins (e.g., hemoglobin in erythrocytes, antibodies in plasma cells, and ovalbumin in the oviduct). Each eukaryotic cell expresses only a small percentage of the genes it contains, and the cells of different tissues express different sets of genes. It is therefore clear that cell differentiation will be explained in molecular terms only when the complex mechanisms of gene regulation in higher cells are understood.

cleavage—rapid cycles of DNA synthesis followed by cell division in which the cytoplasm is partitioned without growth, occurring very early in embryonic development

blastula—early embryonic form consisting of a hollow sphere of cells

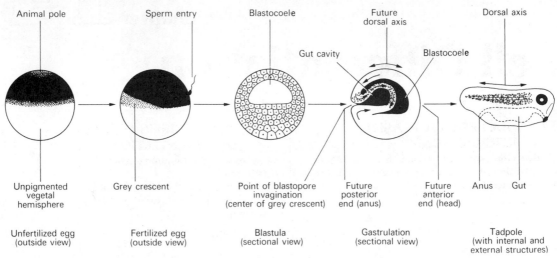

Animal pole · Sperm entry · Blastocoele · Future dorsal axis · Dorsal axis

Gut cavity · Blastocoele

Unpigmented vegetal hemisphere · Grey crescent · Point of blastopore invagination (center of grey crescent) · Future posterior end (anus) · Future anterior end (head) · Anus · Gut

Unfertilized egg (outside view) · Fertilized egg (outside view) · Blastula (sectional view) · Gastrulation (sectional view) · Tadpole (with internal and external structures)

**Figure 14-1**  Development of *Xenopus laevis*. During gastrulation cells invaginate at the blastopore, and extensive morphogenetic movements of cells occur (arrows). The blastopore always forms at the center of a pigmented area called the *grey crescent,* which appears in eggs opposite to the point of sperm entry shortly after fertilization. Thus, the orientation of the dorsal axis of the embryo is determined by the point of sperm entry. (Courtesy of J. B. Gurdon, from *Gene Expression During Cell Differentiation.* Carolina Biology Readers, Vol. 25. Carolina Biological, Durham, North Carolina, 1978.)

In this chapter we will analyze the possible mechanisms by which the initial differences between cells arise in early embryos. The first part of the chapter will be devoted to the general characteristics of cell differentiation and to *nucleocytoplasmic interactions.* This is because a large part of our knowledge of the mechanisms by which differentiation is established and maintained comes from experiments in which nuclei are placed in a foreign cytoplasm. We will then analyze the level at which differential gene expression is controlled in early development, and we will examine the possible role that is played in this process by *determinants,* or molecules located in specific regions of the egg cytoplasm.

## 14-1  NUCLEOCYTOPLASMIC INTERACTIONS

One of the principal characteristics of cell differentiation in higher cells is that once established, the differentiated state is very stable and can persist throughout many cell generations. For example, a neuron will persist as such throughout the lifetime of an individual, and a cell committed to become a skin cell will gradually keratinize and eventually die. These persistent changes are very different from the type of regulation involved in enzyme induction and repression in bacteria, which is specially designed to respond rapidly to changes in the environment.

### Nucleocytoplasmic Interactions in *Acetabularia*

The interdependence of the nucleus and cytoplasm is perhaps best illustrated by experiments with the unicellular marine alga *Aceta-*

*bularia.* The nucleus produces messenger, ribosomal, and transfer RNAs, while the cytoplasm provides the cell with energy in the form of ATP, and synthesizes the proteins and most other structural components. *Acetabularia* has a stem between 3 and 5 cm long and a cap 1 cm in diameter (Fig. 14–2). The nucleus is located in the basal or *rhizoid* end of the cell. The great experimental advantage of *Acetabularia* is that the nucleus can be removed simply by cutting off the rhizoid. The resulting enucleated cells are active in photosynthesis and can survive for many weeks.

Each *Acetabularia* species has a particular cap morphology (Fig. 14–2). If a nucleus from *A. crenulata* is implanted into an enucleated *A. mediterranea*, the resulting cell will eventually develop a cap of the *A. crenulata* type. If two nucleate cells of different species are grafted together, a hybrid having a cap with intermediate morphology is formed, as shown in Figure 14–2. This result clearly demonstrates that the shape of the cap is determined by the nucleus.

Although dependent on the nucleus, the cytoplasm has some degree of functional autonomy. The information required to form the cap passes to the cytoplasm before cap formation, where it can be stored in an inactive form. *A. crenulata* does not form a cap in the darkness, only an elongated stem. If such a stem is enucleated and then exposed to light, a cap will develop. Therefore, the cap formation triggered by light is not dependent on new nuclear activity, and all the necessary information is present in the cytoplasm but is not converted to the final active form until a light stimulus is pres-

*Acetabularia*—genus of unicellular marine alga well suited for the study of nucleocytoplasmic interactions

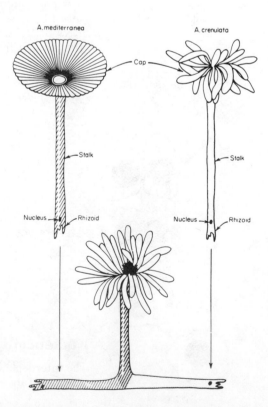

**Figure 14–2**  Experiment in nuclear grafting between two species of the unicellular alga *Acetabularia*. (See description in text.) The type of cap formed appears to depend on substances synthesized by the implanted nucleus. (From J. Hämmerling, *Ann. Rev. Plant Physiol.*, 14:65, 1963.)

ent. This cap formation can be prevented by inhibitors of protein synthesis. The information for cap production is very stable; enucleated stalks maintained in the dark for weeks are still able to produce a cap upon illumination. This experimental result is probably due to the presence of stable mRNA molecules stored in the cytoplasm.

The nucleus itself is dependent on the cytoplasm. *Acetabularia* has a giant nucleus in its rhizoid, but its volume decreases markedly when cytoplasmic energy production is impaired by darkness or by amputation of a large part of the cytoplasm.

## Reactivation of Erythrocyte Nuclei by Cell Fusion

heterokaryon—a single cell containing nuclei of two types, produced by the fusion of two different cells

synkaryon—hybrid cell line generated by a heterokaryon whose nuclei have entered mitosis synchronously and merged

HeLa cell—cell from the undifferentiated line originally derived from the uterine carcinoma cells of a patient named Henrietta Lacks

The fusion of cells using inactivated *Sendai virus* (a parainfluenza virus) makes it possible to place a nucleus in a different cytoplasmic environment. The initial product of the fusion of two different cells is a *heterokaryon* (i.e., a single cell containing nuclei of two types), as shown in Figure 14–3. Eventually, both nuclei might enter mitosis synchronously, form a single metaphase plate, divide, and produce a *hybrid cell line* (also known as a *synkaryon*). The cells of a hybrid cell line have a single nucleus containing chromosomes from both parental nuclei.

In 1965 H. Harris found that chick erythrocyte nuclei are reactivated when fused to HeLa cells (an undifferentiated cell line so named because it was derived from the uterine carcinoma cells of a woman named *Henrietta Lacks*). These heterokaryons are of interest because the erythrocyte nucleus does not normally synthesize RNA or DNA. Chick erythrocytes are terminally differentiated cells that have a highly condensed nucleus and are destined to die (Fig.

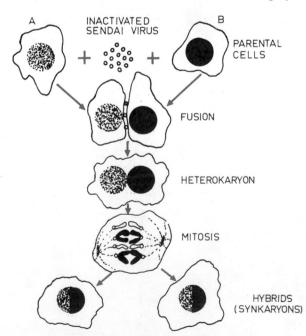

A  INACTIVATED SENDAI VIRUS  B  PARENTAL CELLS

FUSION

HETEROKARYON

MITOSIS

HYBRIDS (SYNKARYONS)

**Figure 14–3** Cell fusion with Sendai virus results first in the production of a heterokaryon with two nuclei, one from each parent. A hybrid cell line (synkaryon) arises when the two nuclei undergo mitosis synchronously. The hybrid cell line has one nucleus which contains chromosomes from both parents, and is capable of multiplying in cell culture. (Modified from N. R. Ringertz and R. E. Savage, *Cell Hybrids.* Academic, New York, 1976.)

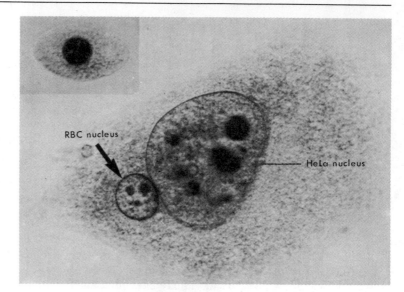

RBC nucleus

HeLa nucleus

Figure 14-4 Heterokaryon formed by fusing a chick erythrocyte with a human HeLa cell. The inset shows a normal chick erythrocyte with condensed inactive chromatin. Three days after fusion the red blood cell nucleus has enlarged, and its chromatin has dispersed. (Courtesy of N. R. Ringertz.)

14-4, inset). (Mammalian erythrocytes normally eliminate their nuclei during red blood cell maturation, whereas the red cells of birds, amphibians, and reptiles retain their nuclei, which become inactivated.) When fused to HeLa cells, the chick erythrocyte nucleus increases 20 times in volume, disperses its chromatin, resumes RNA synthesis, develops a nucleolus, and eventually replicates its DNA. This process is accompanied by the uptake of large amounts of human nuclear proteins which are thought to reactivate the erythrocyte nucleus.

These experiments clearly show that the synthesis of macromolecules in a nucleus is controlled by the cytoplasmic environment. Even though the erythrocytes are terminally differentiated cells, they can resume RNA and DNA synthesis.

Only the cell cytoplasm is required to reactivate the chick erythrocyte nucleus. This was established by fusing erythrocytes to enucleated HeLa cells, which were still able to reactivate the nuclei. Populations of enucleated cultured cells can be obtained by centrifugation after the use of cytochalasin B, a drug that affects mainly actin microfilaments. Cells enucleated by the cytochalasin method are viable for at least two days after enucleation and perform many cell functions normally. For example, cell movements, pinocytosis, and contact inhibition are unaffected, and if cells are detached from the plastic surface of a petri dish, they can reattach and spread on a new dish. It is evident from this that many cytoplasmic functions are independent of the cell nucleus.

Somatic cell hybridization experiments show that the patterns of nucleic acid synthesis and gene expression by a nucleus can be modified by substances present in the cell cytoplasm. Similar conclusions can be drawn from work with nuclei transplanted into *Xenopus laevis* oocytes.

## Transplanted Nuclei in *Xenopus* Oocytes

Frog oocytes are growing egg cells, obtained from the abdominal cavity of frogs, which due to their large size (1.2 mm) and tolerance for micromanipulation have been used in a variety of microinjection experiments. Aside from their hardiness, oocytes interest developmental biologists because they are destined to become eggs and embryos and already contain most of the molecular machinery necessary for early development. Oocytes are active in RNA synthesis but do not synthesize DNA, and if somatic cell nuclei are injected into them, the transplanted nuclei also show this type of synthetic activity. The mature frog egg (which is laid into pond water), in contrast, does not synthesize RNA but actively synthesizes DNA after penetration by sperm or a microinjection needle. Nuclei transplanted into mature eggs also mimic the behavior of the host cell, actively replicating their DNA but not synthesizing RNA.

The oocyte cytoplasm not only affects the pattern of macromolecular synthesis but is also able to *reprogram* the expression of individual genes in transplanted nuclei. Figure 14–5 shows *Xenopus* oocytes that were injected with a suspension of about 200 HeLa nuclei. The nuclei survive for several weeks inside the oocytes and during that period synthesize substantial amounts of RNA. In the first few days after injection the somatic nuclei enlarge to about ten times their original volume. This change is accompanied by chromatin dispersion and a massive uptake of proteins from the surrounding cytoplasm. The injected nuclei resemble the oocyte's own nucleus morphologically, and sometimes the human chromosomes form structures reminiscent of lampbrush chromosomes (Fig. 14–5).

The resemblance between the injected and oocyte nuclei is not only morphological. The oocyte cytoplasm reprograms the gene expression of the injected nuclei in such a way that only those genes which are normally active in oocytes are expressed. This can be determined by taking advantage of the fact that the RNAs synthesized by the injected nuclei in the course of several days accumulate in the cytoplasm, where they code for new proteins.

In one such experiment, *Xenopus* cultured kidney cell nuclei were injected into oocytes of a different amphibian species (the salamander *Pleurodeles*, whose proteins can be readily distinguished from those of *Xenopus* by two-dimensional electrophoresis). Those genes that are normally expressed in kidney cells but not in oocytes became inactive after injection into *Pleurodeles* oocytes. More importantly, some oocyte-active genes which were not expressed by the kidney cell nuclei were activated by the oocyte cytoplasm. It is therefore clear that the oocyte contains components which can determine that a particular spectrum of protein-coding genes will be active and that others will be inactive.

The work with cell fusion and *Xenopus* oocytes indicates that the cytoplasm of all cells contains components which determine the state of activity of nuclear genes. In the next section of this chapter we will see that cytoplasmic gene-controlling substances (or determinants), which sometimes become unequally distributed in the cytoplasm, play a crucial role in establishing the initial steps of cell differentiation in early embryos and adult tissues.

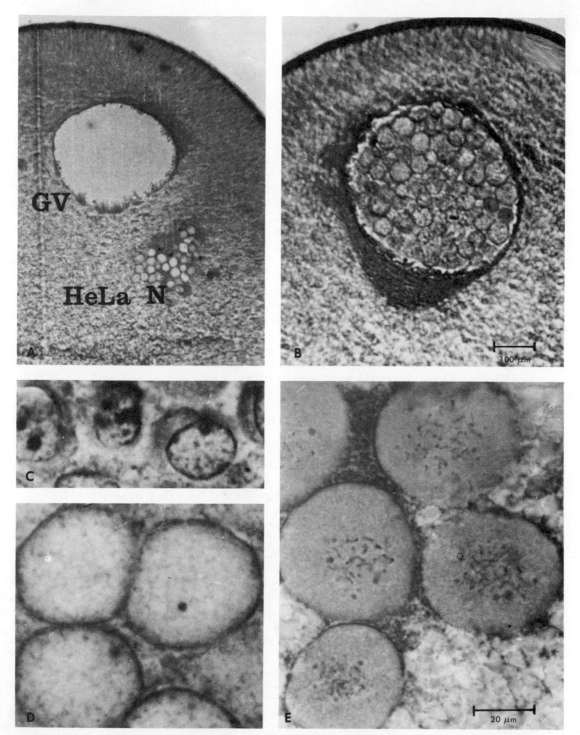

**Figure 14-5** *Xenopus* oocytes injected with human HeLa nuclei. *GV*, germinal vesicle or nucleus of the oocyte. **A,** HeLa nuclei injected into the cytoplasm three days previously. **B,** HeLa nuclei injected into the oocyte nucleus enlarge further. **C,** HeLa nuclei immediately after injection into the cytoplasm. **D,** three days later, the nuclei have swollen more than 10-fold. **E,** one week after injection the HeLa chromosomes become visible, and tend to resemble the nucleus of the oocyte, which is in meiotic prophase (late lampbrush stage). (From E. M. De Robertis.)

## 14-2 MECHANISMS OF CELL DIFFERENTIATION

### Constancy of the Genome during Cell Differentiation

We will now analyze at which level cell differentiation is controlled, and will examine the experimental evidence indicating that the genome remains constant throughout cell differentiation. Specialized cells from different tissues contain the same kind and number of genes, and therefore the differences between specialized cells must be explained in terms of variable rates of gene expression. Later in this chapter we will arrive at the conclusion that cell differentiation is probably controlled at the level of RNA transcription.

Nineteenth century embryologists devoted much attention to the possibility that cell differentiation could be due to the loss of those genes that are not expressed. However, massive loss of chromosomal material cannot be the cause of cell differentiation, since we know that the DNA content of diploid cells is constant in all tissues of an organism (Chapter 7).

Rigorous proof that there is no loss of genetic information during embryonic differentiation came from the classic nuclear transplantation experiments performed by J. B. Gurdon. As shown in Figure 14-6, *Xenopus laevis* unfertilized eggs can be irradiated with ultraviolet light to destroy the endogenous nucleus and can then be injected with a single *Xenopus* diploid nucleus. Nuclei obtained from *Xenopus* tadpole intestinal cells (which are clearly differentiated cells, since they have a "brush border" of microvilli) are able to sustain the development of normal, fertile adult frogs. This demonstrates that the intestinal cells have retained all the genes required for the complete life cycle of a frog. In addition, since many frogs can be obtained from the intestine of the same tadpole, nuclear transplantation makes it possible for a *clone* of genetically identical siblings to be created. Development up to the stage at which the tadpole swims has been obtained using a variety of adult tissues (such as keratinizing skin cells and lymphocytes), thus demonstrating that the genes required to make nerve, blood, muscle, cartilage, and other tadpole tissues were not irreversibly inactivated in the donor nuclei.

Similar conclusions have been obtained using plant cells. It is common knowledge that whole plants can be grown from cuttings. In some cases, such as in carrots, a complete plant can be grown from a single cultured cell.

One possible way of obtaining differential gene activity would be to increase the number of copies of a given gene. Since some differentiated cells produce large amounts of certain gene products (for example, 90 percent of the protein synthesized in a reticulocyte is globin), it seemed conceivable that this could be the case. However, nucleic acid hybridization studies have shown that there is only one copy per haploid genome of globin genes in all tissues, regardless of whether the gene is preferentially expressed. In other words, a

**nuclear transplantation—** technique involving the destruction of an unfertilized egg's endogenous nucleus with ultraviolet light, followed by its replacement with a single diploid nucleus from a differentiated cell

**clone—**line of genetically identical siblings originating from a single individual

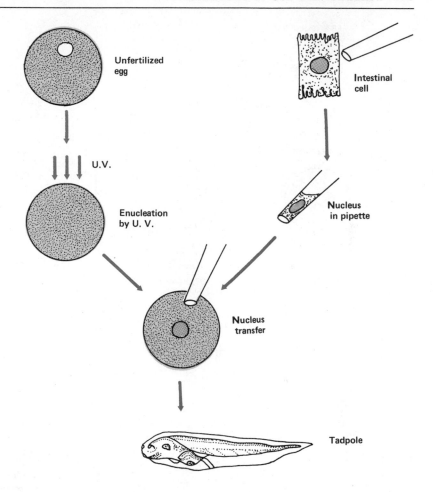

**Figure 14–6** Schematic diagram of the experiment on nuclear transplantation in an unfertilized egg of *Xenopus laevis*. The egg is treated with ultraviolet irradiation to destroy the nucleus. The nucleus of an intestinal cell is drawn into a micropipette and then implanted in the egg. The result is a normal tadpole. (From J. B. Gurdon.)

Unfertilized egg

Intestinal cell

U.V.

Enucleation by U. V.

Nucleus in pipette

Nucleus transfer

Tadpole

single globin gene, when fully activated, can give rise to all the globin required by a red blood cell.

We have seen that differential gene expression does not result from changes in the *kind* or *number* of genes. Another possibility is that it could arise from changes in the *location* of genes in the chromosomes. Gene rearrangements and translocations do occur in the case of antibody-producing cells; however, the DNA sequences adjacent to the globin and ovalbumin genes have been analyzed by cleaving cell DNA with restriction enzymes and were found to be the same regardless of the tissue from which the DNA was extracted. Similarly, no differences were found in the restriction enzyme fragments of *Drosophila* DNA extracted from embryonic and adult tissues.

## Control at the Level of Transcription

Since the genome remains constant throughout development, the differences between cells must be due to differential gene activity;

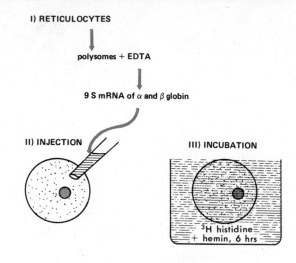

I) RETICULOCYTES

polysomes + EDTA

9 S mRNA of α and β globin

II) INJECTION

III) INCUBATION

³H histidine + hemin, 6 hrs

IV) HOMOGENIZATION AND PROTEIN ANALYSIS ON SEPHADEX G 200

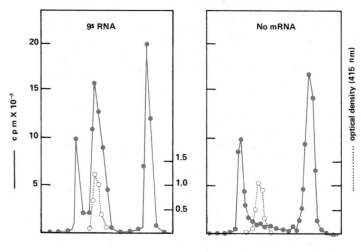

9s RNA

No mRNA

c p m X 10⁻³

optical density (415 nm)

**Figure 14–7** Diagram of the experiment of the injection of hemoglobin messenger RNA into an amphibian egg and demonstration of its translation into hemoglobin molecules. (From J. B. Gurdon.)

that is, some genes must be expressed at a higher rate than others. This could be explained by differential transcription (RNA synthesis or processing) or by differential translation (protein synthesis) of genes. One possibility is that cells may have mechanisms by which some mRNAs are translated in a given cell type but not in others. Experiments in which mRNA is injected into living cells argue strongly against this possibility.

Figure 14–7 shows that when purified mRNA coding for rabbit globin is injected into *Xenopus* oocytes, it is translated into globin protein. A large number of mRNAs of animal and plant origin are efficiently translated in oocytes (including those coding for immunoglobulins, interferon, collagen, viral coat proteins, and many others). This shows that oocytes do not have a mechanism that excludes the translation of certain mRNAs. Thus, protein synthesis does not seem to be regulated in the way required to explain the expression of, for example, hemoglobin in red blood cells but not in other tissues.

Several lines of evidence suggest that eukaryotic gene expression is regulated at the level of RNA synthesis. The clearest example is provided by polytene chromosomes in which transcription of genes can be visualized in the form of puffs. As discussed in Chapter 13, specialized cells have distinct patterns of puffing (and therefore of transcription) which differ in different tissues.

Although many mRNAs are different between tissues, many others are shared by different cell types. The shared sequences represent the so-called *housekeeping genes* required for the survival of all cells, such as, for example, genes coding for membrane proteins, glycolytic enzymes, ribosomal and mitochondrial proteins, and so on. The unshared sequences represent the so-called *luxury functions* which are characteristic of specialized cells, for example, hemoglobin in red blood cells, ovalbumin in the oviduct, and keratin in skin.

The abundance of transcripts for individual genes such as globin, ovalbumin, and so forth can be measured by nucleic acid hybridization studies, and for all differentiated genes that have been examined, the experimental evidence is consistent with the view that production of specialized proteins is due to differential gene transcription.

## Importance of Cytoplasmic Localizations in Development

We will now discuss briefly how the first differences appear between the cells of an early embryo. We mentioned in the first part of this chapter that the cytoplasm of oocytes and other cells contains molecules that influence the activity of the nucleus. Eggs have such substances in their cytoplasm, and in some cases these are localized in specific regions of the egg. As development proceeds, these substances, called *determinants* of development, are unequally distributed in the cytoplasm of certain groups of cells, which then become committed to a particular type of cell differentiation.

Eggs are generally very large cells that stockpile many of the molecules required for early development. For example, a *Xenopus* egg contains about 100,000 times more RNA polymerases, histones, mitochondria, and ribosomes than does a normal adult *Xenopus* somatic cell. One reason for accumulating these ready-made materials rather than making them *de novo* during early embryogenesis is that cell division is extraordinarily rapid during cleavage. At the midblastula stage, one of the first stages of development, *Xenopus* replicates its DNA every 10 minutes, a process that takes 19 hours in the somatic cells of adults. Similarly, *Drosophila* blastulae double their DNA content in 3.4 minutes.

This rapid pace allows little time for new RNA and protein synthesis, but it is during this period that the first differences between

**determinants**—molecular substances that, upon becoming unequally distributed into the cytoplasm of certain groups of cells, commit those cells to a particular course of differentiation

nuclei are established. At least in some cases, these differences result from the presence of the determinants mentioned above. The best example of determinants in development is provided by the *germ plasm* (Fig. 14–8). Amphibian cells contain in their so-called vegetal (yolky) pole a specialized region of cytoplasm which can be recognized morphologically by the presence of special granules. This cytoplasm has the property of inducing germ cell formation; that is, those cells that contain the germ plasm will eventually become the germ cells of the new organism. When the posterior poles of the eggs are irradiated with ultraviolet light, sterile (but otherwise normal) animals are obtained. The effect of ultraviolet treatment can be reversed or *rescued* by injecting cytoplasm containing pole plasm determinants into irradiated eggs.

Although determinants are undoubtedly very important in establishing early differences between cells, they cannot entirely explain development. For example, there is no evidence of cytoplasmic localization in mammalian eggs. Furthermore, as development advances, *cell interactions* become increasingly important. At the stage of gastrulation, extensive cell movements and migrations occur, and different types of cells interact with each other in the phenomenon known as *embryonic induction.* Notochord tissue, for example, induces the overlying ectoderm to become neural tissue, and the optic vesicles (an outgrowth of the brain) induce nearby ectoderm to become the eye lens. These inductions are mediated by diffusible substances, but in spite of numerous attempts to isolate them, their chemical nature remains unknown.

It is possible that the same principles involved in the action of egg determinants could also apply to adult cells. At the beginning of this chapter we saw that all cells contain in their cytoplasm molecules that can reprogram gene expression. Cells continually exchange information between the nucleus and cytoplasm, so that gene products accumulated in the cytoplasm can subsequently modify nuclear activity. If these substances are localized in certain regions of the cytoplasm, upon cell division they can become unequally distributed between daughter cells, giving rise to two different cell types.

Figure 14–8 shows how this might occur in adult and embryonic cells. In the differentiation of adult tissues it is frequently observed that only one of the daughter cells becomes specialized; the other remains as a *stem cell*, which is able to divide again. This occurs in red blood cell differentiation and could occur in skin and intestinal epithelium, in which the dividing cells are located in certain regions of the tissue (attached to the basal membrane or at the bottom of the intestinal crypts).

The hypothetical mechanism shown in Figure 14–8 is supported by experimental evidence. During nerve cell differentiation in grasshoppers, some cell divisions result in the formation of a neuron (ganglion cell) and a stem cell (neuroblast) which are always in the same position and morphologically recognizable. By introducing a needle at mitosis in one experiment, it was possible to rotate the spindle and chromosomes by 180 degrees, but in spite of this manipulation the resulting daughter cells had the neuron and stem cell

embryonic induction— developmental phenomenon in which different types of cells interact with each other to induce specific regional differentiations

stem cell—undifferentiated cell capable of giving rise to specialized daughter cells

(a) Propagation of the determined or specialized state

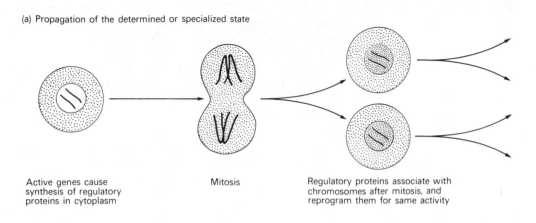

Active genes cause synthesis of regulatory proteins in cytoplasm

Mitosis

Regulatory proteins associate with chromosomes after mitosis, and reprogram them for same activity

(b) Unequal division following unequal distribution of cytoplasm

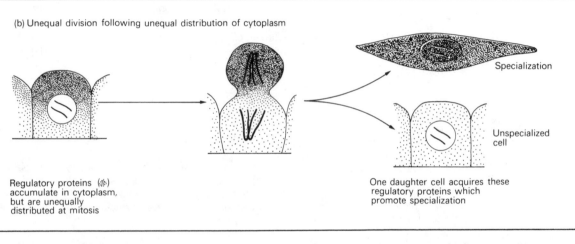

Specialization

Unspecialized cell

Regulatory proteins (⚛) accumulate in cytoplasm, but are unequally distributed at mitosis

One daughter cell acquires these regulatory proteins which promote specialization

(c) Unequal distribution of materials in egg (e.g., germ plasm in Amphibia)

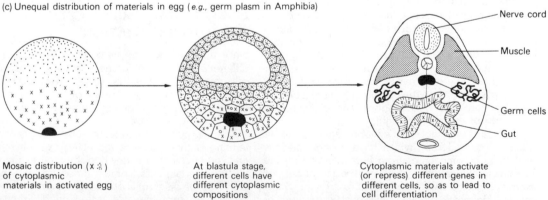

Nerve cord

Muscle

Germ cells

Gut

Mosaic distribution (x ⚛) of cytoplasmic materials in activated egg

At blastula stage, different cells have different cytoplasmic compositions

Cytoplasmic materials activate (or repress) different genes in different cells, so as to lead to cell differentiation

**Figure 14-8** Possible mechanism for the cytoplasmic control of cell differentiation. **a,** under normal conditions the daughter cells have the same differentiated state as the parent cell. **b,** the parent's gene-regulating cytoplasmic molecules become unequally distributed, and one of the daughter cells differentiates, while the other remains as a stem cell. **c,** egg cells have unequally distributed determinants which cause embryonic cells to specialize in different ways, as in the case of the germ plasm, which induces germ cell differentiation. (Courtesy of J. B. Gurdon, from *Gene Expression During Cell Differentiation*. Carolina Biology Readers, Vol. 25. Carolina Biological, Durham, North Carolina, 1978.)

in the normal positions. This shows that the ability to become a neuron does not depend on a particular chromosome set but rather on the type of cytoplasm inherited by the daughter cell.

The idea that the cytoplasm contains determinants that can become unequally distributed in the daughter cells and affect nuclear activity is by no means new. In the 1896 edition of his classic book *The Cell in Development and Heredity*, E. B. Wilson viewed development as follows:

> If chromatin be the idioplasm [i.e., an old term referring to the genes] in which inheres the sum-total of hereditary forces, and if it be equally distributed at every cell-division, how can its mode of action so vary in different cells as to cause diversity of structure [i.e., differentiation]? Through the influence of this idioplasm [i.e., the genes] the cytoplasm of the egg, or of the blastomeres derived from it, undergoes specific and progressive changes, each change reacting upon the nucleus and thus inciting a new change. These changes differ in different regions of the egg because of pre-existing differences, chemical and physical, in the cytoplasmic structure; and these form the conditions under which the nucleus operates.

# SUMMARY

## 14-1 NUCLEOCYTOPLASMIC INTERACTIONS

Once established, the differentiated state in higher cells is very stable and can persist throughout many cell generations.

The interdependence of the nucleus and cytoplasm can be illustrated by experiments with *Acetabularia,* a unicellular marine alga. If a nucleus from one species is implanted in a different, enucleated species, the cap that forms on the cell will have the morphology genetically specified by the new rather than by the original nucleus. The information necessary for cap formation passes first into the cytoplasm, where it can be stored in stable but inactive form and subsequently activated (for example) by light.

The effects of cytoplasmic environment can also be studied in hybrid cells, which have a single nucleus containing chromosomes from both parental nuclei. Terminally differentiated chick erythrocyte nuclei, for example, can be reactivated when fused to undifferentiated HeLa cells, showing that nuclear synthesis of macromolecules is controlled by the cytoplasmic environment. Other experiments have shown that many cytoplasmic functions can occur independently of the cell nucleus.

*Xenopus* oocytes are large and hardy cells that are active in RNA sythesis. The oocyte cytoplasm is able to reprogram the expression of individual genes in transplanted somatic nuclei, in such a way that only those genes which are normally active in oocytes are expressed.

## 14-2 MECHANISMS OF CELL DIFFERENTIATION

Since the genome remains constant throughout cell differentiation, the differences between specialized cells must be explained in terms of variable rates of gene expression. Several lines of evidence suggest that eukaryotic gene expression is regulated at the level of RNA synthesis.

The egg cytoplasm contains substances called determinants that influence the activity of the nucleus and, by becoming unequally distributed in the cytoplasm of certain cells, commit those cells to a particular type of differentiation. Determinants, while important, are not the sole factors in differentiation, however; cell interactions become increasingly important as development progresses. It is possible that the same principles involved in the action of egg determinants might also apply to adult cells, as in the case of stem cells of various kinds.

# STUDY QUESTIONS

- Explain how the experiments performed with *Acetabularia* illustrate the interdependence of the nucleus and cytoplasm.
- What is a synkaryon, and how is it produced?
- Describe Harris's experiment using chick erythrocyte nuclei and HeLa cells. What did it demonstrate?
- Give examples of cytoplasmic functions that seem to be independent of the nucleus, and explain why this is the case.
- What did the nuclear transplantation experiments with amphibian oocytes and eggs demonstrate?
- What are "housekeeping" and "luxury" genes?
- At what level is cell differentiation controlled? What evidence supports your answer?
- What are determinants? What is their function in development?
- What is a stem cell? What distinguishes it from a differentiated cell?

# SUGGESTED READINGS

**Danielli, J. F., and Diberardino, M. A., eds.** (1979) *Nuclear Transplantation.* (International Review of Cytology, Supplement 9) Academic Press, New York.

**Davidson, E. H.** (1977) *Gene Activity in Early Development.* Academic Press, New York.

**De Robertis, E. M., and Gurdon, J. B.** (1977) Gene activation in somatic nuclei after injection into amphibian oocytes. *Proc. Natl. Acad. Sci. USA,* 74:2470.

**De Robertis, E. M., and Gurdon, J. B.** (1979) Gene transplantation and the analysis of development. *Sci. Am.,* 241:74.

**Gurdon, J. B.** (1977) Egg cytoplasm and gene control in development. *Proc. R. Soc. London [Biol.],* 198:211.

**Harris, H.** (1974) *Nucleus and Cytoplasm.* Clarendon, Oxford.

**Ringertz, N. R., and Savage, R. E.** (1976) *Cell Hybrids.* Academic, New York.

**Sublelny, S., and Konigsberg, I. R., eds.** (1979) *Determinants of Spatial Organization.* Academic Press, New York.

**Wilson, E. B.** (1928) *The Cell in Development and Heredity.* 3rd Ed. MacMillan, New York.

# Index

Note: Page numbers in *italics* indicate figures; page numbers followed by "t" indicate tables.